OXFORD MATHEMATICAL MONOGRAPHS

OXFORD MATHEMATICAL MONOGRAPHS

System Control
and Rough Paths

TERRY LYONS

Wallis Professor of Mathematics
University of Oxford and St Anne's College

and

ZHONGMIN QIAN

Chargé de Recherche
Université Paul-Sabatier and CNRS

CLARENDON PRESS · OXFORD

2002

OXFORD
UNIVERSITY PRESS

Great Clarendon Street, Oxford OX2 6DP

Oxford University Press is a department of the University of Oxford.
It furthers the University's objective of excellence in research, scholarship,
and education by publishing worldwide in

Oxford New York

Auckland Bangkok Buenos Aires Cape Town Chennai
Dar es Salaam Delhi Hong Kong Istanbul Karachi Kolkata
Kuala Lumpur Madrid Melbourne Mexico City Mumbai Nairobi
São Paulo Shanghai Taipei Tokyo Toronto

Oxford is a registered trade mark of Oxford University Press
in the UK and in certain other countries

Published in the United States
by Oxford University Press Inc., New York

© Terry Lyons and Zhongmin Qian, 2002

The moral rights of the authors have been asserted

Database right Oxford University Press (maker)

First published 2002

British Library Cataloguing in Publication Data

Data available

Library of Congress Cataloging in Publication Data

ISBN 0 19 850648 1

10 9 8 7 6 5 4 3 2 1

Typeset by Julie Harris using the authors' TEX files

Printed in Great Britain
on acid-free paper by T. J. International Ltd., Padstow, Cornwall

To our parents
To Barbara

PREFACE

Words differently arranged have different meanings and meanings
differently arranged have different effect.

Blaise Pascal (1632–1662)

Pensées 1660

A sentence, a passage, or even a book such as this one can be represented mathematically as a path in a moderate-dimensional vector space. It is capable of
influencing or having an 'effect' on a complex nonlinear system known as human
understanding.

Pascal reminds us of the obvious fact that the order of the words is important
to the effect of the passage or book.

Although the mathematics of understanding is still out of reach, there are
many other systems that can be modelled in mathematically precise terms as
control problems. Classical approaches apply the theory of differential equations,
with the control being a path in a vector space and the state of the system being
represented as a point in a (possibly infinite-dimensional) manifold.

However, the approach is limited by the demands classical (and for proba-
bilists Itô) calculus place on the control. Information about the control over a
small time step is represented by the increment of the control—a vector that ig-
nores the order of small-time events. The approach works very well if the control
is a smooth path, as in this case all the sub-events in a small time interval point
in the same direction and 'essentially' commute.

However, if the control has a lot of structure, even on small time steps (e.g. it
has fractal structure), this approach is inadequate. The order of events becomes
important. We cannot understand a text if every paragraph is summarized into
a count of the occurrence of each word without reference to their order in the
paragraph!

An adequate summary of a rough control must contain a certain amount of
non-commutative structure, even over small time steps. The rougher the control,
the more non-commutative 'order information' is required.

The main purpose of this book is to provide an account of the mathematics
required to model the evolution of a system that is subject to 'rough' external
forcing, and which has now emerged as a body of knowledge loosely called the
'theory of rough paths'. It aims to provide a direct and relatively unsophisticated
route to the basic results of the subject—and particularly to the result that the
differential equation

$$\frac{\mathrm{d}y}{\mathrm{d}t} = \sum_i f^i(y) \frac{\mathrm{d}x^i}{\mathrm{d}t}$$

vii

is uniformly continuous on bounded sets in appropriately chosen metrics, and to present its consequences for stochastic differential equations in finite and infinite dimensions.

The coordinate iterated integrals form a natural family of scalar functions on rough path space—moreover, their linear span is closed under multiplication with pointwise multiplication going to the shuffle product. In this way one sees that rough path space can be viewed formally, up to a simple equivalence relation, as a group, and the tensor algebra can be seen (formally) as the enveloping algebra. Many really concrete and fascinating developments in signal processing are coming from this perspective. They have been omitted to keep a consistent (and basic) approach for the book.

There are two authors associated with this book. The first, in the alphabetical sense, wishes to acknowledge the huge debt he owes to the second—without whose efforts this book would never have existed. Both authors would like to thank OUP for their patience and care.

The subject is developing rapidly. Hopefully this text will provide an initial platform for those interested in going further.

The authors happily acknowledge the assistance of Andy Dickinson for reading substantial parts of the text in draft form and Julie Harris who did a brilliant copy-editing job, and Elizabeth Johnstone who encouraged us to write this book at all. We acknowledge the support provided by students, friends and colleagues who have found some of this interesting and who have played a role in developing the canvas. We regret that this introductory book has to finish without mentioning many of these contributions.

Oxford and Toulouse   T. L.
July 2002   Z. Q.

CONTENTS

1

INTRODUCTION

1.1 Background and general description

The large volumes of data available as a consequence of modern technology have created new and interesting problems which some regard as being at the heart of scientific computation (Hackbusch, 1998). Put simply, it is not clear how one should reduce these large data sets so as to capture the critical information.

While the concept of what constitutes a satisfactory condensation clearly depends on the context, it seems inevitable that, for continuous numerical data streams, any such 'reduction' will involve the discretization of the data stream and a continuity theorem, allowing one to be confident that the reduction accurately represents the original data.

At one level, this book focuses on systems responding to rough external stimuli. In this case, it turns out that the natural reduction of the rough external stimuli approximates it as a sequence of nilpotent elements. The core result of the book is a continuity theorem. The response of the system depends continuously on these nilpotent elements.

Consider the following very simple (illustrative, but completely artificial) example. A sphere is placed on an oriented surface.

Suppose that we have it in our power to successively roll it distances of one (small) unit in any of the four compass directions. Let $X_t = (e_t, n_t)$ represent the path mapped out on $\mathbb{R}^2$ by the point of contact between the sphere and the surface. The path X records our decisions. The orientation Z_t of the sphere is described by an orthogonal transformation, or 3×3 matrix, satisfying

$$Z_t^\top \cdot Z_t = I$$

and the following differential equation relates X with Z:

$$\frac{dZ_t}{dt} = \frac{1}{R} \begin{pmatrix} 0 & 1 & 0 \\ -1 & 0 & 0 \\ 0 & 0 & 0 \end{pmatrix} \frac{de_t}{dt} + \frac{1}{R} \begin{pmatrix} 0 & 0 & 0 \\ 0 & 0 & 1 \\ 0 & -1 & 0 \end{pmatrix} \frac{dn_t}{dt},$$

$$Z_0 = \begin{pmatrix} 1 & 0 & 0 \\ 0 & 1 & 0 \\ 0 & 0 & 1 \end{pmatrix},$$

where R is the radius of the sphere. Since X is piecewise-smooth the solution is uniquely determined.

Now for the catch! Suppose (X_t) was the result of a random walk of a million or so steps and that one would like to store just enough information about X_t to make sure that we could find the value of Z_1. Obviously, the simplest solution is to solve the equation and store the solution! However, what if we do not know in advance the radius of the sphere; suppose we want to find the solution for different and unknown values of R.

We can express Z as a series in iterated integrals of X_t, namely

$$Z_t = I + \sum_{m=1}^{\infty} \sum_{\alpha \in \{e,n\}^m} \int \cdots \int_{0 \leqslant u_1 \leqslant \cdots \leqslant u_m \leqslant 1} R^m A^{\alpha_1} \cdots A^{\alpha_m} \, d\alpha_{1,u_1} \cdots d\alpha_{m,u_m},$$

where

$$A^e = \begin{pmatrix} 0 & 1 & 0 \\ -1 & 0 & 0 \\ 0 & 0 & 0 \end{pmatrix} \quad \text{and} \quad A^n = \begin{pmatrix} 0 & 0 & 0 \\ 0 & 0 & 1 \\ 0 & -1 & 0 \end{pmatrix}.$$

Moreover, one has good uniform estimates for the magnitudes of these integrals. For example, if the path X comprises K steps of unit length, then

$$\left\| \sum_{\alpha \in \{e,n\}^m} \int \cdots \int_{0 \leqslant u_1 \leqslant \cdots \leqslant u_m \leqslant 1} \frac{1}{R^m} A^{\alpha_1} \cdots A^{\alpha_m} \, d\alpha_{1,u_1} \cdots d\alpha_{m,u_m} \right\| < \frac{1}{m!} \left(\frac{K}{R} \right)^m.$$

If we suppose that Z is wanted to a machine precision of 2^{-32} for all paths with 10^6 steps and that $R \geqslant 10^6$ (the scale which ensures that all paths have a length comparable with the diameter of the sphere), then computation shows that it suffices to store the first thirteen or fourteen 3×3 matrices

$$\sum_{m=1}^{\infty} \sum_{\alpha \in \{e,n\}^m} \int \cdots \int_{0 \leqslant u_1 \leqslant \cdots \leqslant u_m \leqslant 1} R^m A^{\alpha_1} \cdots A^{\alpha_m} \, d\alpha_{1,u_1} \cdots d\alpha_{m,u_m},$$

with $m < 14$. Taking into account that the matrices are no longer sparse, one only needs at most a few thousand bits to record all of the necessary information, instead of the 2 million bits we started with.

The approach we set out above is too simple to be optimal, but it captures some important points. Perhaps the most striking (and obvious) is that the relationship between the controlling process X and the response of the sphere Z is far from linear!

None the less, one could ignore this and, since X is a path in a vector space, one could express X as a Fourier Series

$$X_t = \sum_{m \in \mathbb{Z}} a_m e^{2\pi i m t}$$

and consider using $X_t^{(M)} = \sum_{m \in [-M,M]} a_m e^{2\pi i m t}$ as a proxy for x. Unfortunately, this approach fails to give sensible results. For example, suppose that x

is the path that goes one unit north, then one unit east, then one unit south, and then one unit west. It goes around the four edges of the unit square. Let it do this 250 000 times. What happens to our rotation Z_1? In fact, it is easy to compute the answer to be $Z_1 = B^{250\,000}$, where

$$
B = \begin{pmatrix} \cos\left(1/R\right) & \sin\left(1/R\right) & 0 \\ -\sin\left(1/R\right) & \cos\left(1/R\right) & 0 \\ 0 & 0 & 1 \end{pmatrix} \begin{pmatrix} 1 & 0 & 0 \\ 0 & \cos\left(1/R\right) & \sin\left(1/R\right) \\ 0 & -\sin\left(1/R\right) & \cos\left(1/R\right) \end{pmatrix}
$$
$$
\times \begin{pmatrix} \cos\left(1/R\right) & -\sin\left(1/R\right) & 0 \\ \sin\left(1/R\right) & \cos\left(1/R\right) & 0 \\ 0 & 0 & 1 \end{pmatrix} \begin{pmatrix} 1 & 0 & 0 \\ 0 & \cos\left(1/R\right) & -\sin\left(1/R\right) \\ 0 & \sin\left(1/R\right) & \cos\left(1/R\right) \end{pmatrix},
$$

which, for R chosen to be approximately 10^6, gives a rotation that is, on our scale, of non-negligible magnitude. On the other hand, the function X is periodic with period $1/250\,000$ and so will not be approximated at all by the Fourier series until m is approximately $250\,000$. A wavelet basis will not achieve any useful compression either!

We should not be surprised that Fourier and wavelet decompositions do so badly in this rather simple problem. Both assume that it is reasonable to express a path or signal as a sum. There are certainly situations where this is very reasonable (for example, to solve for the evolution of heat, where X represents the initial spatial temperature distribution), but in settings where the process X_t controls the evolution of a second process Z_t this is not reasonable (except in the special case where X is one-dimensional).

As we will explain in the rest of this text, when the context is control theory, and our effort is dedicated to describing a 'rough' path evolving, the natural condensation can be expressed in terms of iterated integrals and their logarithms. The increments of the paths are approximated and represented by nilpotent elements.[1] To help the intuition, we give two other examples, but with the health warning that they are simplistic and are to be taken in the same spirit as examples in books intended to teach computing.

Consider the following problem.[2] An engineer, who has designed and built a computer model of a suspension bridge, wants to test its response to earthquakes. He must collect data and then simulate the earth movements that would be experienced. However, collecting data is expensive, and one obviously needs to decide exactly what data should be collected. Perhaps the surface height $h\left(t\right)$ should be measured? Maybe it is important to measure the behavior of the

---

[1] Coding and Shannon information theory can provide a vital tool, but does not remove the need to decide on the output data that is useful. A bank might process 350M credit card transactions a year. A zip file could be an effective way to reduce the data for storage, but it does not seriously offer the bank help in identifying ways to abstract useful information for creating additional value in its customers.

[2] The oversimplified 'real world' examples here are included to help understanding and aid communication. Theoretical developments of the sort mapped out could lead to advances in the application areas specified—but we do not claim this here.

ground surface on a piece of ground which is the same size as the bridge? If one regards the height as a surface $h(X, Y, t)$, how should one sample it? If an engineering institute wanted to keep a collection of data sets for engineers to stress test their designs against historical scenarios, how could they minimize the size of each data set and standardize its internal structure, while ensuring that they had a rich and non-repetitive collection of scenarios suitable for testing a range of architectural structures. Similar modelling could have resulted in a certain foot-bridge on the Thames being usable!

A second 'example' of a practical kind could be in the reproduction and perception of sound (and perhaps vision). Suppose that one listens to ones favorite stereo sound downloaded from the web to a personal portable. Then bandwidth on the internet, and convenience, as well as the cost of memory on the personal listening device, all push the design engineer to describe the 'sounds' as economically as possible. At the same time, our satisfaction and willingness to listen require that engineer to use this limited description to create sounds of quality and realism.

It is tempting to believe that Shannon's information theory provides a complete answer to this problem. However, Shannon's theory assumes that one already understands the information to be transmitted or recorded. It is invaluable for the design of an efficient coding or compression algorithm after one has the answer to the question: when do two sound streams sound the same? Formally, one needs a discrete alphabet which is rich enough to represent all the possible sounds we can perceive and sparse enough so that no two distinct elements are perceived as the same sound. Answering this question is not so much a problem in information theory, *as a problem in analysis and biology.* One would like to find effective metrics on sound signals relating to how indistinguishable the humans' responses would be. Information that enables one to construct sounds that are indistinguishable to the listener wastes bandwidth.

MPEG3 is a reasonably successful commercial attempt which takes perception into account before applying information-theoretic techniques. It achieves dramatic improvements in the size of sound files, compared with techniques used in standard CD technology, without a severe drop in sound quality.

The methods of MPEG are specific to sound perception. However, we would argue that both examples above, and many others we have not mentioned, are distinguished by the following important common features:

(i) We are interested in the process X_t for the way it affects some more complex system.

(ii) The response of the system is not linear in input data stream.

(iii) The data stream X_t is abruptly changing and generally rich in information ('a rough path'), making it hard to summarize in simple ways and making the nonlinearity of the response more significant.

1.1.1 *Controlled systems*

The examples (the earthquake and the sound) are, at one level of modelling, examples of controlled systems. In each case one has a complex dynamic system (the bridge or our ear drums and perception) subject to an external control (the ground movements or air pressure). Without trying to guess the precise nature of the governing differential equations, it is not unreasonable to hypothesize a model relating the internal state Y_t of the (high-dimensional) system and the (lower-dimensional) external control X_t in the form of a differential system:

$$\frac{\mathrm{d}Y_t}{\mathrm{d}t} = f(Y_t) + \sum_{i=1}^{d} g^i(Y_t)\frac{\mathrm{d}x_t^i}{\mathrm{d}t}. \tag{1.1}$$

However, the rich flow of information in the external control means that the control is not smooth on normal time scales even if it is continuous. The impulse of an earthquake is more like a nonlinear delta function. A spectral analysis of sounds suggests that it is typically much rougher that Brownian motion.

Acknowledging this, and adopting the approach mapped out in Lyons (1994, 1998) and subsequent papers, one can make sense of eqn (1.1) without going to a much finer scale at which X_t is perhaps smooth. Indeed, in making sense of these equations when X is not differentiable, one is forced to answer, in a mathematical way, the question 'when do two signals X and $\tilde{X}$ produce essentially the same response Y?' and analyze various notions of continuity and smoothness for the map[3] $X \to Y$ defined, at least for differentiable inputs X, by

$$\mathrm{d}Y_t = f(Y_t)\,\mathrm{d}t + g(Y_t)\,\mathrm{d}X_t\,,$$
$$Y_0 = a\,.$$

1.1.2 *Vector systems*

In the same way that systems of ordinary differential equations exhibit a richness of behavior not matched by scalars equations, controlled systems exhibit a striking and essential contrast in their behavior according to the dimension of the control variable.

Keeping to our notation, let $Y_t \in M$ denote the state of some evolving system. We may represent its uncontrolled evolution by a differential equation

$$\frac{\mathrm{d}Y_t}{\mathrm{d}t} = f(Y_t)\,,$$

where f is a vector field on M. Suppose that X_t is a smooth function with values in $\mathbb{R}^d$ representing some control influencing the evolution Y_t via the second differential equation

$$\frac{\mathrm{d}Y_t}{\mathrm{d}t} = f(Y_t) + \sum_{i=1}^{d} g^i(Y_t)\frac{\mathrm{d}X_t^i}{\mathrm{d}t}.$$

---

[3]Henceforth we call this map the Itô map.

In general, we will use the more concise notation

$$I : X \to Y \,,$$
$$\mathrm{d}Y_t = f(Y_t)\,\mathrm{d}t + g(Y_t)\,\mathrm{d}X_t \,,$$
$$Y_0 = a$$

to describe the relationship between control and response and will refer to I as the Itô functional.

We have the following elementary, but striking, contrast according to whether $d = 1$ or $d > 1$.[4]

Theorem 1.1.1 *In the case where X is one-dimensional, the Itô functional I is continuous in the topology of uniform convergence. In other words, it is robust to errors in (X_t). On the other hand, if X is multi-dimensional then I is only continuous in the uniform topology in the special case when the vector fields $g^i\,(\cdot)$ commute.*

The condition that the g^i commute is highly non-generic for $d > 1$. The GENERIC behavior is that the mapping I is UNSTABLE if X is multi-dimensional, but STABLE if X is one-dimensional.[5] This instability underlines the essentially nonlinear nature of the problem. Any approach to summarizing a multi-dimensional signal that treats the different coordinate channels independently is missing something essential. It is suggestive to note that a stereo CD uses a sampling depth of 16 bits and a frequency of about 43 kHz applied to a signal bandwidth limited to about 21 kHz, while mono radio can be of reasonable quality with 10 kHz sampling and 8 bit depth.

On the other hand, there is a positive theorem.

Theorem 1.1.2 *The Itô map I is continuous in the topology of convergence in the metric of p-variation of rough paths even in the vector case.*

Our approach to this result, via rough paths, depends on a nonlinear transformation of X using iterated integrals. Consider a path X on $[t_n, t_{n+1}]$. A truncated Fourier series $X \approx \sum a_k e^{2\pi i k t/(t_{n+1}-t_n)}$ might also provide a useful summary of X over short time intervals. However, the intrinsically linear nature of the Fourier approach underlines its weakness and makes it less effective in the more sensitive multi-dimensional setting of the bridge, stereo sound, vision,... The *natural transformation of a path to a set of coefficients*, appropriate when that path is being viewed as a control, is the transformation *taking paths to iterated integral sequences*. Formally, it is a map from path segments to free group elements.

---

[4]For a real world illustration of this dichotomy, contrast the complexity of motion of a train moving backwards and forwards, with only one degree of freedom, with a car having two degrees of freedom.

[5]We say the map $x \to y$ is unstable (for a given metric) if small perturbations in x can lead to big changes in y. Perhaps we should simply use the word discontinuous.

1.1.3 *Iterated integral expansions*

Controlled systems responding to a multi-dimensional stimulus $X_t = (x_t^1, \ldots, x_t^d)$ $\in \mathbb{R}^d$ are NOT stable under small perturbations in the uniform metric. So suppose that X is smooth on some very fine time scale, but 'rough' on observable scales, and that

$$\mathrm{d}Y_t = f(Y_t)\,\mathrm{d}t + g(Y_t)\,\mathrm{d}X_t .$$

What is an appropriate statistic or proxy $\tilde{X}$ for X over a short time interval so that $I(X) \approx I(\tilde{X})$ and the system responses are essentially the same?[6]

Group-like increments

If X is a smooth path at a very fine scale, then we can consider its (higher-order) increments over an interval $[s,t]$, namely

$$X_{s,t}^n := \int \cdots \int_{s \leqslant u_1 \leqslant \cdots \leqslant u_n \leqslant t} \mathrm{d}X_{u_1} \cdots \mathrm{d}X_{u_n} \in \mathbb{R}^{nd} .$$

If the path X has some sort of statistical scale invariance (on normal scales), then one might find a $p > 1$ so that, roughly,

$$X_{s,t}^1 = (X_t - X_s) \sim (t-s)^{1/p} ,$$
$$X_{s,t}^k \sim (t-s)^{k/p} .$$

Then a quantitative sufficient condition for the paths X, $\tilde{X}$ to be close enough to ensure that $I(X)$ and $I(\tilde{X})$ are close together is that

$$\sup_{s=t_0 \leqslant t_1 \leqslant \cdots \leqslant t_{r-1} \leqslant t_r = t} \sum_{i=0}^{r-1} \left| X_{t_i,t_{i+1}}^k - \tilde{X}_{t_i,t_{i+1}}^k \right|^{p/k} \tag{1.2}$$

should be small FOR EVERY $k \leqslant p$. However, delving deeper, one learns that in order to approximately predict the response of the system it is NECESSARY and sufficient to know the values of $X_{t,t+\delta t}^k$ for $k \leqslant p$. The iterated integrals with degree less than p form an asymptotically sufficient statistic.

At this point we can see why there is such a big difference between the one-dimensional case and all of the others. In one dimension,

$$X_{s,t}^n := \int \cdots \int_{s \leqslant u_1 \leqslant \cdots \leqslant u_n \leqslant t} \mathrm{d}X_{u_1} \cdots \mathrm{d}X_{u_n}$$
$$= \frac{1}{n!} (X_t - X_s)^n ,$$

and the increment $X_{s,t}^1 = (X_t - X_s)$ carries full information about $X_{s,t}^n$ for every n. If the path is smooth then knowledge of the increments $X_{u,v}^1$ for all u and v

---

[6]For simplicity of presentation we will drop the expression in f from our further discussions.

still gives full information about $X_{s,t}^n$ in a limiting sense. We can construct $X_{s,t}^n$ as a limit of sums, for example

$$X_{0,1}^n = \lim_{r \to \infty} \sum_{0 \leqslant k_1 < \cdots < k_n < 2^r} \left(X_{(k_1+1)2^{-r}} - X_{k_1 2^{-r}} \right) \cdots \left(X_{(k_n+1)2^{-r}} - X_{k_n 2^{-r}} \right).$$

However, this means that to provide an approximation of $X_{s,t}^n$ on one scale one has to have $X_{u,v}^1$ on a finer scale. If the path fails to be smooth enough (e.g. $p > 2$) then matters get worse. The approximations will fail to converge as n tends to infinity.

The linear increments of a two-dimensional rough path do not in general determine the response! When they do, they describe it in an inefficient way.

In this book we wish to explore mathematically the very reasonable hypothesis that it is more 'information efficient' to record this information[7] directly on a relatively coarse time scale than to record $X_{t,t+\delta t}^1$ on a much finer time scale. Stochastic differential equations give concrete evidence of the benefits of this approach. The numerical approximations to solutions of stochastic differential equations, obtained by recording the increment and Lévy area of the Brownian motion, achieve substantial speed, accuracy, and information gains over traditional approaches, thereby improving the order of approximation sufficiently to ensure that the time intervals used in the numerical simulations need no longer be chosen predictably in relation to the path (Gaines and Lyons, 1997).

1.2   Mathematics of rough paths

This book addresses the mathematics which underpin the applications we have outlined. As we have indicated, we study the differential equation

$$\mathrm{d}Y_t^j = \sum_i F_i^j(t, Y_t) \, \mathrm{d}X_t^i, \tag{1.3}$$

where the driving path $X_t = (X_t^i)$ is a multi-dimensional (perhaps infinite-dimensional) control. An important example concerns the case where X_t is modelled by a multi-dimensional Brownian motion. However, we shall develop a theory which can be applied to far rougher driving paths. The theory shall provide a new setting both for the classical ODE theory and for the theory of stochastic differential equations, going beyond the range of the classical theory and of standard probabilistic techniques.

A dynamical system of input and output is usually described by a differential equation of the form

$$\mathrm{d}Y_t = f(t, Y_t) \, \mathrm{d}X_t. \tag{1.4}$$

As we have explained, we have to consider an eqn (1.4) with an input X containing some noise. Mathematically this means that $t \to X_t$ may be very rough. The first duty we face is to establish a good theory for such a dynamical system.

---

[7]Appropriately rounded.

The main issue is to explore in what sense such a dynamical system is robust, as one is never able to know the precise noise contained in the input X.

The classical ODE theory collapses when the path $t \to X_t$ is very rough. The reason is very simple: the differential dX_t in the usual sense does not make sense. However, on a close observation of the classical approach, one notices that the classical ODE theory can be established by using the increment process $\{X_t - X_s : 0 \leqslant s \leqslant t\}$ of the path X to construct approximate solutions, and one may take the limit to obtain the solution without first identifying the differential dX_t. Therefore, one may regard the whole collection $\{X_t - X_s : 0 \leqslant s \leqslant t\}$ as the 'differential' dX_t if the path $t \to X_t$ possesses finite variation. However, if $t \to X_t$ is very rough, then it is easy to understand that only the increment process $\{X_t - X_s : 0 \leqslant s \leqslant t\}$ is not enough to capture the differential dX_t (which is the impulse propelling our system through the differential equation). One needs higher-order terms. The basic viewpoint of this book lies in the interpretation of the differential dX. The fundamental idea is that the 'full differential' dX is the collection of all iterated path integrals, namely

$$\left\{ \int_{s<t_1<\cdots<t_k<t} dX_{t_1} \otimes \cdots \otimes dX_{t_k} : 0 \leqslant s \leqslant t \right\}.$$

We argue the justification for this as follows. Suppose that X is a continuous path in, for example, a finite-dimensional vector space V such that one is able to solve a differential equation like (1.4), and therefore we at least hope to define the path integral $\int \alpha(X) \, dX$, for any smooth one-form α, and so especially the iterated path integrals

$$X_{s,t}^k \equiv \int_{s<t_1<\cdots<t_k<t} dX_{t_1} \otimes \cdots \otimes dX_{t_k}.$$

(It will be clear later why we use the tensor form.) For example, if X is a path with finite variation then the first-order iterated path integral is nothing more than its increment process $X_{s,t}^1 \equiv X_t - X_s$, and the higher-order integrals can be obtained from

$$X_{s,t}^k = \lim_{m(D)\to 0} \sum_{l=1}^{m} \sum_{i=1}^{k-1} X_{s,t_{l-1}}^i \otimes X_{t_{l-1},t_l}^{k-i},$$

where the limit runs over all finite partitions D of $[s, t]$ as the mesh size tends to zero. This means that the increment process $\{X_{s,t}^1 : 0 \leqslant s \leqslant t\}$ determines all higher-order iterated path integrals, and hence the integrals of one-forms against the path X. This explains why we only need $\{X_{s,t}^1 : 0 \leqslant s \leqslant t\}$ in the classical smoothly-controlled ODE. Moreover, the bounded variation of X leads, for each k, to the following bound:

$$\sup_D \sum_l \left| X_{t_{l-1},t_l}^k \right|^{1/k} < \infty, \quad k = 1, 2, \ldots. \tag{1.5}$$

Now let us explain why we define iterated path integrals as elements in tensor products $V^{\otimes k}$. The advantage is that we can easily express a basic requirement for any reasonable integration theory. That is, the additive property of integrals over different intervals. More precisely, let us set

$$X_{s,t} = \left(1, X_{s,t}^1, \ldots, X_{s,t}^k, \ldots\right)$$

and regard it as an element in the tensor algebra $\oplus V^{\otimes k}$ (direct sum of $V^{\otimes k}$). Then the additive property exactly means that

$$X_{s,t} \otimes X_{t,u} = X_{s,u}, \quad 0 \leqslant s \leqslant t \leqslant u. \tag{1.6}$$

This multiplicative property will play an important role, and eqn (1.6) is called the Chen identity, although K. T. Chen was not the one who discovered this equation. However, it was K. T. Chen, in a series of papers, who exhibited its usefulness in relating analysis and geometry, see Chen (1977).

However, it makes the analysis hard if one is working with an infinite sequence $X_{s,t} = (1, X_{s,t}^1, \ldots, X_{s,t}^k, \ldots)$, though it is not impossible. As we have explained, the condition (1.5) yields that the higher-order integrals X^k are determined uniquely by X^1, but those paths in which we are interested rarely satisfy (1.5). For example, almost all Brownian paths do not satisfy (1.5) even for $k = 1$, but they do satisfy the following weaker condition:

$$\sup_D \sum_l \left| X_{t_{l-1},t_l}^1 \right|^p < \infty, \tag{1.7}$$

for any $p > 2$. Therefore, if X is a such a non-smooth path which satisfies the analytic condition (1.7) and if we are able to define its iterated path integrals $X_{s,t}^k$, then it is reasonable for us to also expect these iterated path integrals to satisfy the following scaling control:

$$\sup_D \sum_l \left| X_{t_{l-1},t_l}^k \right|^{p/k} < \infty, \quad k = 1, 2, \ldots, \tag{1.8}$$

even though (1.5) does not hold.

One of the results we shall prove is that if X satisfies the Chen identity and the analytic condition (1.8) then X^k are uniquely determined by $X^1, \ldots, X^{[p]}$, for all $k \geqslant [p] + 1$. Therefore, in this case we only need to know $X^1, \ldots, X^{[p]}$. On the other hand, $X^2, \ldots, X^{[p]}$ are never unique if we are given $X_{s,t}^1$. In this sense, p relates non-smoothness of the path X—the bigger p is the rougher the path X—and more high-order iterated path integrals are needed in order to integrate one-forms.

2

LIPSCHITZ PATHS

In this chapter we consider continuous paths with finite variation in Banach spaces. Such continuous paths we call Lipschitz paths, as we express the condition of finite variation in terms of Lipschitz-type domination. We will see in later chapters that such a kind of domination expression is especially convenient when we treat rough paths. A complete integration theory (including a theory of differential equations) for continuous paths with finite variation is presented; this can be read independently for those who are only interested in regular paths. We begin with several examples of solving differential equations via integral curves of vector fields, and an example which shows that in general the solution to an ordinary differential equation is not necessarily continuous under the topology of uniform convergence. We then develop an integration theory for continuous paths with finite variation, and give a detailed proof of a continuity theorem for solutions to differential equations with Lipschitz controls.

2.1 Several examples

The simplest case

Differential equations driven by one-dimensional paths are special, and can be solved for all continuous paths. Consider the following differential equation:

$$dY_t = a(Y_t)\,dX_t\,, \quad Y_0 = \xi\,, \tag{2.1}$$

where $X_\bullet\colon [0,T] \to \mathbb{R}$ is a continuous path in $\mathbb{R}$, $\xi \in \mathbb{R}$ is an initial point, and $a \in C_b^\infty(\mathbb{R}, \mathbb{R})$, which we identify as the vector field $V_a(r) = a(r)\,d/dr$ on $\mathbb{R}$.

Suppose that $X_\bullet$ is a smooth path in $\mathbb{R}$. Then eqn (2.1) is equivalent to the following ordinary differential equation:

$$dY_t = a(Y_t)\dot{X}_t\,dt\,, \quad Y_0 = \xi\,, \tag{2.2}$$

and one can solve this equation. In the special case of $X_t = t$ the differential eqn (2.2) reduces to the flow equation of the vector field V_a, namely

$$dZ_t = a(Z_t)\,dt\,, \quad Z_0 = \xi\,, \tag{2.3}$$

or

$$\dot{Z}_t = V_a(Z_t)\,, \quad Z_0 = \xi\,. \tag{2.4}$$

We denote the unique solution Z_t to eqn (2.4) by $\pi(t,\xi)$, in order to emphasize the dependence on the initial point ξ. Then it is well known that $\pi(t,\xi)$ is well defined on $\mathbb{R}$.

Doss (1977) and Sussmann (1978) proposed a procedure to solve eqn (2.1) for any continuous path X which we shall describe in the following.

Observe that $Y_t = \pi(X_t - X_0, \xi)$ is the unique solution of eqn (2.1) if $X_{\bullet}$ is a smooth path in $\mathbb{R}$. In fact, we have

$$
\begin{aligned}
\frac{\mathrm{d}Y_t}{\mathrm{d}t} &= \left.\frac{\mathrm{d}\pi(s,\xi)}{\mathrm{d}s}\right|_{s=X_t-X_0} \frac{\mathrm{d}X_t}{\mathrm{d}t} \\
&= a(\pi(s,\xi))|_{s=X_t-X_0} \frac{\mathrm{d}X_t}{\mathrm{d}t} \\
&= a(\pi(X_t - X_0, \xi)) \frac{\mathrm{d}X_t}{\mathrm{d}t} \\
&= a(Y_t) \frac{\mathrm{d}X_t}{\mathrm{d}t},
\end{aligned}
$$

so that

$$
\mathrm{d}Y_t = a(Y_t)\,\mathrm{d}X_t
$$

and $Y_0 = \pi(X_0 - X_0, \xi) = \pi(0, \xi) = \xi$. Therefore, for any continuous path X, we may regard $Y_t = \pi(X_t - X_0, \xi)$ as a solution of eqn (2.1), although we have not even defined the differential $\mathrm{d}X_t$!

The map $X_{\bullet} \to \pi(X_{\bullet} - X_0, \xi)$ is a continuous extension of the map defined by solving the differential eqn (2.1), and is called the Itô map. We use this name in honour of Itô as the first to regard the solution to a (stochastic) differential equation driven by the Brownian motion as a function of its driving path under the name 'strong solution'.

Let us push the idea a little further. Consider the slightly more complicated differential equation

$$
\mathrm{d}Y_t = a(Y_t)\,\mathrm{d}X_t + b(Y_t)\,\mathrm{d}t, \quad Y_0 = \xi, \tag{2.5}
$$

where $b \in C_b^{\infty}(\mathbb{R}, \mathbb{R})$ is another smooth vector field. The driving path of this equation is a two-dimensional path (X_t, t). One can also solve such an equation even for a continuous path X as the second component t of the path (X_t, t) is very smooth. This point of view will become clear after we establish the whole theory.

We again denote by $\pi(t, \xi)$ the solution flow of the vector field V_a. Let us try to find a path η_t such that $Y_t = \pi(X_t - x_0, \eta_t)$ solves eqn (2.5) when X is smooth—a naive but sometimes efficient trick in the theory of ODEs. Then

$$
\begin{aligned}
\frac{\mathrm{d}Y_t}{\mathrm{d}t} &= \left.\frac{\partial\pi(s,\eta_t)}{\partial s}\right|_{s=X_t-X_0} \frac{\mathrm{d}X_t}{\mathrm{d}t} + D\pi(X_t - X_0, \eta_t)\dot{\eta}_t \\
&= a\left(\pi(X_t - X_0, \eta_t)\right)\frac{\mathrm{d}X_t}{\mathrm{d}t} + D\pi(X_t - X_0, \cdot)(\eta_t)\dot{\eta}_t \\
&= a(Y_t)\dot{X}_t + b(Y_t),
\end{aligned}
$$

where $D\pi(t, \xi)$ denotes the derivative of $\pi(t, \xi)$ in the space variable, that is $D\pi(t, \xi) = \partial\pi(t, \xi)/\partial\xi$. Therefore, η should satisfy the following differential equation:

$$D\pi(X_t - X_0, \cdot)(\eta_t)\dot{\eta}_t = b(\pi(X_t - X_0, \eta_t)),$$

that is

$$\dot{\eta}_t = D\pi(X_t - X_0, \eta_t)^{-1}b(\pi(X_t - X_0, \eta_t)), \quad \eta_0 = \xi.$$

In other words, η_t is the solution flow of the time-dependent vector field

$$C(t, u) = D\pi(X_t - X_0, u)^{-1}b(\pi(X_t - X_0, u)).$$

The vector field $C(t, u)$ is smooth in the space variable u and continuous in the time variable t, and therefore a solution (η_t) with the initial point ξ exists, and this is denoted by $\zeta(t, \xi)$. Therefore, a similar argument as before shows that $Y_t = \pi(X_t - X_0, \zeta(t, \xi))$ is the solution of eqn (2.5), which again makes sense even for a continuous path X.

We may write down explicitly the vector field $C(t, u)$. In fact, since $\pi(t, \xi)$ is the solution flow of the vector field V_a, we have

$$\frac{\mathrm{d}}{\mathrm{d}t}D\pi(t, \xi) = a'(\pi(t, \xi))D\pi(t, \xi),$$

$$D\pi(0, \xi) = 1, \quad \forall \xi \in \mathbb{R}.$$

The initial condition that $D\pi(0, \xi) = 1$ comes from the fact that the solution $\pi(t, \xi)$ starts from a fixed point ξ. Therefore,

$$D\pi(t, \xi) = \exp\left(\int_0^t a'(\pi(s, \xi))\,\mathrm{d}s\right),$$

so that

$$D\pi(t, \xi)^{-1} = \exp\left(-\int_0^t a'(\pi(s, \xi))\,\mathrm{d}s\right).$$

Hence,

$$C(t, u) = \exp\left[-\int_0^{X_t - X_0} a'(\pi(s, u))\,\mathrm{d}s\right]b(\pi(X_t - X_0, u)).$$

Multi-dimensional systems

Consider the following multi-dimensional differential equation:

$$\mathrm{d}Y_t = \sum_{i=1}^m A_i(Y_t)\,\mathrm{d}X_t^i, \quad Y_0 = \xi, \tag{2.6}$$

where $X_\bullet = (X_\bullet^i)_{i=1}^m$ is a continuous path in $\mathbb{R}^m$, the A_i, $i = 1, \ldots, m$, are m vector fields on $\mathbb{R}^d$, and $\xi = (\xi^i)_{i=1}^d \in \mathbb{R}^d$ is an initial value. As we have mentioned, in this book we study multi-dimensional differential equations driven by non-smooth paths, and the main goal is to construct a so-called Itô map $F(t, X_\bullet)$ which solves eqn (2.6).

In a special case when the vector fields $A_1, \ldots, A_m$ are commutative, we may define such a map (or solve eqn (2.6)) by using Doss's method.

We say the vector fields $A_1, \ldots, A_m$ are commutative if their Lie brackets $[A_i, A_j] = 0$, $\forall i \neq j$, where

$$[A_i, A_j]f = A_i(A_j f) - A_j(A_i f).$$

The commutative condition implies that $(A_1, \ldots, A_m)$ is integrable. This means that, if we write

$$A_i(r) = \sum_{j=1}^{d} A_i^j(r) \frac{\partial}{\partial r^j},$$

then the following system of PDEs:

$$\frac{\partial u^i}{\partial r^j} = A_j^i(u(r, \xi)), \quad i, j = 1, \ldots, d,$$

$$u(0, \xi) = \xi,$$

is solvable, where $u = (u^1, \ldots, u^d)$, $r = (r^1, \ldots, r^d)$, and $\xi = (\xi^i) \in \mathbb{R}^d$.

We then call $u = (u^1, \ldots, u^d)$ the solution flow of the commutative system $(A_1, \ldots, A_m)$. Define $Y_t^i = u^i(X_t - X_0, \xi)$, $Y_t = (Y_t^i)_{i=1}^d$. Then Y solves eqn (2.6).

The case where A_i, $i = 1, \ldots, m$, are not commutative (and therefore $m \geqslant 2$) is quite different. In this case the input path X is a multi-dimensional path, and the key fact that is revealed is that a multi-dimensional path actually is not determined by the 'horizontal shape' of the path, and indeed involves the mixed structure of intervening paths—they can be expressed as a sequence of 'iterated path integrals'.

A counter-example

To exhibit where we may go wrong for such a differential equation, let us consider the following simple differential equation with two vector fields A_1, A_2 on $\mathbb{R}$:

$$dY_t = A_1(Y_t) \, dX_t^1 + A_2(Y_t) \, dX_t^2,$$
$$Y_0 = \xi \in \mathbb{R}, \tag{2.7}$$

where $X_t = (X_t^1, X_t^2)$ is a two-dimensional path (a path in $\mathbb{R}^2$) and $A_1(r) = r \, d/dr$, $A_2(r) = d/dr$. We may easily see that $[A_1, A_2] = -[A_2, A_1] = -A_2$. Therefore,

$$[\ldots [\ldots [A_2, A_1], A_1], \ldots, A_1] = A_2$$

and any Lie bracket containing two or more A_2 factors is zero. Hence, the Lie algebra $\mathrm{Lie}(A_1, A_2)$ generated by A_1, A_2 is not nilpotent.

Equation (2.7) may be written as

$$\mathrm{d}Y_t = Y_t\,\mathrm{d}X_t^1 + \mathrm{d}X_t^2, \quad Y_0 = \xi, \tag{2.8}$$

and the unique solution of this equation is given by the following formula:

$$Y_t = e^{X_t^1}\left(\int_0^t e^{-X_s^1}\,\mathrm{d}X_s^2 + \xi_0\,e^{-X_0^1}\right).$$

Denote the map which sends a path $X_\bullet = (X_\bullet^1, X_\bullet^2)$ to the solution $Y_\bullet$ by $F(X_\bullet, \xi_0)$. What we are going to show is that $X_\bullet \to F(X_\bullet, \xi)$ is not continuous under the topology of uniform convergence.

It is easily seen that

$$Y_t = e^{X_t^1 - X_0^1}\left(\xi_0 + \int_0^t \mathrm{d}X_{t_1}^2 + \int_{0<t_1<t_2<t} \mathrm{d}X_{t_1}^1\,\mathrm{d}X_{t_2}^2\right.$$
$$\left. + \sum_{k=2}^\infty (-1)^k \int_{0<t_1<\cdots<t_k<s<t} \mathrm{d}X_{t_1}^1 \cdots \mathrm{d}X_{t_k}^1\,\mathrm{d}X_s^2\right).$$

Let $X(n)_t^1 = (1/n)\cos n^2 t$ and $X(n)_t^2 = (1/n)\sin n^2 t$. Then $X(n)_\bullet^i \to 0$ uniformly on any finite interval as $n \to \infty$, and therefore if $X_\bullet \to F(X_\bullet, \xi)$ were continuous we would have that $F(X(n)_\bullet, \xi)_\bullet \to \xi$. An elementary calculation shows that

$$e^{X(n)_\bullet^1 - X(n)_0^1} \to 1, \quad \text{as} \quad n \to \infty,$$
$$\int_0^t \mathrm{d}X(n)_{t_1}^2 = \frac{1}{n}\sin n^2 t,$$

and, however,

$$\int_{0<t_1<\cdots<t_k<s<t} \mathrm{d}X(n)_{t_1}^1 \cdots \mathrm{d}X(n)_{t_k}^1\,\mathrm{d}X(n)_s^2$$
$$= \frac{1}{k!n^{k-1}}\int_0^t (\cos n^2 s - 1)^k \cos n^2 s\,\mathrm{d}s.$$

Hence,

$$\sum_{k=2}^\infty (-1)^k \int_{0<t_1<\cdots<t_k<s<t} \mathrm{d}X(n)_{t_1}^1 \cdots \mathrm{d}X(n)_{t_k}^1\,\mathrm{d}X(n)_s^2 \to 0,$$

and

$$\int_{0<t_1<t_2<t} \mathrm{d}X(n)_{t_1}^1\,\mathrm{d}X(n)_{t_2}^2 = \frac{t}{2} + \frac{1}{4n^2}\sin 2n^2 t \to \frac{t}{2}$$

as $n \to \infty$. Therefore,

$$F(X(n)_\bullet, \xi_0)_t \to \xi_0 + \frac{t}{2}, \quad \text{as} \quad n \to \infty.$$

The map $X_\bullet \to F(X_\bullet, \xi_0)$ is not continuous.

Moreover, as we have seen, the troublesome term in this example is the second iterated path integral

$$\int_{0<t_1<t_2<t} dX^1_{t_1}\, dX^2_{t_2}\,,$$

which relates to the area enclosed by the curve $X_\bullet = (X^1_\bullet, X^2_\bullet)$.

2.2 Integration theory

Let $[0, T]$ be a finite interval and let Δ_T denote the simplex $\{(s, t) : 0 \leqslant s \leqslant t \leqslant T\}$. A control[8] (function) ω is a non-negative continuous function on Δ_T which is super-additive, namely

$$\omega(s, t) + \omega(t, u) \leqslant \omega(s, u)\,,$$

for all $0 \leqslant s \leqslant t \leqslant u \leqslant T$, and for which $\omega(t, t) = 0$, for all $t \in [0, T]$. For simplicity, we will drop the subscript T if no confusion may arise.

A Lipschitz path in a Banach space V is a map X from a time interval $[0, T]$ into V which satisfies the following condition:

$$|X_t - X_s| \leqslant \omega(s, t)\,, \quad \forall\, 0 \leqslant s \leqslant t \leqslant T\,,$$

for some control $\omega : \Delta \to \mathbb{R}^+$. In this case, we also call X a Lipschitz path in V controlled by ω, or a Lipschitz path with control ω. In this section the running time interval $[0, T]$ will be fixed, except when otherwise specified. A simple observation is that (we will prove a general claim similar to this in the next chapter) a continuous path X in V is a Lipschitz path if and only if it possesses finite variation, namely

$$\sup_D \sum_l \left| X_{t_l} - X_{t_{l-1}} \right| < \infty\,,$$

where the supremum is taken over all finite partitions D of $[0, T]$. Indeed, in this case

$$\omega(s, t) = \sup_{D_{[s,t]}} \sum_l \left| X_{t_l} - X_{t_{l-1}} \right|$$

is a control of X, where $D_{[s,t]}$ denotes all finite partitions of $[s, t]$.

Given a Lipschitz path $X : [0, T] \to V$ with control ω_1, we may define its increment function $X^1 : \Delta \to V$ by $X^1_{s,t} = X_t - X_s$, for all $(s, t) \in \Delta$.

Suppose now that $Y : [0, T] \to W$ is another Lipschitz path in a Banach space W with a control ω_2, and suppose that $f : W \to \boldsymbol{L}(V, U)$ is a continuous map, where $\boldsymbol{L}(V, U)$ denotes the Banach space of all bounded linear operators from V into U. Then we may define the integral $\int_s^t f(Y)\, dX$ as the limit of

---

[8]Added in the final proofs: Unfortunately, the word control is used in two senses in the literature: (a) for a function ω regulating the irregularity of X and (b) for X itself, regarding it as controlling Y.

Riemannian sums. The integral $\int_s^t f(Y)\,\mathrm{d}X$ exists as a continuous path in W under some conditions on f. Since our main concern is the continuity of the integral $\int f(Y)\,\mathrm{d}X$ in X and Y, we will therefore assume that f is Lipschitz. More precisely, f satisfies the following condition: there is a non-negative constant M_1 such that, for any $\xi, \eta \in W$,

$$|f(\xi) - f(\eta)| \leqslant M_1|\xi - \eta|,$$
$$|f(\xi)| \leqslant M_1(1 + |\xi|).$$

Then the integral

$$\int_s^t f(Y)\,\mathrm{d}X = \lim_{m(D)\to 0} \sum_{l=1}^{m} f(Y_{t_{l-1}})(X^1_{t_{l-1},t_l}) \tag{2.9}$$

exists, where the limit is taken over all finite divisions $D = \{s = t_0 < t_1 < \cdots < t_m = t\}$ of $[s,t]$, and $m(D) = \max_l(t_l - t_{l-1})$. Therefore, $\int_s^t f(Y)\,\mathrm{d}X$ is simply the Lebesgue–Stieltjes integral.

We are going to study the continuity of the map $(X,Y) \to \int f(Y)\,\mathrm{d}X$. Suppose that we are given two Lipschitz paths $X, \hat{X}$ in V, and two Lipschitz paths $Y, \hat{Y}$ in W, and they are controlled by ω_1 and ω_2, respectively. Suppose that, for some $\varepsilon, \delta > 0$,

$$\left|X^1_{s,t} - \hat{X}^1_{s,t}\right| \leqslant \varepsilon\omega_1(s,t), \quad \left|Y^1_{s,t} - \hat{Y}^1_{s,t}\right| \leqslant \delta\omega_2(s,t), \tag{2.10}$$

for all $(s,t) \in \Delta$. Then we have the following theorem.

Theorem 2.2.1 *Under the above conditions on X, $\hat{X}$, Y, $\hat{Y}$, and if f is a Lipschitz one-form with Lipschitz constant M_1, then*

$$\left|\int_s^t f(Y)\,\mathrm{d}X - \int_s^t f(\hat{Y})\,\mathrm{d}\hat{X}\right| \leqslant M_2\left(\varepsilon, \delta, |\xi - \hat{\xi}|\right)\omega_1(s,t), \tag{2.11}$$

for all $(s,t) \in \Delta$, where $\xi = Y_0$, $\hat{\xi} = \hat{Y}_0$, and

$$M_2\left(\varepsilon, \delta, |\xi - \hat{\xi}|\right) = M_1|\xi - \hat{\xi}| + \delta\omega_2(0,T) + \varepsilon M_1\left(1 + |\hat{\xi}| + \omega_2(0,T)\right).$$

Proof For simplicity, we introduce the following notation. If D is a finite partition of the interval $[s,t]$, namely

$$D = \{s = t_0 < t_1 < \cdots < t_m = t\},$$

then we set

$$I(D) = \sum_{l=1}^{m} f(Y_{t_{l-1}})(X^1_{t_{l-1},t_l}) \quad \text{and} \quad \hat{I}(D) = \sum_{l=1}^{m} f(\hat{Y}_{t_{l-1}})(\hat{X}^1_{t_{l-1},t_l}). \tag{2.12}$$

By definition,

$$\int_s^t f(Y)\,\mathrm{d}X = \lim_{m(D)\to 0} I(D) \quad \text{and} \quad \int_s^t f(\hat{Y})\,\mathrm{d}\hat{X} = \lim_{m(D)\to 0} \hat{I}(D). \qquad (2.13)$$

Consider the difference $I(D) - \hat{I}(D)$. It is easily seen that

$$I(D) - \hat{I}(D) = \sum_{l=1}^m \left[f(Y_{t_{l-1}})(X^1_{t_{l-1},t_l}) - f(\hat{Y}_{t_{l-1}})(\hat{X}^1_{t_{l-1},t_l}) \right]$$

$$= \sum_{l=1}^m \left[f(Y_{t_{l-1}}) - f(\hat{Y}_{t_{l-1}}) \right] (X^1_{t_{l-1},t_l})$$

$$+ \sum_{l=1}^m f(\hat{Y}_{t_{l-1}})(X^1_{t_{l-1},t_l} - \hat{X}^1_{t_{l-1},t_l}),$$

and therefore

$$|I(D) - \hat{I}(D)| \leqslant \sum_{l=1}^m |f(Y_{t_{l-1}}) - f(\hat{Y}_{t_{l-1}})||X^1_{t_{l-1},t_l}|$$

$$+ \sum_{l=1}^m |f(\hat{Y}_{t_{l-1}})||X^1_{t_{l-1},t_l} - \hat{X}^1_{t_{l-1},t_l}|$$

$$\leqslant M_1 \sum_{l=1}^m |Y_{t_{l-1}} - \hat{Y}_{t_{l-1}}|\omega_1(t_{l-1},t_l)$$

$$+ \varepsilon M_1 \sum_{l=1}^m \left(1 + |\hat{Y}_{t_{l-1}}| \right) \omega_1(t_{l-1},t_l).$$

However, by assumption,

$$|Y_{t_{l-1}} - \hat{Y}_{t_{l-1}}| \leqslant |Y_0 - \hat{Y}_0| + |Y_{0,t_{l-1}} - \hat{Y}_{0,t_{l-1}}|$$

$$\leqslant |Y_0 - \hat{Y}_0| + \delta\omega_2(0,t_{l-1})$$

$$\leqslant |\xi - \hat{\xi}| + \delta\omega_2(0,T),$$

and

$$|\hat{Y}_{t_{l-1}}| \leqslant |\hat{Y}_0| + |\hat{Y}_{0,t_{l-1}}|$$

$$\leqslant |\hat{Y}_0| + \omega_2(0,t_{l-1})$$

$$\leqslant |\hat{\xi}| + \omega_2(0,T).$$

Inserting these two estimates into the previous inequality, we obtain

$$|I(D) - \hat{I}(D)|$$

$$\leqslant M_1 \left(|\xi - \hat{\xi}| + \delta\omega_2(0,T) \right) \sum_{l=1}^{m} \omega_1(t_{l-1}, t_l)$$

$$+ M_1 \varepsilon \left(1 + |\hat{\xi}| + \omega_2(0,T) \right) \sum_{l=1}^{m} \omega_1(t_{l-1}, t_l)$$

$$\leqslant M_1 \left[|\xi - \hat{\xi}| + \delta\omega_2(0,T) + \varepsilon \left(1 + |\hat{\xi}| + \omega_2(0,T) \right) \right] \omega_1(s,t).$$

By letting $m(D) \to 0$, we obtain the conclusion. ∎

The above theorem shows that the integral $(X, Y) \to \int f(Y) \, dX$ is continuous. In particular,

$$\left| \int_s^t f(Y) \, dX \right| \leqslant M_1 \left(1 + |\xi| + 2\omega_2(0,T) \right) \omega_1(s,t), \qquad (2.14)$$

for all $(s, t) \in \Delta$, so that the continuous path $t \to \int_0^t f(Y) \, dX$ is a Lipschitz path.

2.3 Equations driven by Lipschitz paths

In this section, we consider a differential equation driven by a Lipschitz path.

Let $X : [0, T] \to V$ be a Lipschitz path in a Banach space V controlled by ω_1, and let $f : W \to L(V, W)$ be a Lipschitz function on W valued in the Banach space $L(V, W)$ of all bounded linear operators from V to W. The function f may be regarded as a map which sends elements of V linearly to vector fields on W. The differential equation which we are going to study is

$$dY_t = f(Y_t) \, dX_t, \quad Y_0 = \xi, \qquad (2.15)$$

where $\xi \in W$ is an initial point. By a solution to eqn (2.15) we mean a Lipschitz path $Y : [0, T] \to W$ such that

$$\int_s^t f(Y) \, dX = Y_{s,t}^1, \quad \forall (s,t) \in \Delta \quad \text{and} \quad Y_0 = \xi. \qquad (2.16)$$

2.3.1 *Existence of solutions*

We shall use the Picard iteration to solve eqn (2.15). Set

$$Y(0)_t = \xi,$$

$$Y(n+1)_t = \xi + \int_0^t f(Y(n)) \, dX, \quad n = 1, 2, \dots.$$

Then, by definition,

$$Y(n+1)_{s,t}^1 = \int_s^t f(Y(n)) \, dX, \quad \forall (s,t) \in \Delta. \qquad (2.17)$$

We will prove that the sequence of Lipschitz paths $\{Y(n) : n \in \mathbb{N}\}$ converges (in a relatively strong sense) to a unique Lipschitz path Y, which will be the unique solution of eqn (2.15). This solution will be denoted by $F(X, \xi)$, to indicate the dependence on the path X and on the initial point ξ as well.

The main result we are going to show is that the map (recall that we call it an Itô map) $X \to F(X, \xi)$ is continuous in a sense that will be made precise in what follows.

Let us first show that the sequence $Y(n)$ does converge. We have

$$Y(1)_t = \xi + \int_0^t f(\xi)\, dX = \xi + f(\xi)(X_{0,t}^1)$$

and therefore

$$\begin{aligned} \left| Y(1)_{s,t}^1 \right| &\leqslant \left| f(\xi)(X_{s,t}^1) \right| \\ &\leqslant M_1(1 + |\xi|)\omega_1(s,t)\,, \quad \forall (s,t) \in \Delta\,. \end{aligned} \tag{2.18}$$

Since

$$Y(2)_t = \xi + \int_0^t f(Y(1))\, dX\,,$$

by (2.14) we have

$$\begin{aligned} \left| Y(2)_{s,t}^1 \right| &\leqslant \left| \int_s^t f(Y(1))\, dX \right| \\ &\leqslant M_1 \left[(1 + |\xi|) + 2M_1(1 + |\xi|)\omega_1(0, T_1) \right] \omega_1(s,t)\,, \end{aligned}$$

for all $0 \leqslant s \leqslant t \leqslant T_1$. Therefore, if we choose T_1 small enough such that $M_1\omega_1(0, T_1) \leqslant 1/2$, then

$$\left| Y(2)_{s,t}^1 \right| \leqslant 2M_1(1 + |\xi|)\omega_1(s,t)\,, \quad \forall 0 \leqslant s \leqslant t \leqslant T_1\,. \tag{2.19}$$

Proposition 2.3.1 *Let $T_1 > 0$ be such that $M_1\omega_1(0, T_1) \leqslant 1/2$. Then*

$$\left| Y(n)_{s,t}^1 \right| \leqslant 2M_1(1 + |\xi|)\omega_1(s,t)\,, \quad \forall 0 \leqslant s \leqslant t \leqslant T\,. \tag{2.20}$$

Proof We prove the estimate by induction. It is true for $n = 1$ by eqn (2.18). Suppose now that it is true for n. Then, by (2.14),

$$\begin{aligned} \left| Y(n+1)_{s,t}^1 \right| &\leqslant M_1 \left[1 + |\xi| + 2M_1(1 + |\xi|)\omega_1(0, t) \right] \omega_1(s,t) \\ &\leqslant 2M_1(1 + |\xi|)\omega_1(s,t)\,, \quad \forall 0 \leqslant s \leqslant t \leqslant T_1\,, \end{aligned}$$

and the induction completes the proof. ∎

Next we show that $Y(n)$ converges on $[0, T_1]$. It is clear that

$$\begin{aligned} \left| Y(1)_{s,t}^1 - Y(0)_{s,t}^1 \right| &= \left| \int_s^t f(\xi)\, dX \right| \\ &= \left| f(\xi)(X_{s,t}^1) \right| \\ &\leqslant M_1(1 + |\xi|)\omega_1(s,t)\,, \quad \forall 0 \leqslant s \leqslant t \leqslant T_1\,, \end{aligned}$$

and in general we have the following proposition.

Proposition 2.3.2 *Let $T_1 > 0$ such that $M_1 \omega_1(0, T_1) \leqslant 1/2$. Then*

$$\left| Y(n+1)^1_{s,t} - Y(n)^1_{s,t} \right| \leqslant M_1(1 + |\xi|) \left(M_1 \omega_1(0, T_1) \right)^n \omega_1(s, t), \qquad (2.21)$$

for all $0 \leqslant s \leqslant t \leqslant T_1$.

Proof Use induction. We have already seen that the inequality holds for $n = 0$. Suppose that it is true for $n - 1$. Then, by Theorem 2.2.1 and Proposition 2.3.1, we have

$$
\begin{aligned}
\left| Y(n+1)^1_{s,t} - Y(n)^1_{s,t} \right| \\
= \left| \int_s^t f(Y(n)) \, \mathrm{d}X - \int_s^t f(Y(n-1)) \, \mathrm{d}X \right| \\
\leqslant M_1 \left[\frac{1}{2}(1 + |\xi|)(M_1 \omega_1(0, T_1))^{n-1} 2M_1 \omega_1(0, T_1) \right] \omega_1(s, t) \\
\leqslant M_1(1 + |\xi|) \left(M_1 \omega_1(0, T_1) \right)^n \omega_1(s, t),
\end{aligned}
$$

which implies inequality (2.21). ∎

Corollary 2.3.1 *Choose $T_1 > 0$ such that $M_1 \omega_1(0, T_1) \leqslant 1/2$. Then*

$$\lim_{n \to \infty} Y(n)_t = Y_t, \quad \forall 0 \leqslant t \leqslant T_1,$$

exists. Moreover, the convergence is uniform on the time interval $[0, T_1]$, and Y satisfies the integration equation

$$Y_t = \xi + \int_0^t f(Y) \, \mathrm{d}X, \quad \forall 0 \leqslant t \leqslant T_1,$$

and

$$\left| Y^1_{s,t} \right| \leqslant 2M_1(1 + |\xi|) \omega_1(s, t), \quad \forall 0 \leqslant s \leqslant t \leqslant T_1. \qquad (2.22)$$

2.3.2 *Uniqueness*

Suppose that Z is a Lipschitz path in W controlled by some ω_2, which satisfies the same equation, namely

$$\mathrm{d}Z_t = f(Z_t) \, \mathrm{d}X_t, \quad Z_0 = \xi$$

on some interval $[0, T_2]$. Then we claim that, for some $T_3 > 0$ depending only on $\max \omega_1$, $\max \omega_2$, and M_1 (the Lipschitz constant for f), $\lim_{n \to \infty} Y(n)_t = Z_t$ on $[0, T_3]$, where $Y(n)$ is the Picard iteration sequence defined in the previous subsection.

By Proposition 2.3.1 we know that there is a $0 < T_1 \leqslant T_2$ such that

$$\left|Y(n)^1_{s,t}\right| \leqslant 2M_1(1 + |\xi|)\omega_1(s,t), \quad \forall 0 \leqslant s \leqslant t \leqslant T_1. \tag{2.23}$$

Let

$$\omega_3(s,t) = [2M_1(1 + |\xi|) \vee 1]\,\omega_1(s,t) + \omega_2(s,t). \tag{2.24}$$

Then ω_3 is a control, and $|Y(n)^1_{s,t}| \leqslant \omega_3(s,t)$, $|X^1_{s,t}| \leqslant \omega_3(s,t)$, and $|Z^1_{s,t}| \leqslant \omega_3(s,t)$, for all $0 \leqslant s \leqslant t \leqslant T_1$. By definition,

$$\left|Y(1)^1_{s,t} - Z^1_{s,t}\right| \leqslant \omega_3(s,t), \quad \forall 0 \leqslant s \leqslant t \leqslant T_1, \tag{2.25}$$

and therefore, by Theorem 2.2.1,

$$\left|Y(1)^1_{s,t} - Z^1_{s,t}\right| = \left|\int_s^t f(Y(0))\,\mathrm{d}X - \int_s^t f(Z)\,\mathrm{d}X\right|$$
$$\leqslant M_1\omega_3(0,t)\omega_1(s,t),$$

for all $0 \leqslant s \leqslant t \leqslant T_1$. In general, one can show by induction that

$$\left|Y(n)^1_{s,t} - Z^1_{s,t}\right| \leqslant (M_1\omega_3(0,t))^n\,\omega_1(s,t). \tag{2.26}$$

Therefore, any $T_3 > 0$ such that $M_1\omega_3(0,T_3) < 1$ will do. This completes the proof of the uniqueness.

2.3.3 *Existence of solutions revisited*

In this subsection we show the existence of the solution to eqn (2.15) on the whole time range $[0,T]$.

To this end, we define a sequence

$$0 < T_1 < T_2 < \cdots < T_k \leqslant T$$

such that $M_1\omega_1(T_{l-1}, T_l) = 1/2$, for $l = 1, \ldots, k$, and $M_1\omega_1(T_k, T) < 1/2$. Then k is uniquely determined by M_1 and ω_1, and $k \leqslant \omega_1(0,T)/2M_1$. Define $X(l)_t = X_{t+T_{l-1}}$, for $t \in [0, T_l - T_{l-1}]$, and

$$\omega_1^l(s,t) = \omega_1(s + T_{l-1}, t + T_{l-1}).$$

Then

$$\left|X(l)^1_{s,t}\right| \leqslant \omega_1^l(s,t), \quad \forall 0 \leqslant s \leqslant t \leqslant T_l - T_{l-1},$$

and therefore we can solve the differential equation recursively as follows:

$$\mathrm{d}Y_t^l = f(Y_t^l)\,\mathrm{d}X(l)_t, \quad Y_0^l = Y_{T_{l-1}}.$$

Let $Y_t = Y_{t-T_{l-1}}^l$, for $t \in [T_{l-1}, T_l]$.

Theorem 2.3.1 *Under the above notation, Y defined on the whole interval $[0, T]$ is a Lipschitz path and is the unique solution to the differential equation*

$$\mathrm{d}Y_t = f(Y_t)\,\mathrm{d}X_t, \quad Y_0 = \xi.$$

Moreover, if the Lipschitz path X is controlled by ω_1, namely $|X^1_{s,t}| \leqslant \omega_1(s,t)$, then

$$\left|Y^1_{s,t}\right| \leqslant 2M_1(1+|\xi|)\left[1+2M_1\omega_1(0,T)\right]^k \omega_1(s,t), \tag{2.27}$$

for all $(s,t) \in \Delta$.

Proof It is clear that, by definition, Y is a solution of eqn (2.15). Uniqueness follows immediately from the above subsection. To prove (2.27), we prove the following claim. If $0 \leqslant s \leqslant t \leqslant T_l$, then

$$\left|Y^1_{s,t}\right| \leqslant 2M_1(1+|\xi|)[1+2M_1\omega_1(0,T)]^{l-1}\omega_1(s,t). \tag{2.28}$$

Certainly (2.28) is true for $l = 1$. We now use induction. Suppose that it is true for l. Then especially

$$|Y_{T_l}| \leqslant |\xi| + 2M_1(1+|\xi|)[1+2M_1\omega_1(0,T)]^{l-1}\omega_1(0,T).$$

Hence, if $T_l \leqslant s < t \leqslant T_{l+1}$, we have

$$\left|Y(l+1)^1_{s,t}\right| \leqslant 2M_1\left(1+|Y_{T_l}|\right)\omega_1(s,t),$$

so that (2.28) holds. ∎

2.3.4 *Continuity of the Itô map*

We next explore the continuity of the map $(X, \xi) \to F_t(X, \xi)$ defined by solving the differential equation

$$\mathrm{d}Y_t = f(Y_t)\,\mathrm{d}X_t, \quad Y_0 = \xi. \tag{2.29}$$

Suppose that X and $\hat{X}$ are two Lipschitz paths in V on $[0, T]$ which are both controlled by ω_1, namely

$$\left|X^1_{s,t}\right|, \left|\hat{X}^1_{s,t}\right| \leqslant \omega_1(s,t), \quad \forall (s,t) \in \Delta.$$

Suppose that

$$\left|X^1_{s,t} - \hat{X}^1_{s,t}\right| \leqslant \varepsilon\omega_1(s,t), \quad \forall (s,t) \in \Delta. \tag{2.30}$$

Let $\xi, \hat{\xi} \in W$, and define

$$Y(n)_t = \xi + \int_0^t f(Y(n-1))\,\mathrm{d}X, \quad Y(0) = \xi$$

and

$$\hat{Y}(n)_t = \hat{\xi} + \int_0^t f(\hat{Y}(n-1))\,\mathrm{d}\hat{X}, \quad \hat{Y}(0) = \hat{\xi}.$$

Then, as we have proved,

$$F_t(X, \xi) = \lim_{n \to \infty} Y(n)_t \quad \text{and} \quad F_t(\hat{X}, \hat{\xi}) = \lim_{n \to \infty} \hat{Y}(n), \quad \text{on} \quad [0, T_1].$$

Proposition 2.3.3 *Let $T_1 > 0$ such that $4M_1\omega_1(0, T_1) \leqslant 1$. Then we have*

$$\left|Y(n)^1_{s,t} - \hat{Y}(n)^1_{s,t}\right| \leqslant 2M_1\big[|\xi - \hat{\xi}| + \varepsilon\big(1 + |\xi| \vee |\hat{\xi}|\big)\big]\omega_1(s,t),\qquad(2.31)$$

for all $0 \leqslant s \leqslant t \leqslant T_1$.

Proof It is easily seen that

$$
\begin{aligned}
\left|Y(1)^1_{s,t} - \hat{Y}(1)^1_{s,t}\right| &\leqslant \left|\int_s^t f(Y(0))\,\mathrm{d}X - \int_s^t f(\hat{Y}(0))\,\mathrm{d}\hat{X}\right| \\
&\leqslant \left|f(\xi)(X^1_{s,t}) - f(\hat{\xi})(\hat{X}^1_{s,t})\right| \\
&\leqslant \left|f(\xi) - f(\hat{\xi})\right|\left|X^1_{s,t}\right| + \left|f(\hat{\xi})\right|\left|X^1_{s,t} - \hat{X}^1_{s,t}\right| \\
&\leqslant M_1\big[|\xi - \hat{\xi}| + \varepsilon(1 + |\hat{\xi}|)\big]\omega_1(s,t).
\end{aligned}
$$

Therefore (2.31) holds for $n = 1$. Notice that

$$\left|Y(n)^1_{s,t}\right| \leqslant 2M_1\left(1 + |\xi|\right)\omega_1(s,t),\quad \forall\, 0 \leqslant s \leqslant t \leqslant T_1$$

and

$$\left|\hat{Y}(n)^1_{s,t}\right| \leqslant 2M_1\left(1 + |\hat{\xi}|\right)\omega_1(s,t),\quad \forall\, 0 \leqslant s \leqslant t \leqslant T_1.$$

Hence, if (2.31) is true for n, then by Theorem 2.2.1 we have

$$
\begin{aligned}
\big|Y(n&+1)^1_{s,t} - \hat{Y}(n+1)^1_{s,t}\big| \\
&\leqslant M_1\Big\{|\xi - \hat{\xi}| + 2M_1\big[|\xi - \hat{\xi}| + \varepsilon\big(1 + |\xi| \vee |\hat{\xi}|\big)\big]\omega_1(0, T_1) \\
&\qquad + \varepsilon\big[1 + |\hat{\xi}| + 2M_1\big(1 + |\xi| \vee |\hat{\xi}|\big)\big]\omega_1(0, T_1)\Big\}\omega_1(s,t) \\
&\leqslant M_1\Big\{|\xi - \hat{\xi}|\big[1 + 2M_1\omega_1(0, T_1)\big] \\
&\qquad + \varepsilon\big(1 + |\xi| \vee |\hat{\xi}|\big)\big[1 + 4M_1\omega_1(0, T_1)\big]\Big\}\omega_1(s,t) \\
&\leqslant 2M_1\Big\{|\xi - \hat{\xi}| + \varepsilon\big(1 + |\xi| \vee |\hat{\xi}|\big)\Big\}\omega_1(s,t).
\end{aligned}
$$

The proof may be completed by induction. ∎

Theorem 2.3.2 *Let $Y = F(X, \xi)$ and $\hat{Y} = F(\hat{X}, \hat{\xi})$. Set*

$$M_3 = 2M_1\big(1 + |\xi| \vee |\hat{\xi}|\big)\big[1 + 2M_1\omega_1(0, T)\big]^k\omega_1(0, T) + |\xi|,$$

where k is defined by choosing a sequence

$$0 < T_1 < T_2 < \cdots < T_k \leqslant T$$

such that $4M_1\omega_1(T_{l-1}, T_l) = 1$, for $l = 1, \ldots, k$, and $4M_1\omega_1(T_k, T) < 1$. Then

$$\left|Y^1_{s,t} - \hat{Y}^1_{s,t}\right| \leqslant 2M_1\big[1 + 2M_1\omega_1(0, T)\big]^k\big(|\xi - \hat{\xi}| + \varepsilon M_3\big)\omega_1(s,t),\qquad(2.32)$$

for all $0 \leqslant s \leqslant t \leqslant T$. Moreover, $k \leqslant \omega_1(0, T)/4M_1$.

Proof The super-additive property of ω ensures that $k \leqslant \omega_1(0, T)/4M_1$. For $0 \leqslant l \leqslant k$, let Z and $\hat{Z}$ be the solutions of the equations

$$dZ_t = f(Z_t)\, dX(l)_t, \quad Z_0 = Y_{T_{l-1}}$$

and

$$d\hat{Z}_t = f(\hat{Z}_t)\, d\hat{X}(l)_t, \quad \hat{Z}_0 = \hat{Y}_{T_{l-1}},$$

respectively, where $X(l)_t = X_{T_{l-1}+t}$ and $\hat{X}(l)_t = \hat{X}_{T_{l-1}+t}$. Let

$$\omega_1^l(s, t) = \omega_1(T_{l-1} + s, T_{l-1} + t).$$

Then

$$\left|Z_{s,t}^1 - \hat{Z}_{s,t}^1\right| \leqslant 2M_1 \left(\left|Y_{T_{l-1}} - \hat{Y}_{T_{l-1}}\right| + M_3\varepsilon\right)\omega_1^l(s, t),$$

for all $0 \leqslant s \leqslant t \leqslant T_l - T_{l-1}$. We claim that

$$\left|Z_{s,t}^1 - \hat{Z}_{s,t}^1\right| \leqslant 2M_1 \left[1 + 2M_1\omega_1(0, T)\right]^l \left(|\xi - \hat{\xi}| + \varepsilon M_3\right)\omega_1^l(s, t).$$

By Proposition 2.3.3, the above estimate clearly holds for $l = 1$. Suppose that it is true for $l - 1$, and therefore

$$\left|Y_{T_{l-1}} - \hat{Y}_{T_{l-1}}\right| \leqslant |\xi - \hat{\xi}| + 2M_1 \left[1 + 2M_1\omega_1(0, T)\right]^{l-1} \left(|\xi - \hat{\xi}| + M_3\varepsilon\right)\omega_1(0, T),$$

so that

$$\left|Z_{s,t}^1 - \hat{Z}_{s,t}^1\right|$$
$$\leqslant 2M_1\Big\{|\xi - \hat{\xi}| + M_3\varepsilon$$
$$\qquad + 2M_1\left[1 + 2M_1\omega_1(0, T)\right]^{l-1}\left(|\xi - \hat{\xi}| + \varepsilon M_3\right)\omega_1(0, T)\Big\}\omega_1(s, t)$$
$$= 2M_1\left[1 + 2M_1\omega_1(0, T)\right]^l\left(|\xi - \hat{\xi}| + M_3\varepsilon\right)\omega_1^l(s, t),$$

for any $0 \leqslant s \leqslant t \leqslant T_l - T_{l-1}$. Thus we have proved the theorem. ∎

2.4 Comments and notes on Chapter 2

All of the results in this chapter are more or less well known in the theory of ordinary differential equations, although the form in which we present them is perhaps new. Our approach here closely follows Lyons (1994, 1998).

There are several ways to look at a system of differential equations at different levels. In ODE theory, one of the main objects is the following system:

$$dY_t^i = f^i(Y_t^1, \ldots, Y_t^d)\, dt, \quad i = 1, \ldots, d, \tag{2.33}$$

with the initial condition that $Y_0^i = y^i$, $i = 1, \ldots, d$. Such a system was studied as an initial problem. Another viewpoint for looking at this system is the following. Let $Y_t = (Y_t^1, \ldots, Y_t^d)$ and define a vector field

$$A(x) = \sum_{i=1}^{d} f^i(x^1, \ldots, x^d) \frac{\partial}{\partial x^i} \,.$$

Then the above system can be rewritten as

$$\frac{\mathrm{d}Y_t}{\mathrm{d}t} = A(Y_t) \,, \quad Y_0 = y \,,$$

and the solution Y is called the integral curve starting at y of the vector field A. The advantage with this view is that eqn (2.33) is coordinate free—no coordinate chart is needed in proposing such a differential equation, and it makes sense perfectly well on smooth manifolds. A deep point of view, although it does not reduce the difficulty in solving such a differential equation, but it is helpful in the study of such an equation, we describe as follows. Denote the integral curve with initial point y by $Y(t, y)$. Then the map $\zeta_t : x \to Y(t, x)$ is a diffeomorphism of the manifold M which supports the vector field A. In fact (ζ_t) is a group, that is,

$$\zeta_{t+s} = \zeta_t \circ \zeta_s \,, \quad \zeta_0 = \mathrm{id} \,, \quad \forall s, t \in \mathbb{R} \,.$$

This is why we may regard (2.33) as the flow equation

$$\frac{\mathrm{d}\zeta_t}{\mathrm{d}t} = A(\zeta_t) \,, \quad \zeta_0 = \mathrm{id} \,.$$

The solution to the above flow equation is a one-parameter sub-group of $\mathrm{Diff}(M)$, the 'Lie' group of all diffeomorphisms of M.

A nonlinear dynamical system can commonly be described by a differential equation evolving in t, which takes the following form in many cases:

$$\mathrm{d}Y_t^i = \sum_{i=1}^{d} f^i(Y_t, t) \, \mathrm{d}X_t^i \,. \tag{2.34}$$

In the classical theory, such an equation was studied as an ODE by writing $\mathrm{d}X_t^i$ as $\dot{X}_t^i \, \mathrm{d}t$. The pathwise point of view—which says that eqn (2.34) is the dynamic governing the input X and the output Y of a dynamical system—seems to have first appeared in engineering. In the 1940s Itô initiated and developed his stochastic integration theory for Brownian motion paths, which is one of the main achievements of the twentieth century in mathematics. Itô has shown how to solve an equation like (2.34) even when X contains white noise. Moreover, it was shown that the solution to (2.34) is a measurable function of the path X. Partial continuity results for solutions to stochastic differential equations (SDEs) were later provided by various authors, notably in the Wong–Zakai-type approximation theorem, see Wong and Zakai (1965 a, 1965 b). For a description of Itô's theory, see Ikeda and Watanabe (1981) and Malliavin (1993), for example.

Section 2.1. For a pathwise study of one-dimensional (stochastic) differential equations, see McShane (1974). For Doss–Sussmann's formula, see Doss (1977) and Sussmann (1978). For multi-dimensional systems which we may solve via integral curves of given vector fields under a certain commutative condition, see Yamato (1979) and Ikeda and Watanabe (1981). It is known that the area enclosed by a planar curve is not continuous in curve under the topology of uniform convergence.

Section 2.2. The integration theory for paths with bounded variations is classical, although the result of the continuity in variation of solutions to ODEs seems new. However, it turns out to be a special case in Lyons (1998).

3

ROUGH PATHS

The main notion of this book is the concept of rough paths, which will be introduced in this chapter, and the main machinery of the analysis for a class of non-smooth continuous paths will be established as well. The chapter is organized as follows. The first section contains the basic definitions about rough paths. Roughly speaking, a rough path is such a continuous path for which we have an integration theory, and therefore from which a sequence of iterated path integrals may be constructed. Also in the first section, the continuity theorem for extensions of rough paths to higher-order iterated path integrals is proved. In Section 3.2, we introduce the concept of almost rough paths, and give a method to construct rough paths from almost rough ones. The continuity result about this construction is also proved in this section. In Section 3.3, we consider the p-variation topology on the spaces of rough paths, especially the p-variation distance. We also discuss several basic operations on rough paths, and Young's integration theory.

3.1 Basic definitions and properties

We begin with several notations which will be used intensively in this book. The setting is the following. We are given a Banach space V with norm $|\cdot|$ together with a sequence of tensor norms $|\cdot|_k$ on the algebraic tensor products $V^{\otimes_a k} \equiv V \otimes_a \cdots \otimes_a V$ (of k copies of V) satisfying the following compatibility condition:[9]

$$|\xi \otimes \eta|_{k+l} \leqslant |\xi|_k \, |\eta|_l \,, \quad \forall \xi \in V^{\otimes_a k}, \quad \forall \eta \in V^{\otimes_a l} \,. \tag{3.1}$$

When $k = 1$, $|\cdot|_1$ coincides with $|\cdot|$. The completion of the algebraic tensor product $V^{\otimes_a k}$ under the norm $|\cdot|_k$ is denoted by $(V^{\otimes k}, |\cdot|_k)$, and the lower index will be suppressed from the norms if no confusion may arise. By the compatibility condition, the algebraic tensor product $V^{\otimes k} \otimes_a V^{\otimes l}$ of two Banach spaces $V^{\otimes k}$ and $V^{\otimes l}$ is embedded into $V^{\otimes(k+l)}$ in an obvious sense, and therefore it will be regarded as a subspace of $V^{\otimes(k+l)}$. Moreover, the compatibility condition is preserved, that is

$$|\xi \otimes \eta| \leqslant |\xi| \, |\eta| \,, \quad \forall \xi \in V^{\otimes k}, \quad \forall \eta \in V^{\otimes l} \,. \tag{3.2}$$

---

[9]If σ is a permutation of $\{1, \ldots, k\}$ then σ induces a linear transformation of $V^{\otimes_a k}$. We will sometimes require this transformation to be bounded. In most concrete settings it is an isometry.

For each $n \in \mathbb{N}$, we then build the following (truncated) tensor algebra $T^{(n)}(V)$:

$$T^{(n)}(V) = \sum_{k=0}^{n} \oplus V^{\otimes k}, \quad V^{\otimes 0} = \mathbb{R}. \tag{3.3}$$

Its multiplication (also called tensor product) is the usual multiplication as polynomials, except that the higher-order (than degree n) terms are omitted. In other words, if $\xi = (\xi^0, \xi^1, \ldots, \xi^n)$, $\eta = (\eta^0, \eta^1, \ldots, \eta^n)$ are two vectors in $T^{(n)}(V)$, then $\zeta = \xi \otimes \eta \in T^{(n)}(V)$, where its kth component is

$$\zeta^k = \sum_{j=0}^{k} \xi^j \otimes \eta^{k-j}, \quad k = 0, 1, \ldots, n. \tag{3.4}$$

The norm $|\cdot|$ on $T^{(n)}(V)$ is defined by

$$|\xi| = \sum_{i=0}^{n} |\xi^i|, \quad \text{if} \quad \xi = (\xi^0, \ldots, \xi^n),$$

though different, but equivalent, norms may be used from time to time in our next development. The pair $(T^{(n)}(V), |\cdot|)$ is a tensor algebra with identity element $e = (1, 0, \ldots, 0)$, and, for $\xi, \eta \in T^{(n)}(V)$,

$$\begin{aligned}
|\xi \otimes \eta| &= \sum_{k=0}^{n} \left| \sum_{j=0}^{k} \xi^j \otimes \eta^{k-j} \right| \\
&\leqslant \sum_{k=0}^{n} \sum_{i+j=k} |\xi^i \otimes \eta^j| \\
&\leqslant \sum_{k=0}^{n} \sum_{i+j=k} |\xi^i| |\eta^j| \\
&= |\xi| |\eta|.
\end{aligned}$$

The running time interval I of paths considered in this book will be a finite interval $[0, T]$, although any bounded sub-interval of $\mathbb{R}$ is allowed. In the most part, $[0, T]$ will be $[0, 1]$. We use Δ or Δ_T to denote the simplex $\{(s, t) : 0 \leqslant s \leqslant t \leqslant T\}$. Recall that a control ω is then a continuous, super-additive function on Δ with values in $[0, +\infty)$ such that $\omega(t, t) = 0$. Therefore,

$$\omega(s, t) + \omega(t, u) \leqslant \omega(s, u), \quad \forall (s, t), (t, u) \in \Delta.$$

In this chapter, $p \geqslant 1$ is a constant.

Definition 3.1.1 *A continuous map X from the simplex $\triangle$ into a truncated tensor algebra $T^{(n)}(V)$, and written as*

$$X_{s,t} = (X^0_{s,t}, X^1_{s,t}, \dots, X^n_{s,t}), \quad \text{with} \quad X^k_{s,t} \in V^{\otimes k}, \quad \text{for any} \quad (s,t) \in \triangle,$$

is called a multiplicative functional of degree n ($n \in \mathbb{N}$, $n \geqslant 1$) if $X^0_{s,t} \equiv 1$ (for all $(s,t) \in \triangle$) and

$$X_{s,t} \otimes X_{t,u} = X_{s,u}, \quad \forall (s,t), (t,u) \in \triangle, \tag{3.5}$$

where the tensor product $\otimes$ is taken in $T^{(n)}(V)$.

Equality (3.5) is called the Chen identity, although it appears long before Chen's fundamental works in which a connection is made from iterated path integrals along smooth paths to a class of differential forms on a space of loops on a manifold.

Example 3.1.1 Let $X : [0,T] \to V$ be a continuous path. Then its increment process $X : \triangle \to T^{(1)}(V)$ (we use the same notation), defined by $X_{s,t} = (1, X^1_{s,t})$ and $X^1_{s,t} = X_t - X_s$, is a multiplicative functional of degree 1. In this case, Chen's identity is equivalent to the additive property of increments over different intervals. If, in addition, X is a Lipschitz path, then we may build a sequence of iterated path integrals

$$X^k_{s,t} = \int_{s < t_1 < \cdots < t_k < t} \mathrm{d}X_{t_1} \otimes \cdots \otimes \mathrm{d}X_{t_k}, \tag{3.6}$$

as defined in the previous chapter. That is, $X^k_{s,t}$ ($k \geqslant 2$) is given inductively by

$$X^k_{s,t} = \lim_{m(D) \to 0} \sum_l \sum_{i=1}^{k-1} X^i_{s,t_{l-1}} \otimes X^{k-i}_{t_{l-1}, t_l}.$$

Then, for any $n \in \mathbb{N}$,

$$X_{s,t} = (1, X^1_{s,t}, \dots, X^n_{s,t}), \quad \forall (s,t) \in \triangle \tag{3.7}$$

is a multiplicative functional of degree n. In this case, Chen's identity is equivalent to the additive property of iterated path integrals over different domains.

Remark 3.1.1 We have seen from the above example that Chen's identity represents a basic requirement on any 'continuous path' in $T^{(n)}(V)$ which has an integration theory. Namely, it is equivalent to the additive property of integrals over different intervals.

Remark 3.1.2 Let $D = \{s = t_0 \leqslant t_1 \leqslant \cdots \leqslant t_r = t\}$ be a finite partition of $[s,t]$, and let X be a multiplicative functional in $T^{(n)}(V)$. Then, by the repeated use of Chen's identity, we have

$$X_{s,t} = X_{s,t_1} \otimes X_{t_1,t_2} \otimes \cdots \otimes X_{t_{r-1},t}\,, \tag{3.8}$$

and therefore, for $1 \leqslant k \leqslant n$,

$$X_{s,t}^k = \sum_{l=1}^{r} X_{t_{l-1},t_l}^k + \sum_{i=1}^{k-1} \sum_{l=1}^{r} X_{s,t_{l-1}}^i \otimes X_{t_{l-1},t}^{k-i}\,. \tag{3.9}$$

Chen's identity is purely algebraic. We next turn to an analytic condition required on a multiplicative functional, which on one hand allows us to develop an integration theory and on the other hand relates the 'roughness' of a continuous path.

Definition 3.1.2 *Let $p \geqslant 1$ be a constant. We say that a map $X : \Delta \to T^{(n)}(V)$ possesses finite p-variation if*

$$|X_{s,t}^i| \leqslant \omega(s,t)^{i/p}\,, \quad \forall i = 1,\ldots,n\,, \quad \forall(s,t) \in \Delta\,, \tag{3.10}$$

for some control ω.

Thus, for a Lipschitz path X in V, the associated multiplicative functional of any degree, built via its iterated path integrals, is of finite one-variation.

Before we can prove several basic results, we need a generalized binomial (in)equality.

3.1.1 *The binomial inequality*

The Gamma function is defined by the formula

$$\Gamma(z) = \int_0^\infty u^{z-1} e^{-u}\,\mathrm{d}u\,, \quad \forall \operatorname{Re} z > 0\,,$$

which is extended to an entire function with simple poles $0, -1, -2, \ldots$, and

$$\Gamma(z) = \frac{e^{-\gamma z}}{z} \prod_{n=1}^{\infty} \left(1 + \frac{z}{n}\right)^{-1} e^{z/n}\,,$$

where γ is the Euler constant, namely

$$\gamma = \lim_{n \to \infty} \left(\sum_{k=1}^{n} \frac{1}{k} - \ln n\right).$$

Therefore, $\Gamma(z+1) = z\Gamma(z)$, for all complex z such that $\operatorname{Re} z > 0$. Thus, by an obvious reason, we denote $\Gamma(\lambda + 1)$ by $\lambda!$. By applying a change of variables in the integral formula for $\Gamma(z)$, it is easily seen that

$$\int_0^1 u^{\lambda-1}(1-u)^{\mu-1}\,\mathrm{d}u = \frac{\Gamma(\lambda)\Gamma(\mu)}{\Gamma(\lambda+\mu)}\,, \quad \forall\lambda,\mu \in (0,+\infty)\,, \qquad (3.11)$$

and we may therefore rewrite the binomial formula as

$$\sum_{i=0}^n \frac{n!}{i!(n-i)!}a^i b^{n-i} = (a+b)^n\,,$$

for all $a,b \geqslant 0$ and the natural number $n \in \mathbb{N}$. Let $x = a/(a+b)$. Then the binomial law becomes

$$\sum_{i=0}^n \frac{n!}{i!(n-i)!}x^i(1-x)^{n-i} = 1\,,$$

and, by (3.11), it can be written as

$$\sum_{i=0}^n \frac{x^i(1-x)^{n-i}}{\int_0^1 u^i(1-u)^{n-i}\,\mathrm{d}u} = n+1\,. \qquad (3.12)$$

For some applications presented below, we need to bound (from above) the following expression:

$$\sum_{i=0}^n \frac{x^{i/p}(1-x)^{(n-i)/p}}{\int_0^1 u^{i/p}(1-u)^{(n-i)/p}\,\mathrm{d}u}\,, \quad x \in [0,1]\,, \quad p \geqslant 1\,. \qquad (3.13)$$

Let $A_\theta(t,x) = x^{\theta/t}(1-x)^{(1-\theta)/t}$, for $\theta \in [0,1]$ and $t > 0$. Let $\theta_i = i/n$, for $i = 0,\dots,n$, and $A_i = A_{\theta_i}$. Then (3.13) can be rewritten as

$$\sum_{i=0}^n \frac{A_i(p/n,x)}{\int_0^1 A_i(p/n,u)\,\mathrm{d}u}\,,$$

and (3.12) can be stated as

$$\sum_{i=0}^n \frac{A_i(1/n,x)}{\int_0^1 A_i(1/n,u)\,\mathrm{d}u} = n+1\,, \quad \text{for all} \quad x \in [0,1]\,. \qquad (3.14)$$

Therefore, we consider the following function:

$$\sum_{i=0}^n \frac{A_i(t,x)}{\int_0^1 A_i(t,u)\,\mathrm{d}u}\,, \quad t \geqslant \frac{1}{n}\,, \quad x \in [0,1]\,,$$

and we will use the maximum principle to bound this function. We mention that (3.14) means that on the boundary $t = 1/n$ the above function equals the constant $n+1$.

Let $\varphi(t)$ be a positive function to be chosen later, and let

$$F_\theta(t,x) = \varphi(t) \frac{A_\theta(t,x)}{\int_0^1 A_\theta(t,u)\,du} \, .$$

Taking logarithms, we obtain

$$\ln F_\theta(t,x) = \ln \varphi(t) + \frac{\theta}{t}\ln x + \frac{1-\theta}{t}\ln(1-x) - \ln \int_0^1 A_\theta(t,u)\,du \, ,$$

and then, taking the derivative in x, we obtain

$$\partial_x \ln F_\theta = \frac{1}{t}\frac{(\theta - x)}{x(1-x)} \, .$$

We thus establish that

$$\partial_x \left(x(1-x)\partial_x \ln F_\theta \right) = -\frac{1}{t} \, . \tag{3.15}$$

Similarly, taking the derivative in time t, we have

$$\partial_t \ln F_\theta = \frac{\varphi'}{\varphi} - \frac{1}{t^2}\left[\theta \ln x + (1-\theta)\ln(1-x) \right] - \frac{\int_0^1 \partial_t A_\theta(t,u)\,du}{\int_0^1 A_\theta(t,u)\,du}$$

$$= \frac{\varphi'}{\varphi} + \frac{1}{t^2}\frac{\int_0^1 A_\theta(t,u)\left\{ \theta \ln(u/x) + (1-\theta)\ln[(1-u)/(1-x)] \right\}du}{\int_0^1 A_\theta(t,u)\,du} \, .$$

$$\tag{3.16}$$

Define the following elliptic operator of second order:

$$L = \partial_x \left(x(1-x)\partial_x \right) .$$

Then, using (3.15) and (3.16), and the fact that

$$L \ln F = \frac{1}{F}LF - x(1-x)\left|\partial_x \ln F\right|^2 \, ,$$

we have

$$\frac{1}{F_\theta}(L - \partial_t)F_\theta = (L - \partial_t)\ln F_\theta + x(1-x)\left|\partial_x \ln F_\theta\right|^2$$

$$= -\frac{1}{t} - \frac{\varphi'}{\varphi} + \frac{1}{t^2}\frac{(\theta - x)^2}{x(1-x)}$$

$$- \frac{1}{t^2}\frac{\int_0^1 A_\theta(t,u)\left\{ \theta \ln(u/x) + (1-\theta)\ln[(1-u)/(1-x)] \right\}du}{\int_0^1 A_\theta(t,u)\,du} \, .$$

Set

$$g_\theta(u) \equiv \theta \ln \frac{u}{\theta} + (1-\theta) \ln \frac{1-u}{1-\theta}, \quad \forall u, \theta \in [0,1].$$

Then $g_\theta(\theta) = 0$ and $g_\theta'(u) = (\theta - u)[u(1-u)]$, so that $g_\theta(u) \leqslant 0$. Using g_θ, we may rewrite the previous equation for F_θ as

$$\frac{1}{F_\theta}(L - \partial_t)F_\theta = -\frac{1}{t} - \frac{\varphi'(t)}{\varphi(t)} - \frac{1}{t^2}\frac{\int_0^1 Ag_\theta \, du}{\int_0^t A \, du}$$

$$- \frac{1}{t^2}\left(\theta \ln \frac{\theta}{x} + (1-\theta)\ln \frac{1-\theta}{1-x}\right)$$

$$+ \frac{1}{t^2}\left(\theta^2 \frac{1-x}{x} + (1-\theta)^2 \frac{x}{1-x} - 2\theta(1-\theta)\right).$$

However,

$$f_\theta(x) \equiv \theta \ln \frac{\theta}{x} + (1-\theta)\ln \frac{1-\theta}{1-x} - \theta^2 \frac{1-x}{x} - (1-\theta)^2 \frac{x}{1-x} + 2\theta(1-\theta)$$

$$\geqslant 0,$$

for any $\theta, x \in (0,1)$. Indeed, $f_\theta(\theta) = 0$ and

$$f_\theta'(x) = (\theta - x)\left(\frac{\theta}{x^2} + \frac{1-\theta}{(1-x)^2}\right),$$

so that $f_\theta(x) \geqslant f_\theta(\theta) = 0$. Therefore,

$$\frac{1}{F_\theta}(L - \partial_t)F_\theta \geqslant -\frac{1}{t} - \frac{\varphi'(t)}{\varphi(t)} - \frac{1}{t^2}\frac{\int_0^1 Ag_\theta \, du}{\int_0^1 A \, du}$$

$$\geqslant -\frac{1}{t} - \frac{\varphi'(t)}{\varphi(t)}.$$

Choose $\varphi(t) = t^{-1}$. We have thus established the following lemma.

Lemma 3.1.1 *For $\theta \in [0,1]$, let*

$$F_\theta(t,x) = \frac{1}{t}\frac{A_\theta(t,x)}{\int_0^1 A_\theta(t,u)\, du}, \quad t \geqslant 1/n \quad and \quad x \in [0,1].$$

Then $(L - \partial_t)F_\theta \geqslant 0$, where $L = \partial_x(x(1-x)\partial_x)$.

Let $F = \sum_{i=0}^n F_{i/n}(t,x)$. Lemma 3.1.1 implies that

$$(L - \partial_t)F \geqslant 0.$$

Applying the maximum principle to F, and using the fact that $F(1/n, x) = n(n+1)$, we may conclude that $F(p/n, x) \leqslant n(n+1)$, for all $p \geqslant 1$. Therefore,

$$\frac{n}{p} \sum_{i=0}^{n} \frac{x^{i/p}(1-x)^{(n-i)/p}}{\int_0^1 x^{i/p}(1-x)^{(n-i)/p}\,\mathrm{d}x} \leqslant n(n+1)\,.$$

By (3.11), the above inequality can be written as

$$\sum_{i=0}^{n} \frac{(n/p)!\,x^{i/p}(1-x)^{(n-i)/p}}{(i/p)!\,((n-i)/p)!} \leqslant \frac{n+1}{n+p}p^2$$

$$\leqslant p^2\,.$$

Let $a, b \geqslant 0$ and set $x = a/(a+b)$ in the above inequality. Then we have established the following theorem.

Theorem 3.1.1 *For any $p \in [1, +\infty)$, $n \in \mathbb{N}$, and any $a, b \geqslant 0$, we have*

$$\sum_{i=0}^{n} \frac{a^{i/p}b^{(n-i)/p}}{(i/p)!\,((n-i)/p)!} \leqslant p^2 \frac{(a+b)^{n/p}}{(n/p)!}\,. \tag{3.17}$$

3.1.2 *Several basic results*

Theorem 3.1.2 *Let $p \geqslant 1$, and let $X : \Delta \to T^{(n)}(V)$ be a multiplicative functional with finite p-variation so that*

$$|X_{s,t}^i| \leqslant \omega(s,t)^{i/p}\,, \quad \forall i = 1, \ldots, n \quad and \quad \forall (s,t) \in \Delta\,,$$

for some control ω. If $n \geqslant [p]$, then we may uniquely extend X to be a multiplicative functional in $T^{(\infty)}(V)$ which possesses finite p-variation. More precisely, for any $m \geqslant [p] + 1$, there is a unique continuous function $X^m : \Delta \to V^{\otimes m}$ such that

$$X = (1, X^1, \ldots, X^{[p]}, \ldots, X^m, \ldots)$$

is a multiplicative functional in $T^{(\infty)}(V)$ with finite p-variation. Moreover, if ω is a control such that

$$|X_{s,t}^i| \leqslant \frac{\omega(s,t)^{i/p}}{\beta\,(i/p)!}\,, \quad \forall i = 1, \ldots, [p] \quad and \quad \forall (s,t) \in \Delta\,, \tag{3.18}$$

where β is a constant such that

$$\beta \geqslant p^2 \left(1 + \sum_{r=3}^{\infty} \left(\frac{2}{r-2}\right)^{([p]+1)/p}\right)\,, \tag{3.19}$$

then (3.18) remains true for all $i > [p]$.

The method of proving this theorem is to construct higher (than $[p]$) order 'iterated path integrals' by using limits of Riemannian sums. We describe the idea briefly before the proof. Suppose that we have constructed an extension of X to $T^{(k)}(V)$, where $k \geqslant [p]$, denoted by X. Then, in order to construct

$X^{k+1} : \Delta \to V^{\otimes(k+1)}$, we regard X as a map from Δ into $T^{(k+1)}(V)$ (and denote it by $\hat{X}$) by inserting a final component of zero, namely

$$(s,t) \to (1, X^1_{s,t}, \dots, X^k_{s,t}, 0) \, .$$

Of course, $\hat{X}$ is no longer a multiplicative functional in $T^{(k+1)}(V)$. However, we may define a new multiplicative functional $\tilde{X}$ in $T^{(k+1)}(V)$ as the limit (if the limit exists!)

$$\tilde{X}_{s,t} = \lim_{m(D) \to 0} \hat{X}_{s,t_1} \otimes \cdots \otimes \hat{X}_{t_{r-1},t} \, , \tag{3.20}$$

where the limit is taken over all finite divisions D of $[s,t]$, namely $\{s = t_0 \leqslant t_1 \leqslant \cdots \leqslant t_r = t\}$, and the tensor product $\otimes$ is multiplication in $T^{(k+1)}(V)$ rather than in $T^{(k)}(V)$. Since $X = (1, X^1, \dots, X^k)$ is a multiplicative functional in $T^{(k)}(V)$, we therefore have $\tilde{X}^i = X^i$, for $i \leqslant k$, and

$$\tilde{X}^{k+1}_{s,t} = \lim_{m(D) \to 0} \sum_{l=1}^r \sum_{i=1}^k X^i_{s,t_{l-1}} \otimes X^{k+1-i}_{t_{l-1},t_l} \, . \tag{3.21}$$

Proof of Theorem 3.1.2 By the above analysis, we define $X^k : \Delta \to V^{\otimes k}$, for $k = [p] + 1, \dots$, inductively by

$$X^k_{s,t} = \lim_{m(D) \to 0} \sum_{i=1}^{k-1} \sum_{l=1}^r X^i_{s,t_{l-1}} \otimes X^{k-i}_{t_{l-1},t_l} \, , \quad \forall (s,t) \in \Delta \, . \tag{3.22}$$

We are going to show that (3.22) is well defined, and therefore

$$(1, X^1, \dots, X^k, \dots)$$

is a multiplicative functional with finite p-variation.

For reasons that will be clear through our proof, we may choose a control ω such that

$$|X^i_{s,t}| \leqslant \frac{\omega(s,t)^{i/p}}{\beta \, (i/p)!} \, , \quad \forall i = 1, \dots, [p] \quad \text{and} \quad \forall (s,t) \in \Delta \, , \tag{3.23}$$

and choose β such that (3.19) holds. This can be done by scaling any control since $[p]$ is finite. By the definition of X^k for $k \geqslant [p] + 1$, we only have to prove that X^k is well defined, that is, the limit of (3.22) exists and that it is of finite p-variation. Indeed, we will prove that (3.23) holds for all i.

Let us now prove these claims together. We use induction on $k \geqslant [p]$. If $k = [p]$, there is nothing to prove. So we suppose that X^i, for $i = 1, \dots, k$ and $k \geqslant [p]$, are well defined,

$$(1, X^1, \dots, X^k)$$

is a multiplicative functional in $T^{(k)}(V)$, and (3.23) holds for all $i \leqslant k$. Now we define, for any finite division

$$D = \{s = t_0 \leqslant t_1 \leqslant \cdots \leqslant t_r = t\},$$

a functional

$$X(D)_{s,t} = \hat{X}_{s,t_1} \otimes \cdots \otimes \hat{X}_{t_{r-1},t},$$

which is valued in $T^{(k+1)}(V)$. A simple computation shows that

$$X(D)_{s,t}^{k+1} = \sum_{i=1}^{k} \sum_{l=1}^{r} X_{s,t_{l-1}}^{i} \otimes X_{t_{l-1},t_l}^{k-i}. \tag{3.24}$$

In order to show that the limit $\lim_{m(D)\to 0} X(D)_{s,t}^{k+1}$ exists, we shall use the same trick which we have used in the case of Lipschitz paths. If $r \geqslant 3$, then we may find a point t_l in the partition D such that

$$\omega(t_{l-1}, t_{l+1}) \leqslant \frac{2\omega(s,t)}{r-2}, \tag{3.25}$$

while, if $r = 2$, there is only one point inside D other than s, t. Therefore, for $r = 2$ and $l = 1$, we have

$$\omega(t_{l-1}, t_{l+1}) = \omega(s,t). \tag{3.26}$$

Let $D' = D - \{t_l\}$ be the new partition of $[s,t]$ formed by deleting the point t_l from D. Then

$$X(D)_{s,t}^{k+1} - X(D')_{s,t}^{k+1} = \sum_{i=1}^{k} \Big(X_{s,t_{l-1}}^{i} \otimes X_{t_{l-1},t_l}^{k+1-i}$$
$$+ X_{s,t_l}^{i} \otimes X_{t_l,t_{l+1}}^{k+1-i} - X_{s,t_{l-1}}^{i} \otimes X_{t_{l-1},t_{l+1}}^{k+1-i} \Big). \tag{3.27}$$

Since $i, k + 1 - i \leqslant k$, we may apply Chen's identity to X_{s,t_l}^{i} and $X_{t_{l-1},t_{l+1}}^{k+1-i}$, and we therefore obtain

$$X_{s,t_l}^{i} = \sum_{\lambda+\mu=i} X_{s,t_{l-1}}^{\lambda} \otimes X_{t_{l-1},t_l}^{\mu}$$
$$= X_{s,t_{l-1}}^{i} + X_{t_{l-1},t_l}^{i} + \sum_{\lambda=1}^{i-1} X_{s,t_{l-1}}^{\lambda} \otimes X_{t_{l-1},t_l}^{i-\lambda}. \tag{3.28}$$

Similarly, for $j = k + 1 - i$,

$$X_{t_{l-1},t_{l+1}}^{j} = X_{t_{l-1},t_l}^{j} + X_{t_l,t_{l+1}}^{j} + \sum_{\lambda=1}^{j-1} X_{t_{l-1},t_l}^{\lambda} \otimes X_{t_l,t_{l+1}}^{j-\lambda}. \tag{3.29}$$

Inserting (3.28) and (3.29) into (3.27), we have

$$X(D)_{s,t}^{k+1} - X(D')_{s,t}^{k+1} = \sum_{i=1}^{k} X_{t_{l-1},t_l}^i \otimes X_{t_l,t_{l+1}}^{k+1-i}. \tag{3.30}$$

Hence, by the binomial inequality,

$$\left| X(D)_{s,t}^{k+1} - X(D')_{s,t}^{k+1} \right| \leqslant \sum_{i=1}^{k} \left| X_{t_{l-1},t_l}^i \otimes X_{t_l,t_{l+1}}^{k+1-i} \right|$$

$$\leqslant \sum_{i=1}^{k} \frac{\omega(t_{l-1},t_l)^{i/p}}{\beta\,(i/p)!} \frac{\omega(t_l,t_{l+1})^{(k+1-i)/p}}{\beta\,((k+1-i)/p)!}$$

$$\leqslant \frac{p^2}{\beta^2} \frac{[\omega(t_{l-1},t_l) + \omega(t_l,t_{l+1})]^{(k+1)/p}}{((k+1)/p)!}$$

$$\leqslant \frac{p^2}{\beta} \left(\frac{2}{r-2} \right)^{(k+1)/p} \frac{\omega(s,t)^{(k+1)/p}}{\beta\,((k+1)/p)!}. \tag{3.31}$$

Since $(k+1)/p > 1$, (3.31) thus yields that $X(D)_{s,t}^{k+1}$ satisfies the Cauchy condition, and therefore

$$\lim_{m(D)\to 0} X(D)_{s,t}^{k+1} = X_{s,t}^{k+1}$$

exists. Moreover, by repeating the above argument until all of the inside points t_l of the partition D other than s,t have been deleted, we obtain

$$\left| X(D)_{s,t}^{k+1} \right| \leqslant \frac{p^2}{\beta} \left(1 + \sum_{r=3}^{\infty} \left(\frac{2}{r-2} \right)^{(k+1)/p} \right) \frac{\omega(s,t)^{(k+1)/p}}{\beta\,((k+1)/p)!}$$

$$\leqslant \frac{p^2}{\beta} \left[1 + \sum_{r=3}^{\infty} \left(\frac{2}{r-2} \right)^{([p]+1)/p} \right] \frac{\omega(s,t)^{(k+1)/p}}{\beta\,((k+1)/p)!}$$

$$\leqslant \frac{\omega(s,t)^{(k+1)/p}}{\beta\,((k+1)/p)!}.$$

The above estimate is true for any finite partition D, and therefore

$$\left| X_{s,t}^{k+1} \right| \leqslant \frac{\omega(s,t)^{(k+1)/p}}{\beta\,((k+1)/p)!}.$$

Since we have seen that

$$X^{k+1} = \lim_{m(D)\to 0} \left(\hat{X}_{s,t_1} \otimes \cdots \otimes \hat{X}_{t_{r-1},t} \right)^{k+1}, \tag{3.32}$$

where $\hat{X} = (1, X^1, \ldots, X^k, 0)$, then by a simple induction on $k \geqslant [p]$ one can see that, for all k, $(1, X^1, \ldots, X^k)$ is a multiplicative functional in $T^{(k)}(V)$. Thus we have completed the proof. ∎

Theorem 3.1.2 shows that the higher (than $[p]$) order terms X^k ($k > [p]$) are determined uniquely by X^i ($i \leqslant [p]$) among all possible extensions to a multiplicative functional which possess finite p-variations. On the other hand, those terms X^i, for $i \leqslant [p]$, are never uniquely determined by the terms with orders lower than $[p]$.

Example 3.1.2 Let $p = 2$, and let $X = (1, X^1, X^2)$ be any multiplicative functional with finite p-variation. Then, for any bounded linear operator A,

$$(1, X^1_{s,t}, X^2_{s,t} + (t - s)A)$$

is still a multiplicative functional with finite p-variation.

Therefore, we may give the following definition.

Definition 3.1.3 *A multiplicative functional with finite p-variation in $T^{([p])}(V)$ is called a rough path (of roughness p). We say that a rough path (of roughness p) X in $T^{([p])}(V)$ is controlled by ω if*

$$|X^i_{s,t}| \leqslant \omega(s,t)^{i/p}, \quad \forall i = 1, \ldots, [p] \quad and \quad \forall (s,t) \in \triangle. \qquad (3.33)$$

The set of all rough paths with roughness p in $T^{([p])}(V)$ will be denoted by $\Omega_p(V)$. Thus, any rough path with roughness p in $T^{([p])}(V)$ has a unique, canonical extension to a multiplicative functional in $T^{(\infty)}(V)$ with finite p-variation.

Theorem 3.1.3 *Let X and Y be two rough paths of roughness p, and let β be a constant such that*

$$\beta \geqslant 2p^2 \left(1 + \sum_{r=3}^{\infty} \left(\frac{2}{r-2} \right)^{([p]+1)/p} \right). \qquad (3.34)$$

If ω is a control such that

$$|X^i_{s,t}|, |Y^i_{s,t}| \leqslant \frac{\omega(s,t)^{i/p}}{\beta \, (i/p)!}, \quad \forall i = 1, \ldots, [p] \quad and \quad \forall (s,t) \in \triangle \qquad (3.35)$$

and

$$|X^i_{s,t} - Y^i_{s,t}| \leqslant \varepsilon \frac{\omega(s,t)^{i/p}}{\beta \, (i/p)!}, \quad \forall i = 1, \ldots, [p] \quad and \quad \forall (s,t) \in \triangle, \qquad (3.36)$$

then (3.36) holds for all i.

Therefore, the extension of a rough path X to a higher-order multiplicative functional is continuous in p-variation distance.

Proof Use induction on $k \geqslant [p]$. We shall continue to use the notation established in the proof of Theorem 3.1.2, and shall apply them to both X and Y. By (3.35), which holds for all i, we have

$$\left| X(D)_{s,t}^{k+1} - X(D')_{s,t}^{k+1} - \left[Y(D)_{s,t}^{k+1} - Y(D')_{s,t}^{k+1} \right] \right|$$

$$\leqslant \sum_{i=1}^{k} \left| X_{t_{l-1},t_l}^i \otimes X_{t_l,t_{l+1}}^{k+1-i} - Y_{t_{l-1},t_l}^i \otimes Y_{t_l,t_{l+1}}^{k+1-i} \right|$$

$$\leqslant \varepsilon \sum_{i=1}^{k} \frac{2}{\beta^2} \frac{\omega(t_{l-1},t_l)^{i/p} \omega(t_l,t_{l+1})^{(k+1-i)/p}}{(i/p)! \, ((k+1-i)/p)!}$$

$$\leqslant \varepsilon \frac{2p^2}{\beta^2} \frac{[\omega(t_{l-1},t_l) + \omega(t_l,t_{l+1})]^{(k+1)/p}}{((k+1)/p)!}$$

$$\leqslant \varepsilon \frac{2p^2}{\beta} \left(\frac{2}{r-2} \right)^{(k+1)/p} \frac{\omega(s,t)^{(k+1)/p}}{\beta \, ((k+1)/p)!}, \qquad (3.37)$$

where, in the third inequality, we have used the binomial inequality. By repeating the same procedure until $D' = [s,t]$, we obtain

$$\left| X(D)_{s,t}^{k+1} - Y(D)_{s,t}^{k+1} \right| \leqslant \varepsilon \frac{2p^2}{\beta} \left(1 + \sum_{r=3}^{\infty} \left(\frac{2}{r-2} \right)^{(k+1)/p} \right) \frac{\omega(s,t)^{(k+1)/p}}{\beta \, ((k+1)/p)!}$$

$$\leqslant \varepsilon \frac{\omega(s,t)^{(k+1)/p}}{\beta \, ((k+1)/p)!},$$

for any finite partition D of $\{s,t\}$. Letting $m(D) \to 0$, we obtain

$$\left| X_{s,t}^{k+1} - Y_{s,t}^{k+1} \right| \leqslant \varepsilon \frac{\omega(s,t)^{(k+1)/p}}{\beta \, ((k+1)/p)!}.$$

This completes the proof.   ∎

3.2   Almost rough paths

In this section we give a method of constructing rough paths.

Definition 3.2.1 *Let $p \geqslant 1$ be a constant. A function $X : \Delta \to T^{([p])}(V)$ is called an almost rough path (of roughness p) if it is of finite p-variation, $X_{s,t}^0 = 1$, and, for some control ω and some constant $\theta > 1$,*

$$\left| (X_{s,t} \otimes X_{t,u})^i - X_{s,u}^i \right| \leqslant \omega(s,u)^\theta, \qquad (3.38)$$

for all $(s,t), (t,u) \in \Delta$ and $i = 1, \ldots, [p]$.

The following theorem justifies the name of an almost rough path.

Theorem 3.2.1 *If $X : \triangle \to T^{([p])}(V)$ is an almost rough path of roughness p, then there is a unique rough path $\hat{X}$ (with roughness p) in $T^{([p])}(V)$ such that*

$$|\hat{X}_{s,t}^i - X_{s,t}^i| \leqslant \omega(s,t)^\theta , \quad \forall 1 \leqslant i \leqslant [p] \quad and \quad \forall (s,t) \in \triangle , \tag{3.39}$$

for some control ω and constant $\theta > 1$.

The idea for proving this theorem is similar to the proof of Theorem 3.1.2. Let us describe it as follows by again using induction. Suppose that we have constructed $\hat{X}^i : \triangle \to V^{\otimes i}$, for $i \leqslant k$, such that $(1, \hat{X}^1, \ldots, \hat{X}^k)$ is a multiplicative functional with finite p-variation in $T^{(k)}(V)$. Then $\hat{X}_{s,t}^{k+1}$ is to be defined as the last component of the following limit:

$$\lim_{m(D) \to 0} \bar{X}_{s,t_1} \otimes \cdots \otimes \bar{X}_{t_{r-1},t} , \tag{3.40}$$

where

$$\bar{X} = (1, \hat{X}^1, \ldots, \hat{X}^k, X^{k+1}) , \tag{3.41}$$

and the tensor product $\otimes$ is computed in $T^{(k+1)}(V)$. By a simple calculation, one can easily see that

$$\hat{X}_{s,t}^{k+1} = \lim_{m(D) \to 0} \sum_{l=1}^r \left(X_{t_{l-1},t_l}^{k+1} + \sum_{i=1}^k \hat{X}_{s,t_{l-1}}^i \otimes \hat{X}_{t_{l-1},t_l}^{k+1-i} \right) , \quad \forall (s,t) \in \triangle . \tag{3.42}$$

Proof of Theorem 3.2.1 Choose a control ω such that

$$|X_{s,t}^i| \leqslant \omega(s,t)^{i/p}$$

and

$$\left| (X_{s,t} \otimes X_{t,u})^i - X_{s,u}^i \right| \leqslant \omega(s,u)^\theta ,$$

for $i = 1, \ldots, [p]$ and all $(s,t), (t,u) \in \triangle$. Let us show that the limit in (3.42) exists by induction on k, and

$$|\hat{X}_{s,t}^k - X_{s,t}^k| \leqslant K_k \omega(s,t)^\theta , \quad \forall (s,t) \in \triangle , \tag{3.43}$$

for some constants K_k, which will be made precise later. First we consider the case $k = 1$. Clearly,

$$\hat{X}_{s,t}^1 = \lim_{m(D) \to 0} \sum_{l=1}^r X_{t_{l-1},t_l}^1$$

$$= \lim_{m(D) \to 0} X(D)_{s,t}^1 . \tag{3.44}$$

Using the same trick as that used in the proof of Theorem 3.1.2, we choose $t_l \in D$ such that

$$\omega(t_{l-1}, t_{l+1}) \leqslant \frac{2\omega(s,t)}{r-2}, \quad \text{if} \quad r \geqslant 3, \tag{3.45}$$

and $\omega(t_{l-1}, t_l) = \omega(s,t)$, if $r = 2$. Let $D' = D - \{t_l\}$. Then

$$X(D)^1_{s,t} - X(D')^1_{s,t} = X^1_{t_{l-1},t_l} + X^1_{t_l,t_{l+1}} - X^1_{t_{l-1},t_{l+1}}.$$

Hence, by (3.38),

$$\left| X(D)^1_{s,t} - X(D')^1_{s,t} \right| \leqslant \omega(t_{l-1}, t_{l+1})^\theta$$

$$\leqslant \left(\frac{2}{r-2} \right)^\theta \omega(s,t)^\theta, \quad \forall (s,t) \in \Delta, \tag{3.46}$$

which implies that $\lim_{m(D) \to 0} X(D)^1_{s,t}$ exists. Repeating the above argument until $D' = [s,t]$, we obtain

$$\left| X(D)^1_{s,t} - X^1_{s,t} \right| \leqslant \left(1 + \sum_{r=3}^{\infty} \left(\frac{2}{r-2} \right)^\theta \right) \omega(s,t)^\theta, \quad \forall (s,t) \in \Delta, \tag{3.47}$$

which is true for any finite partition D, so that

$$\left| \hat{X}^1_{s,t} - X^1_{s,t} \right| \leqslant K_1 \omega(s,t)^\theta, \quad \forall (s,t) \in \Delta, \tag{3.48}$$

where

$$K_1 = 1 + \sum_{r=3}^{\infty} \left(\frac{2}{r-2} \right)^\theta. \tag{3.49}$$

Next we assume that for $i \leqslant k$ we have defined $\hat{X}^i$ using (3.42). Let us consider $i = k+1$. For a finite division $D = \{s = t_0 \leqslant t_1 \leqslant \cdots \leqslant t_r = t\}$, let

$$X(D)^{k+1}_{s,t} = \sum_{l=1}^{r} \left(X^{k+1}_{t_{l-1},t_l} + \sum_{i=1}^{k} \hat{X}^i_{s,t_{l-1}} \otimes \hat{X}^{k+1-i}_{t_{l-1},t_l} \right). \tag{3.50}$$

We use the same trick as before to choose t_l, and we let $D' = D - \{t_l\}$. Then

$$X(D)^{k+1}_{s,t} - X(D')^{k+1}_{s,t} = X^{k+1}_{t_{l-1},t_l} + X^{k+1}_{t_l,t_{l+1}} - X^{k+1}_{t_{l-1},t_{l+1}}$$

$$+ \sum_{i=1}^{k} \left(\hat{X}^i_{s,t_{l-1}} \otimes \hat{X}^{k+1-i}_{t_{l-1},t_l} + \hat{X}^i_{s,t_l} \otimes \hat{X}^{k+1-i}_{t_l,t_{l+1}} \right.$$

$$\left. - \hat{X}^i_{s,t_{l-1}} \otimes \hat{X}^{k+1-i}_{t_{l-1},t_{l+1}} \right). \tag{3.51}$$

Next, using the identity

$$X_{t_{l-1},t_{l+1}}^{k+1} - \left(X_{t_{l-1},t_l} \otimes X_{t_l,t_{l+1}}\right)^{k+1}$$

$$= X_{t_{l-1},t_{l+1}}^{k+1} - X_{t_{l-1},t_l}^{k+1} - X_{t_l,t_{l+1}}^{k+1} - \sum_{i=1}^{k} X_{t_{l-1},t_l}^i \otimes X_{t_l,t_{l+1}}^{k+1-i}$$

and the fact that $(1, \hat{X}^1, \ldots, \hat{X}^k)$ is a multiplicative functional in $T^{(k)}(V)$, we have

$$X(D)_{s,t}^{k+1} - X(D')_{s,t}^{k+1} = \left(X_{t_{l-1},t_l} \otimes X_{t_l,t_{l+1}}\right)^{k+1} - X_{t_{l-1},t_{l+1}}^{k+1}$$

$$+ \sum_{i=1}^{k} \left(\hat{X}_{t_{l-1},t_l}^i \otimes \hat{X}_{t_l,t_{l+1}}^{k+1-i} - X_{t_{l-1},t_l}^i \otimes X_{t_l,t_{l+1}}^{k+1-i}\right),$$

$$(3.52)$$

and therefore

$$\left|X(D)_{s,t}^{k+1} - X(D')_{s,t}^{k+1}\right|$$

$$\leqslant \omega(t_{l-1}, t_{l+1})^\theta$$

$$+ \sum_{i=1}^{k} K_i \omega(t_{l-1}, t_l)^\theta \left(\omega(t_l, t_{l+1})^{(k+1-i)/p} + K_{k+1-i}\omega(t_{l-1}, t_{l+1})^\theta\right)$$

$$+ \sum_{i=1}^{k} \omega(t_{l-1}, t_l)^{i/p} K_{k+1-i}\omega(t_{l-1}, t_{l+1})^\theta .$$

Set $K_0 = \max \omega \vee 1$. Then the above estimate yields that

$$\left|X(D)_{s,t}^{k+1} - X(D')_{s,t}^{k+1}\right|$$

$$\leqslant \left(\frac{2}{r-2}\right)^\theta \omega(s,t)^\theta$$

$$+ \sum_{i=1}^{k} \left(\frac{2}{r-2}\right)^\theta K_i \omega(s,t)^\theta \left(K_0^{(k+1-i)/p} + K_{k+1-i}K_0^\theta\right)$$

$$+ \sum_{i=1}^{k} \left(\frac{2}{r-2}\right)^\theta K_{k+1-i}\omega(s,t)^\theta K_0^{i/p}$$

$$= \left(\frac{2}{r-2}\right)^\theta \omega(s,t)^\theta \left(1 + \sum_{i=1}^{k} 2K_{k+1-i}K_0^{i/p} + K_i K_{k+1-i}K_0^\theta\right),$$

and therefore $\lim_{m(D) \to 0} X(D)_{s,t}^{k+1}$ exists. By deleting all of the t_l other than s, t according to the above procedure, we may thus establish that

$$\left|X(D)_{s,t}^{k+1} - X_{s,t}^{k+1}\right| \leqslant K_{k+1}\omega(s,t)^\theta, \quad \forall (s,t) \in \Delta, \qquad (3.53)$$

where the constants K_k are defined inductively by

$$K_{k+1} = K_1 \left[1 + \sum_{i=1}^{k} \left(2K_0^{i/p} K_{k+1-i} + K_i K_{k+1-i} K_0^{\theta} \right) \right]. \qquad (3.54)$$

The inequality (3.53) yields that

$$\left| \hat{X}_{s,t}^{k+1} - X_{s,t}^{k+1} \right| \leqslant K_{k+1} \omega(s,t)^{\theta}, \quad \forall (s,t) \in \Delta.$$

The proof is completed.   ∎

Corollary 3.2.1 *Let X be an almost rough path with roughness p, let ω be a control, and let $\theta > 1$. If*

$$\left| X_{s,t}^i \right| \leqslant \omega(s,t)^{i/p}, \quad \forall i = 1, \dots, [p], \quad \forall (s,t) \in \Delta$$

and

$$\left| (X_{s,t} \otimes X_{t,u})^i - X_{s,u}^i \right| \leqslant \omega(s,u)^{\theta}, \quad \forall 1 \leqslant i \leqslant [p],$$

for all $(s,t), (t,u) \in \Delta$, then

$$\left| \hat{X}_{s,t}^i \right| \leqslant K \omega(s,t)^{i/p}, \quad \forall i = 1, \dots, [p], \quad \forall (s,t) \in \Delta,$$

where $\hat{X}$ is the unique rough path associated to X and K is a constant depending only on $\max\{\omega, p, \theta\}$.

The following theorem shows that in fact the map $X \to \hat{X}$ established in Theorem 3.2.1 is continuous.

Theorem 3.2.2 *Let X, Y be two almost rough paths of roughness p in $T^{([p])}(V)$, both of which are controlled by a control ω, namely*

$$\left| X_{s,t}^i \right|, \left| Y_{s,t}^i \right| \leqslant \omega(s,t)^{i/p}, \quad \forall i = 1, \dots, [p], \quad \forall (s,t) \in \Delta, \qquad (3.55)$$

and, for some $\theta > 1$,

$$\left| (X_{s,t} \otimes X_{t,u})^i - X_{s,u}^i \right| \leqslant \omega(s,u)^{\theta}, \qquad (3.56)$$

for all $(s,t), (t,u) \in \Delta$, $i = 1, \dots, [p]$, with the same inequality (3.56) also holding for Y. Suppose that

$$\left| X_{s,t}^i - Y_{s,t}^i \right| \leqslant \varepsilon \omega(s,t)^{i/p}, \qquad \forall i = 1, \dots, [p], \quad \forall (s,t) \in \Delta, \qquad (3.57)$$

then

$$\left| \hat{X}_{s,t}^i - \hat{Y}_{s,t}^i \right| \leqslant B_i(\varepsilon) \omega(s,t)^{i/p}, \quad \forall i = 1, \dots, [p], \quad \forall (s,t) \in \Delta, \qquad (3.58)$$

where

$$B_1(\varepsilon) = \varepsilon + 3 \left\{ \sum_{r=3}^{\infty} \left[\varepsilon \left(\frac{2}{r-2} \right)^{1/p} \right] \wedge \left[\left(\frac{2}{r-2} \right)^{\theta} K_0^{\theta} \right] + \varepsilon \wedge K_0^{\theta} \right\}$$

and $B_{k+1}(\varepsilon)$ is defined inductively by (3.63) (see below) for $k \geqslant 1$.

Proof Let us use the same notation as in the proof of Theorem 3.2.1. Use induction on $k \leqslant [p]$. For $k = 1$, then

$$\left| X(D)_{s,t}^1 - Y(D)_{s,t}^1 - (X(D')_{s,t}^1 - Y(D')_{s,t}^1) \right|$$
$$= \left| X_{t_{l-1},t_l}^1 + X_{t_l,t_{l+1}}^1 - X_{t_{l-1},t_{l+1}}^1 - (Y_{t_{l-1},t_l}^1 + Y_{t_l,t_{l+1}}^1 - Y_{t_{l-1},t_{l+1}}^1) \right|, \tag{3.59}$$

which on one hand is smaller than

$$2\omega(t_{l-1}, t_{l+1})^\theta \leqslant 2 \left(\frac{2}{r-2} \right)^\theta \omega(s,t)^\theta,$$

and on the other hand, by (3.57), it is not bigger than

$$3\varepsilon\omega(t_{l-1}, t_{l+1})^{1/p} \leqslant 3\varepsilon \left(\frac{2}{r-2} \right)^{1/p} \omega(s,t)^{1/p}.$$

Therefore, the right-hand side of (3.59) is less than

$$3\omega(s,t)^{1/p} \left\{ \left[\varepsilon \left(\frac{2}{r-2} \right)^{1/p} \right] \wedge \left(\frac{2}{r-2} \right)^\theta K_0^\theta \right\}.$$

By repeating this procedure we obtain

$$\left| X(D)_{s,t}^1 - Y(D)_{s,t}^1 - (X_{s,t}^1 - Y_{s,t}^1) \right|$$
$$\leqslant 3 \left\{ \sum_{r=2}^\infty \left[\varepsilon \left(\frac{2}{r-2} \right)^{1/p} \right] \wedge \left[\left(\frac{2}{r-2} \right)^\theta K_0^\theta \right] + \varepsilon \wedge K_0^\theta \right\} \omega(s,t)^{1/p},$$

where we have used the convention that the term $2/(r-2)$ should be read as 1 when $r = 2$. This inequality is true for any finite division, and therefore

$$\left| \hat{X}_{s,t}^1 - \hat{Y}_{s,t}^1 \right| \leqslant B_1(\varepsilon)\omega(s,t)^{1/p}, \quad \forall (s,t) \in \Delta. \tag{3.60}$$

Suppose now that (3.58) holds for all $i \leqslant k < [p]$. Consider the case of the $(k+1)$th components. Again, for a finite partition D, delete t_l from D to obtain D'. Here t_l is chosen so that

$$\omega(t_{l-1}, t_{l+1}) \leqslant \left(\frac{2}{r-2} \right) \omega(s,t).$$

Then

$$\left| X_{t_{l-1},t_l}^{k+1} + X_{t_l,t_{l+1}}^{k+1} - X_{t_{l-1},t_{l+1}}^{k+1} - (Y_{t_{l-1},t_l}^{k+1} + Y_{t_l,t_{l+1}}^{k+1} - Y_{t_{l-1},t_{l+1}}^{k+1}) \right|$$
$$\leqslant 3\omega(s,t)^{(k+1)/p} \left[K_0^\theta \left(\frac{2}{r-2} \right)^\theta \wedge \varepsilon \left(\frac{2}{r-2} \right)^{(k+1)/p} \right]$$

and

$$\left| \sum_{i=1}^{k} (\hat{X}_{t_{l-1},t_l}^i \otimes \hat{X}_{t_l,t_{l+1}}^{k+1-i} - X_{t_{l-1},t_l}^i \otimes X_{t_l,t_{l+1}}^{k+1-i}) \right.$$

$$\left. - \sum_{i=1}^{k} (\hat{Y}_{t_{l-1},t_l}^i \otimes \hat{Y}_{t_l,t_{l+1}}^{k+1-i} - Y_{t_{l-1},t_l}^i \otimes Y_{t_l,t_{l+1}}^{k+1-i}) \right|$$

$$\leqslant \sum_{i=1}^{k} |\hat{X}_{t_{l-1},t_l}^i \otimes \hat{X}_{t_l,t_{l+1}}^{k+1-i} - \hat{Y}_{t_{l-1},t_l}^i \otimes \hat{Y}_{t_l,t_{l+1}}^{k+1-i}|$$

$$+ \sum_{i=1}^{k} |X_{t_{l-1},t_l}^i \otimes X_{t_l,t_{l+1}}^{k+1-i} - Y_{t_{l-1},t_l}^i \otimes Y_{t_l,t_{l+1}}^{k+1-i}|$$

$$\leqslant \sum_{i=1}^{k} B_i(\varepsilon)(1 + K_{k+1-i})\omega(t_{l-1},t_l)^{i/p}\omega(t_l,t_{l+1})^{(k+1-i)/p}$$

$$+ \sum_{i=1}^{k} B_{k+1-i}(\varepsilon)(1 + K_i)\omega(t_{l-1},t_l)^{i/p}\omega(t_l,t_{l+1})^{(k+1-i)/p}$$

$$+ 2\varepsilon \sum_{i=1}^{k} K\omega(t_{l-1},t_l)^{i/p}\omega(t_l,t_{l+1})^{(k+1-i)/p}$$

$$\leqslant 2\omega(s,t)^{(k+1)/p} \left(\frac{2}{r-2}\right)^{(k+1)/p} \sum_{i=1}^{k} [(1 + K_{k+1-i})B_i(\varepsilon) + K\varepsilon].$$

Hence

$$\left| X(D)_{s,t}^{k+1} - X(D')_{s,t}^{k+1} - [Y(D)_{s,t}^{k+1} - Y(D')_{s,t}^{k+1}] \right|$$

$$\leqslant 3 \left(\frac{2}{r-2}\right)^{(k+1)/p} \omega(s,t)^{(k+1)/p} \left\{ \sum_{i=1}^{k} [B_i(\varepsilon)(1 + K_{k+1-i}) + \varepsilon K] + \varepsilon \right\}.$$

$$(3.61)$$

Here K is a constant depending only on $\max \omega$ and p. On the other hand, we have by (3.56) that the left-hand side of (3.61) is smaller than

$$2 \left(\frac{r}{r-2}\right)^{\theta} \omega(s,t)^{\theta} \left(1 + 2\sum_{i=1}^{k} K_{k+1-i}K_0^{i/p} + K_i K_{k+1-i}K_0^{\theta}\right)$$

$$\leqslant 2K_0^{\theta} \left(\frac{r}{r-2}\right)^{\theta} \omega(s,t)^{(k+1)/p} \left(1 + 2\sum_{i=1}^{k} K_{k+1-i}K_0^{i/p} + K_i K_{k+1-i}K_0^{\theta}\right).$$

Therefore,

$$\left| X(D)^{k+1}_{s,t} - X(D')^{k+1}_{s,t} - \left[Y(D)^{k+1}_{s,t} - Y(D')^{k+1}_{s,t} \right] \right|$$
$$\leqslant 3A_k(r,\varepsilon)\omega(s,t)^{(k+1)/p}, \quad \forall (s,t) \in \Delta,$$

where

$$A_k(r,\varepsilon) = \min \left\{ K_0^\theta \left(\frac{2}{r-2} \right)^\theta \left(1 + 2\sum_{i=1}^{k} K_{k+1-i} K_0^{i/p} + K_i K_{k+1-i} K_0^\theta \right), \right.$$
$$\left. \left(\frac{2}{r-2} \right)^{(k+1)/p} \left[\sum_{i=1}^{k} [B_i(\varepsilon)(1 + K_{k+1-i}) + K\varepsilon] + \varepsilon \right] \right\}.$$

$$(3.62)$$

By repeating the argument until all of the t_l have been deleted, we obtain

$$\left| X(D)^{k+1}_{s,t} - Y(D)^{k+1}_{s,t} - \left[X^{k+1}_{s,t} - Y^{k+1}_{s,t} \right] \right| \leqslant 3 \sum_{r=3}^{\infty} A_k(r,\varepsilon)\omega(s,t)^{(k+1)/p},$$

for all finite division D of $[s,t]$. Therefore,

$$\left| X(D)^{k+1}_{s,t} - Y(D)^{k+1}_{s,t} \right| \leqslant \left(\varepsilon + 3\sum_{r=2}^{\infty} A_k(r,\varepsilon) \right) \omega(s,t)^{(k+1)/p}.$$

Letting $m(D) \to 0$,

$$\left| \hat{X}^{k+1}_{s,t} - \hat{Y}^{k+1}_{s,t} \right| \leqslant B_{k+1}(\varepsilon)\omega(s,t)^{(k+1)/p}, \quad \forall (s,t) \in \Delta,$$

where

$$B_{k+1}(\varepsilon) = \varepsilon + 3\sum_{r=2}^{\infty} A_k(r,\varepsilon). \tag{3.63}$$

The use of induction then ends the proof. ∎

Remark 3.2.1 The precise values of $B_k(\varepsilon)$ are not important. The essential point here is that $B_k(\varepsilon)$ depend only on $\max\{\omega, p, \theta\}$, and that $\lim_{\varepsilon \to 0} B_k(\varepsilon) = 0$.

3.3 Spaces of rough paths

3.3.1 *Variation distances and variation topology*

Initially, a Lipschitz path $X : [0,T] \to V$ in a Banach space V is a continuous path with finite variation (or finite one-variation as we have called it) in the sense that

$$\sup_D \sum_l |X^1_{t_{l-1},t_l}| < +\infty, \tag{3.64}$$

where $\sup_D$ runs over all finite partitions of $[0,T]$, namely

$$D = \{0 = t_0 \leqslant \cdots \leqslant t_r = T\}.$$

It is clear that, if a function $X : [0,T] \to V$ is a continuous Lipschitz path in V as defined in Chapter 2, then X does have finite variation on $[0,T]$, that is (3.64) holds. Conversely, we have the following proposition.

Proposition 3.3.1 *Suppose that $X : [0,T] \to V$ is a continuous path which has finite variation on $[0,T]$, that is*

$$\sup_D \sum_l |X_{t_l} - X_{t_{l-1}}| < +\infty. \tag{3.65}$$

Then X is a Lipschitz path controlled by ω, so that

$$\omega(s,t) = \sup_{D_{[s,t]}} \sum_l |X_{t_l} - X_{t_{l-1}}|, \quad \forall (s,t) \in \Delta, \tag{3.66}$$

where $\sup_{D_{[s,t]}}$ runs over finite divisions of the interval $[s,t]$.

Actually, a similar conclusion is true for any rough path.

Let $C(\Delta, T^{(n)}(V))$ denote the set of all continuous functions from the simplex Δ into the truncated tensor algebra $T^{(n)}(V)$, with an appropriate norm, as we mentioned at the beginning of this chapter. If $X \in C(\Delta, T^{(n)}(V))$, then we may write

$$X_{s,t} = (X_{s,t}^0, X_{s,t}^1, \ldots, X_{s,t}^n), \quad \forall (s,t) \in \Delta,$$

where $X_{s,t}^i \in V^{\otimes i}$ is the ith component of X (also called the ith level path of X). Of course, we are only interested in those functions X in $C(\Delta, T^{(n)}(V))$ such that $X_{s,t}^0 \equiv 1$, and the subset of these functions will be denoted by $C_0(\Delta, T^{(n)}(V))$. Therefore, any almost rough path with roughness p belongs to $C_0(\Delta, T^{([p])}(V))$.

Definition 3.3.1 *A function $X \in C_0(\Delta, T^{(n)}(V))$ is said to have finite total p-variation if*

$$\sup_D \sum_l |X_{t_{l-1},t_l}^i|^{p/i} < +\infty, \quad i = 1, \ldots, n, \tag{3.67}$$

where $\sup_D$ runs over all finite divisions of $[0,T]$.

It is clear that if $X \in C_0(\Delta, T^{(n)}(V))$ is of finite p-variation, i.e. there is a control ω such that

$$|X_{s,t}^i| \leqslant \omega(s,t)^{i/p}, \quad \forall i = 1, \ldots, n, \quad \forall (s,t) \in \Delta,$$

then X has finite total p-variation. Conversely we have the following proposition.

Proposition 3.3.2 *Let $p \geqslant 1$ be a constant, and let $X \in C_0(\Delta, T^{(n)}(V))$ satisfy Chen's identity (i.e. X is a multiplicative functional in $T^{(n)}(V)$ of order n). If X has finite total p-variation, then*

$$\omega(s,t) = \sum_{i=1}^n \sup_{D_{[s,t]}} \sum_l |X_{t_{l-1},t_l}^i|^{p/i}, \quad \forall (s,t) \in \Delta \tag{3.68}$$

is a control function, and

$$|X_{s,t}^i| \leqslant \omega(s,t)^{i/p}, \quad \forall i = 1, \ldots, n, \quad \forall (s,t) \in \Delta. \tag{3.69}$$

Proposition 3.3.2 follows from the following lemma.

Lemma 3.3.1 *Let $p \geqslant 1$, and let $X, Y \in C_0(\Delta, T^{(n)}(V))$ satisfy Chen's identity. Suppose that X, Y have finite total p-variations. Then, for any $1 \leqslant i \leqslant n$, the function*

$$\omega_i(s,t) = \sup_{D_{[s,t]}} \sum_l \left| X^i_{t_{l-1}, t_l} - Y^i_{t_{l-1}, t_l} \right|^{p/i}$$

is a control function.

Proof By definition,

$$\omega_i(s,t) + \omega_i(t,u) \leqslant \omega_i(s,u), \quad \forall (s,t), (t,u) \in \Delta.$$

Next we show that ω is continuous. First observe that $X^i_{s,s}, Y^i_{s,s} = 0$, for $i \geqslant 1$, due to Chen's identity. For any s, the function $t \to \omega_i(s,t)$ is increasing and bounded, so that

$$\omega_i(s, t-) \equiv \lim_{t' \uparrow t} \omega_i(s, t')$$

exists. Suppose that $D_{[s,t]} = \{s = t_0 < t_1 < \cdots < t_{r-1} < t_r = t\}$ is any finite division and $t_{r-1} < t_p < t$ is any point. Then, by Chen's identity,

$$X^i_{t_{r-1}, t} = X^i_{t_{r-1}, t_p} + X^i_{t_p, t} + \sum_{l=1}^{i-1} X^l_{t_{r-1}, t_p} \otimes X^{i-l}_{t_p, t},$$

and the same identity holds for Y. Therefore,

$$\left| X^i_{t_{r-1}, t} - Y^i_{t_{r-1}, t} \right| \leqslant \left| X^i_{t_{r-1}, t_p} - Y^i_{t_{r-1}, t_p} \right| + \left| X^i_{t_p, t} \right| + \left| Y^i_{t_p, t} \right|$$

$$+ \sum_{l=1}^{i-1} \left(\left| X^l_{t_{r-1}, t_p} \right| \left| X^{i-l}_{t_p, t} \right| + \left| Y^l_{t_{r-1}, t_p} \right| \left| Y^{i-l}_{t_p, t} \right| \right),$$

so that

$$\sum_{l=1}^{r} \left| X^i_{t_{l-1}, t_l} - Y^i_{t_{l-1}, t_l} \right|^{p/i}$$

$$= \sum_{l=1}^{r-1} \left| X^i_{t_{l-1}, t_l} - Y^i_{t_{l-1}, t_l} \right|^{p/i} + \left| X^i_{t_{r-1}, t} - Y^i_{t_{r-1}, t} \right|^{p/i}$$

$$\leqslant \sum_{l=1}^{r-1} \left| X^i_{t_{l-1}, t_l} - Y^i_{t_{l-1}, t_l} \right|^{p/i}$$

$$+ \left[\left| X^i_{t_{r-1}, t_p} - Y^i_{t_{r-1}, t_p} \right| + \left| X^i_{t_p, t} \right| + \left| Y^i_{t_p, t} \right| \right.$$

$$\left. + \sum_{k=1}^{i-1} \left(\left| X^k_{t_{r-1}, t_p} \right| \left| X^{i-k}_{t_p, t} \right| + \left| Y^k_{t_{r-1}, t_p} \right| \left| Y^{i-k}_{t_p, t} \right| \right) \right]^{p/i}$$

$$- \left| X^i_{t_{r-1}, t_p} - Y^i_{t_{r-1}, t_p} \right|^{p/i}$$

$$\leqslant \omega_i(s, t-) + \Bigg[\Big| X^i_{t_{r-1},t_p} - Y^i_{t_{r-1},t_p} \Big| + \Big| X^i_{t_p,t} \Big| + \Big| Y^i_{t_p,t} \Big|$$

$$+ \sum_{k=1}^{i-1} \Big(\Big| X^k_{t_{r-1},t_p} \Big| \Big| X^{i-k}_{t_p,t} \Big| + \Big| Y^k_{t_{r-1},t_p} \Big| \Big| Y^{i-k}_{t_p,t} \Big| \Big) \Bigg]^{p/i}$$

$$- \Big| X^i_{t_{r-1},t_p} - Y^i_{t_{r-1},t_p} \Big|^{p/i}.$$

Letting $t_p \uparrow t$, we obtain

$$\sum_{l=1}^{r} \Big| X^i_{t_{l-1},t_l} - Y^i_{t_{l-1},t_l} \Big|^{p/i} \leqslant \omega_i(s, t-),$$

for any finite division D of $[s, t]$, where we have used the fact that $\lim_{s \to t} X^i_{s,t} = 0$ and $\lim_{s \to t} Y^i_{s,t} = 0$. Therefore,

$$\omega_i(s, t) \leqslant \omega_i(s, t-),$$

so that $\omega_i(s, t) = \omega_i(s, t-)$. Similarly, it is right-continuous as well. Hence ω_i is continuous, so that it is a control function. ∎

Let $C_{0,p}(\triangle, T^{(n)}(V))$ denote the subspace of all $X \in C_0(\triangle, T^{(n)}(V))$ with finite total p-variation. It is clear that $C_{0,p}(\triangle, T^{(n)}(V))$ is a linear space.

The p-variation metric d_p on $C_{0,p}(\triangle, T^{([p])}(V))$ is defined by

$$d_p(X, Y) = \max_{1 \leqslant i \leqslant [p]} \sup_D \left(\sum_l \Big| X^i_{t_{l-1},t_l} - Y^i_{t_{l-1},t_l} \Big|^{p/i} \right)^{i/p}. \qquad (3.70)$$

In fact, d_p makes sense even on the space $C_0(\triangle, T^{([p])}(V))$, but it may well be possible that $d_p(X, Y) = \infty$. Indeed, $X \in C_{0,p}(\triangle, T^{([p])}(V))$ if and only if $X \in C_0(\triangle, T^{([p])}(V))$ and

$$d_p(X, \mathbf{1}) < +\infty, \qquad (3.71)$$

where

$$\mathbf{1} = (1, 0, \ldots, 0) \in C_{0,p}(\triangle, T^{([p])}(V)).$$

The space $\Omega_p(V)$ of rough paths in $T^{([p])}(V)$ is endowed with the p-variation distance d_p as a metric subspace of $C_{0,p}(\triangle, T^{([p])}(V))$.

Lemma 3.3.2   (i) *If* $\{X(n)\}$ *is a Cauchy sequence in* d_p *then there is a subsequence* $\{X(n_k)\}$ *with a uniform control* ω. *The* $X(n_k)$ *are convergent with respect to* ω.

(ii) *A sequence* $X(n) \to X$ *in* $(\Omega_p(V), d_p)$ *if and only if, for all* $1 \leqslant i \leqslant [p]$,

$$\lim_{n \to \infty} \sup_D \sum_l \Big| X(n)^i_{s,t} - X^i_{s,t} \Big|^{p/i} = 0.$$

This is obvious by definition.

Lemma 3.3.3 $(\Omega_p(V), d_p)$ *is a complete metric space.*

Proof The proof is very standard. Suppose that $X(n)$ is a Cauchy sequence in the metric space $(\Omega_p(V), d_p)$. We may choose a subsequence $X(n_k)$ such that

$$\sum_{i=1}^{[p]} \sup_D \sum_l \left| X(n_k)^i_{t_{l-1}, t_l} - X(n_{k-1})^i_{t_{l-1}, t_l} \right|^{p/i} \leq \frac{1}{2^{2k}}, \quad \forall k.$$

Let

$$\omega(s,t) = \sup_{k \geq 0} \sum_{i=1}^{[p]} \sup_{D_{[s,t]}} \sum_l \left| X(n_k)^i_{t_{l-1}, t_l} \right|^{p/i}$$

$$+ \sum_{k=1}^{\infty} \sum_{i=1}^{[p]} \sup_{D_{[s,t]}} \sum_l \left| X(n_k)^i_{t_{l-1}, t_l} - X(n_{k-1})^i_{t_{l-1}, t_l} \right|^{p/i}.$$

Then, by Lemma 3.3.1 and the Dini theorem, we know that ω is a control function. Clearly

$$\left| X(n_k)^i_{s,t} \right| \leq \omega(s,t)^{i/p}, \quad i = 1, \ldots, [p], \quad \forall (s,t) \in \Delta \tag{3.72}$$

and

$$\left| X(n_k)^i_{s,t} - X(n_{k-1})^i_{s,t} \right| \leq \frac{1}{2^{ik/p}} \omega(s,t)^{i/p}, \quad \forall i \leq [p], \quad \forall k \geq 1, \quad \forall (s,t) \in \Delta.$$

Therefore, $\lim_{k \to \infty} X(n_k)^i_{s,t} = X^i_{s,t}$ exists uniformly, so that X is a multiplicative functional. Also, by the above uniform bound (3.72), X satisfies the same inequality (3.72), so that X is a rough path. It is easily seen that

$$\lim_{n \to \infty} \sup_D \sum_l \left| X(n)^i_{s,t} - X^i_{s,t} \right|^{p/i} = 0,$$

and therefore $d_p(X(n), X) = 0$. ∎

By a similar argument, we have the following proposition.

Proposition 3.3.3 *Let* $X(n), X \in \Omega_p(V)$ *such that* $d_p(X(n), X) \to 0$ *as* $n \to \infty$. *Then there is a subsequence* $X(n_k)$ *and a control* ω *such that*

$$\left| X(n_k)^i_{s,t} \right|, \left| X^i_{s,t} \right| \leq \omega(s,t)^{i/p}$$

and

$$\left| X(n_k)^i_{s,t} - X^i_{s,t} \right| \leq \frac{1}{2^k} \omega(s,t)^{i/p},$$

for all $1 \leq i \leq [p]$, $k \in \mathbb{N}$ *and all* $(s,t) \in \Delta$.

However, the distance function d_p is difficult to use in practice. Therefore we need the following definition.

Definition 3.3.2   *(i) In view of the above proposition, we shall say that a sequence $\{X(n)\}$ of $C_{0,p}(\triangle, T^{(N)}(V))$ converges to $X \in C_{0,p}(\triangle, T^{(N)}(V))$ in p-variation topology if there is a control ω such that*

$$\left|X(n)_{s,t}^i\right|, \left|X_{s,t}^i\right| \leqslant \omega(s,t)^{i/p}, \quad i = 1, \ldots, N, \quad \forall (s,t) \in \triangle, \qquad (3.73)$$

for any $n = 1, 2, \ldots$, and

$$\left|X(n)_{s,t}^i - X_{s,t}^i\right| \leqslant a(n)\omega(s,t)^{i/p}, \quad i = 1, \ldots, N, \quad \forall (s,t) \in \triangle, \quad (3.74)$$

for some function $a(n)$ (which may depend on the sequence $X(n)$, X, and the control ω) such that $\lim_{n\to\infty} a(n) = 0$.

(ii) Let $p, q \geqslant 1$ be two constants. We say that a map $F : C_{0,p}(\triangle, T^{(N)}(V)) \to C_{0,q}(\triangle, T^{(N')}(W))$ is continuous in (p, q)-variation topology if, for any control ω, there is a control ω_1 and a function $\alpha : \mathbb{R}^+ \to \mathbb{R}^+$ satisfying the condition $\lim_{\varepsilon \downarrow 0} \alpha(\varepsilon) = 0$, such that, if $X, Y \in C_{0,p}(\triangle, T^{(N)}(V))$ and

$$\left|X_{s,t}^i\right|, \left|Y_{s,t}^i\right| \leqslant \omega(s,t)^{i/p}, \quad i = 1, \ldots, N, \quad \forall (s,t) \in \triangle, \qquad (3.75)$$

$$\left|X_{s,t}^i - Y_{s,t}^i\right| \leqslant \varepsilon\omega(s,t)^{i/p}, \quad i = 1, \ldots, N, \quad \forall (s,t) \in \triangle, \qquad (3.76)$$

then

$$\left|F(X)_{s,t}^j - F(Y)_{s,t}^j\right| \leqslant \alpha(\varepsilon)\,\omega_1(s,t)^{j/q}, \quad j = 1, \ldots, N', \quad \forall (s,t) \in \triangle. \tag{3.77}$$

Remark 3.3.1 We are mainly interested in the rough path space $\Omega_p(V)$, which is a closed subspace of $C_{0,p}(\triangle, T^{([p])}(V))$ under the convergence in p-variation topology. Therefore, the above definition about continuity can be applied to functions on $\Omega_p(V)$. Thus, we say that a map $F : \Omega_p(V) \to C_{0,q}(\triangle, T^{(N')}(W))$ is (uniformly) continuous if, for any control ω, there is a control ω_1 and a function $\alpha : \mathbb{R}^+ \to \mathbb{R}^+$, where $\lim_{\varepsilon \downarrow 0} \alpha(\varepsilon) = 0$, such that, if $X, Y \in \Omega_p(V)$ and

$$\begin{aligned} \left|X_{s,t}^i\right|, \left|Y_{s,t}^i\right| &\leqslant \omega(s,t)^{i/p}, \\ \left|X_{s,t}^i - Y_{s,t}^i\right| &\leqslant \varepsilon\omega(s,t)^{i/p}, \end{aligned} \quad i = 1, \ldots, [p], \quad \forall (s,t) \in \triangle, \qquad (3.78)$$

then

$$\left|F(X)_{s,t}^j - F(Y)_{s,t}^j\right| \leqslant \alpha(\varepsilon)\omega_1(s,t)^{j/q}, \quad j = 1, \ldots, N', \quad \forall (s,t) \in \triangle. \quad (3.79)$$

Remark 3.3.2 The same principle applies to the subset $A\Omega_p(V)$ of all almost rough paths in $C_{0,p}(\triangle, T^{([p])}(V))$. Thus, we say that a map $F : A\Omega_p(V) \to C_{0,q}(\triangle, T^{(N')}(W))$ is (uniformly) continuous in (p, q)-variation topology if, for any control ω and a constant $\theta > 1$, there exists a control ω_1 and a function $\alpha : \mathbb{R}^+ \to \mathbb{R}^+$, where $\lim_{\varepsilon \downarrow 0} \alpha(\varepsilon) = 0$, such that, for any $X, Y \in A\Omega_p(V)$, if

$$
\begin{aligned}
\left| (X_{s,t} \otimes X_{t,u})^i - X_{s,u}^i \right| &\leqslant \omega(s,u)^\theta , \\
\left| (Y_{s,t} \otimes Y_{t,u})^i - Y_{s,u}^i \right| &\leqslant \omega(s,u)^\theta , \\
\left| X_{s,t}^i \right|, \left| Y_{s,t}^i \right| &\leqslant \omega(s,t)^{i/p} ,
\end{aligned}
\tag{3.80}
$$

and

$$
\left| X_{s,t}^i - Y_{s,t}^i \right| \leqslant \varepsilon \omega(s,t)^{i/p} ,
$$

for all $i = 1, \ldots, [p]$ and all $(s,t), (t,u) \in \Delta$, then

$$
\left| F(X)_{s,t}^j - F(Y)_{s,t}^j \right| \leqslant \alpha(\varepsilon) \omega_1(s,t)^{j/p'} ,
\tag{3.81}
$$

for all $j = 1, \ldots, N'$ and all $(s,t) \in \Delta$.

The reason that we put an extra condition in the definition of the continuity of a map on the almost rough path space $A\Omega_p(V)$ is that this subset is not closed under the convergence in p-variation topology. The uniform conditions (3.80) ensure that limits still possess the almost multiplicative property.

Finally, we introduce a class of rough paths called geometric rough paths. A rough path $X \in \Omega_p(V)$ is called a smooth rough path if $t \to X_t \equiv X_{0,t}^1$ is a continuous path with finite variation and $X_{s,t}^i$ is the ith iterated path integral of the path X_t over the interval $[s,t]$ (for $i = 1, \ldots, [p]$), that is

$$
X_{s,t}^i = \int_{s < t_1 < \cdots < t_i < t} dX_{t_1} \otimes \cdots \otimes dX_{t_i} , \quad \forall (s,t) \in \Delta .
$$

Definition 3.3.3 *Geometric rough paths with roughness p are the rough paths in the closure of smooth rough paths under p-variation distance (or, equivalently, under p-variation topology). Thus, a rough path $X \in \Omega_p(V)$ is a geometric rough path if there is a sequence $X(n)$ of smooth rough paths in $\Omega_p(V)$ such that*

$$
d_p(X(n), X) \to 0, \quad as \quad n \to \infty .
$$

Therefore, X is a geometric rough path of roughness p if and only if there is a sequence $X(n)$ of smooth rough paths in $\Omega_p(V)$ and a control ω such that

$$
\left| X(n)_{s,t}^i \right|, \left| X_{s,t}^i \right| \leqslant \omega(s,t)^{i/p}
$$

and

$$
\left| X(n)_{s,t}^i - X_{s,t}^i \right| \leqslant \frac{1}{n} \omega(s,t)^{i/p} ,
$$

for all $i = 1, \ldots, [p]$ and $(s,t) \in \Delta$.

The space of all geometric rough paths with roughness p is denoted by $G\Omega_p(V)$.

3.3.2 Young's integration theory

In Young (1936), L. C. Young established an integration theory for some non-smooth continuous paths. Young showed that path integrals such as $\int f(Y_t)\,\mathrm{d}X_t$ can make sense as the limits of Riemannian sums even if the path X is rougher than paths with finite variation, provided that Y and the one-form f are smooth enough to compensate for the roughness of X.

Let V, H, W be three Banach spaces, and let X (Y, respectively) be a continuous path in V (in W, respectively) with finite p-variation (with finite q-variation, respectively). Therefore, we may choose a control ω such that

$$\left|X_{s,t}^1\right| \leqslant \omega(s,t)^{1/p} \quad \text{and} \quad \left|Y_{s,t}^1\right| \leqslant \omega(s,t)^{1/q}, \quad \forall (s,t) \in \Delta. \tag{3.82}$$

Suppose that $p \geqslant 1$, $q \geqslant 1$, and $1/p + 1/q \equiv \theta > 1$. Let $f : W \to \boldsymbol{L}(V, H)$ be a Lipschitz one-form, namely

$$|f(\xi_1) - f(\xi_2)| \leqslant M\,|\xi_1 - \xi_2|, \quad \forall \xi_1, \xi_2 \in W. \tag{3.83}$$

The Young integral is defined by

$$\int_s^t f(Y)\,\mathrm{d}X = \lim_{m(D)\to 0} \sum_l f(Y_{t_{l-1}})(X_{t_{l-1},t_l}^1), \tag{3.84}$$

for all $(s,t) \in \Delta$, where the limit is taken over finite partitions of $[s,t]$.

Example 3.3.1 Let V and W be two Banach spaces and let $H = W \otimes V$ be a Banach tensor product. Define $f : W \to \boldsymbol{L}(V, W \otimes V)$ by $f(\xi)(\eta) = \xi \otimes \eta$, for all $\xi \in W$, $\eta \in V$. Then f is a Lipschitz one-form with Lipschitz constant $M = 1$.

We may use the same method as in Chapter 2 to show that the Young integral (3.84) exists under the above conditions.

Theorem 3.3.1 *Under the above conditions, we have the following conclusions:*

(i) The integral

$$Z_{s,t}^1 \equiv \int_s^t f(Y)\,\mathrm{d}X = \lim_{m(D)\to 0} \sum_l f(Y_{t_{l-1}})(X_{t_{l-1},t_l}^1)$$

is well defined, and $Z \in C_{0,p}(\Delta, T^{(1)}(H))$, where $Z_{s,t} = (1, Z_{s,t}^1)$. Moreover, Z satisfies the Chen identity, that is $Z_{s,t}^1 = Z_t - Z_s$, where $Z_t = Z_{0,t}^1$.

(ii) With the above control ω, we have

$$\left|\int_s^t f(Y)\,\mathrm{d}X\right| \leqslant M_1\omega(s,t)^{1/p}, \quad \forall (s,t) \in \Delta, \tag{3.85}$$

for some constant M_1.

(iii) The map $(X, Y) \to Z$ is a continuous map from $C_{0,p}(\Delta, T^{(1)}(V)) \times C_{0,q}(\Delta, T^{(1)}(W))$ into $C_{0,p}(\Delta, T^{(1)}(H))$.

Proof The procedure to prove the existence of the Young integral (3.84) is similar to those used in Chapter 2. For any partition D of $[s,t]$, let

$$Z(D)_{s,t} = \sum_{l=1}^{r} f(Y_{t_{l-1}})(X_{t_l} - X_{t_{l-1}}) = \sum_{l=1}^{r} f(Y_{t_{l-1}})(X^1_{t_{l-1},t_l}), \qquad (3.86)$$

where $D = \{s = t_0 < t_1 < \cdots < t_r = t\}$. Using the same trick as in Section 1.2, we choose a point t_l in D such that

$$\omega(t_{l-1}, t_{l+1}) \leqslant \frac{2\omega(s,t)}{r-2}, \quad \text{if} \quad r > 2, \qquad (3.87)$$

and when $r = 2$ (this means that there is no other point in D other than s,t) we just replace the term $2/(r-2)$ in the above inequality by 1. Now delete the point t_l from D to get a new division D'. Then we have

$$Z(D)_{s,t} - Z(D')_{s,t} = f(Y_{t_{l-1}})(X^1_{t_{l-1},t_l}) + f(Y_{t_l})(X^1_{t_l,t_{l+1}}) - f(Y_{t_{l-1}})(X^1_{t_{l-1},t_{l+1}})$$
$$= f(Y_{t_l})(X^1_{t_l,t_{l+1}}) - f(Y_{t_{l-1}})(X^1_{t_l,t_{l+1}})$$
$$= \left(f(Y_{t_l}) - f(Y_{t_{l-1}})\right)(X^1_{t_l,t_{l+1}}),$$

and therefore

$$|Z(D)_{s,t} - Z(D')_{s,t}| \leqslant M\omega(t_{l-1}, t_l)^{1/q}\omega(t_l, t_{l+1})^{1/p}$$
$$\leqslant M\left(\frac{2}{r-2}\right)^{\theta}\omega(s,t)^{\theta}. \qquad (3.88)$$

The estimate (3.88) implies the existence of the limit in (3.84). Moreover, by repeating the above argument we have

$$|Z(D)_{s,t} - f(y_0)(X^1_{s,t})| \leqslant M\left[1 + \sum_{r=3}^{\infty}\left(\frac{2}{r-2}\right)^{\theta}\right]\omega(s,t)^{\theta}, \qquad (3.89)$$

for any finite partition D of $[s,t]$. Therefore,

$$\left|\int_s^t f(Y)\,\mathrm{d}X\right| \leqslant |f(y_0)|\,\omega(s,t)^{1/p} + M\left[1 + \sum_{r=3}^{\infty}\left(\frac{2}{r-2}\right)^{\theta}\right]\omega(s,t)^{\theta}.$$

Thus we have proved the first two conclusions. We next prove the continuity. Let $\hat{X}, \hat{Y}$ be two continuous paths which satisfy the previous conditions and also

$$\left|X^1_{s,t} - \hat{X}^1_{s,t}\right| \leqslant \varepsilon\omega(s,t)^{1/p}, \quad \left|Y^1_{s,t} - \hat{Y}^1_{s,t}\right| \leqslant \varepsilon\omega(s,t)^{1/q}, \qquad (3.90)$$

for all $(s,t) \in \Delta$. Then, by using the same trick as before, we may control the norm of the difference

$$Z(D)_{s,t} - Z(D')_{s,t} - \left(\hat{Z}(D)_{s,t} - \hat{Z}(D')_{s,t}\right)$$

from above by

$$\text{constant} \times \left(\frac{2}{r-2}\right)^{\theta} \omega(s,t)^{1/p}\,.$$

Therefore, by eliminating all of the partition points t_l, we obtain the conclusion (iii) of the theorem. ■

3.3.3 *Elementary operations on rough paths*

The rough path space $\Omega_p(V)$ is not a linear space (it is a closed subset of $C_{0,p}(\triangle, T^{([p])}(V))$ under p-variation topology) due to the nonlinearity of Chen's identity. There are several operations on $\Omega_p(V)$ which we describe in the following.

Scalar multiplication

This is a map from $\mathbb{R} \times \Omega_p(V) \to \Omega_p(V)$. For any $\alpha \in \mathbb{R}$, $X \in \Omega_p(V)$, we define $\alpha X \in \Omega_p(V)$ (we also denote this rough path by $\Gamma(\alpha)X$) to be the rough path

$$\left(1, \alpha X_{s,t}^1, \alpha^2 X_{s,t}^2, \ldots, \alpha^{[p]} X_{s,t}^{[p]}\right), \quad \forall (s,t) \in \triangle\,. \tag{3.91}$$

It is clear that the map $\Gamma(\alpha) : (\alpha, X) \to \Gamma(\alpha)X$ is a continuous map from $\mathbb{R} \times \Omega_p(V)$ to $\Omega_p(V)$ in p-variation topology. Clearly, $G\Omega_p(V)$ is invariant under $\Gamma(\alpha)$.

Remark 3.3.3 The operator $\Gamma(\alpha)$ is called the second quantization of the operator $\alpha : V \to V$, $\xi \to \alpha\xi$ (regarded as a bounded linear operator on V), which is in fact well defined as an operator from $T^{(n)}(V)$ into $T^{(m)}(V)$, in an obvious sense. In general, if V, W are two Banach spaces and $A \in \boldsymbol{L}(V, W)$, then, for any $n \geqslant m$, we define the second quantization $\Gamma(A)$ of A by

$$\Gamma(A) : T^{(n)}(V) \to T^{(m)}(W)\,,$$
$$(1, \xi^1, \xi^2, \ldots, \xi^n) \to (1, A\xi^1, A^{\otimes 2}\xi^2, \ldots, A^{\otimes m}\xi^m)\,,$$

where $A^{\otimes k} : V^{\otimes k} \to W^{\otimes k}$ is the linear operator such that

$$A^{\otimes k}(\xi_1 \otimes \cdots \otimes \xi_k) = A\xi_1 \otimes \cdots \otimes A\xi_k\,, \tag{3.92}$$

for $\xi_i \in V$ and $i = 1, \ldots, k$.

Translations

Let us first look at a simple case. Since the material in this subsection is only interesting for really rough paths, we therefore assume that $p \geqslant 2$. Suppose that $1 \leqslant q < 2$ such that $\theta \equiv 1/p + 1/q > 1$. Let $X \in \Omega_p(V)$ and $H \in \Omega_q(V)$. Notice that $[q] = 1$, and therefore $H_{s,t} = (1, H_{s,t}^1)$. For simplicity, we denote $H_{0,t}^1$ by H_t, which is a continuous path in V with finite q-variation. We define a new rough path $X^H \in \Omega_p(V)$. Formally, it is defined by

$$\left(X^H\right)^k_{s,t} \equiv \int_{s<t_1<\cdots<t_k<t} \mathrm{d}(X+H)_{t_1} \otimes \cdots \otimes \mathrm{d}(X+H)_{t_k}. \tag{3.93}$$

For simplicity we look at (3.93) carefully when $2 \leqslant p < 3$. In this case, $[p] = 2$ and, of course,

$$\left(X^H\right)^1_{s,t} = X^1_{s,t} + H^1_{s,t}, \quad \forall (s,t) \in \Delta, \tag{3.94}$$

$$\left(X^H\right)^2_{s,t} = X^2_{s,t} + H^2_{s,t} + \int_s^t H^1_{s,u} \otimes \mathrm{d}X^1_u + \int_s^t X^1_{s,u} \otimes \mathrm{d}H_u, \tag{3.95}$$

where $H^2_{s,t}$ certainly makes sense by Theorem 3.1.2, and the integrals appearing in the right-hand side of (3.95) are the Young integrals. More precisely,

$$\int_s^t H^1_{s,u} \otimes \mathrm{d}X_u = \lim_{m(D)\to 0} \sum_l H^1_{s,t_{l-1}} \otimes X^1_{t_{l-1},t_l},$$

and

$$\int_s^t X^1_{s,u} \otimes \mathrm{d}H_u = \lim_{m(D)\to 0} \sum_l X^1_{s,t_{l-1}} \otimes H^1_{t_{l-1},t_l}.$$

Then $X^H \in \Omega_p(V)$, and the map $(X,H) \to X^H$ is a continuous map from $\Omega_p(V) \times \Omega_q(V)$ to $\Omega_p(V)$ in variation topology.

In general, if $p \geqslant 2$ and $1 \leqslant q_i < 2$ such that $\theta_i \equiv 1/p + 1/q_i > 1$, if $X \in \Omega_p(V)$, and if $H^i \in C_{0,q_i}(\Delta, T^{(1)}(V^{\otimes i}))$, for $i = 1, \ldots, [p]$, then we have the following theorem.

Theorem 3.3.2 *Under the above assumptions, $X + H \in C_0(\Delta, T^{([p])}(V))$, defined as the sum of two (vector-valued) continuous functions by*

$$(X+H)_{s,t} \equiv \left(1, X^1_{s,t} + H^1_{s,t}, \ldots, X^{[p]}_{s,t} + H^{[p]}_{s,t}\right), \quad \forall (s,t) \in \Delta, \tag{3.96}$$

is an almost rough path with finite p-variation. Denote by X^H the rough path associated with the almost rough path $X + H$ as defined in Theorem 3.2.1. Then the map $(X,H) \to X^H$ is a continuous map from $\Omega_p(V) \times \prod_{i=1}^{[p]} C_{0,q_i}(\Delta, T^{(1)}(V^{\otimes i}))$ $\to \Omega_p(V)$ in variation topology.

Proof Clearly the map $(X,H) \to X + H$ is a continuous map from $\Omega_p(V) \times \prod_{i=1}^{[p]} C_{0,q_i}(\Delta, T^{(1)}(V^{\otimes i}))$ to $C_{0,p}(\Delta, T^{([p])}(V))$ in variation topology. Therefore, we only need to prove that $X + H$ is an almost multiplicative functional, since $X + H$ possesses finite p-variation. However, it is easily seen that

$$\left((X+H)_{s,t} \otimes (X+H)_{t,u} \right)^k = X^k_{s,u} + H^k_{s,t} + H^k_{t,u} + \sum_{i=1}^{k-1} H^i_{s,t} \otimes X^{k-i}_{t,u}$$

$$+ \sum_{i=1}^{k-1} X^i_{s,t} \otimes H^{k-i}_{t,u} + \sum_{i=1}^{k-1} H^i_{s,t} \otimes H^{k-i}_{t,u},$$

and therefore the difference

$$\left((X+H)_{s,t} \otimes (X+H)_{t,u} \right)^k - (X+H)^k_{s,u}$$

$$= H^k_{s,t} + H^k_{t,u} - H^k_{s,u}$$

$$+ \sum_{i=1}^{k-1} \left(H^i_{s,t} \otimes X^{k-i}_{t,u} + X^i_{s,t} \otimes H^{k-i}_{t,u} + H^i_{s,t} \otimes H^{k-i}_{t,u} \right).$$

Hence, the norm

$$\left| \left((X+H)_{s,t} \otimes (X+H)_{t,u} \right)^k - (X+H)^k_{s,u} \right|$$

is dominated by $K\omega(s,t)^\theta$, for some constant K depending only on $\max \omega$, p, and θ. Thus we have completed the proof. $\blacksquare$

Time-reversal of a rough path

For a geometric rough path, we can define its time-reversal at a fixed time. Let us begin with a smooth rough path $X \in \Omega_p(V)$. Its projection $X_t \equiv X^1_{0,t}$ in V is a continuous path with finite variation. The time-reversal of the path X_t at time T is a new path $Y_t = X_{T-t}$, which is again a Lipschitz path, and it generates a smooth rough path, which we again denote by Y. By definition, the ith component of the smooth rough path Y is

$$Y^i_{s,t} = \int_{s<u_1<\cdots<u_i<t} dY_{u_1} \otimes \cdots \otimes dY_{u_i}$$

$$= \int_{s<u_1<\cdots<u_i<t} \dot{Y}_{u_1} \otimes \cdots \otimes \dot{Y}_{u_i} \, du_1 \cdots du_i$$

$$= \int_{T-t<t_i<\cdots<t_1<T-s} (-1)^i \dot{X}_{t_1} \otimes \cdots \otimes \dot{X}_{t_i} \, dt_1 \cdots dt_i$$

$$= (-1)^i \pi_i \int_{T-t<t_i<\cdots<t_1<T-s} \dot{X}_{t_i} \otimes \cdots \otimes \dot{X}_{t_1} \, dt_i \cdots dt_1$$

$$= (-1)^i \pi_i X^i_{T-t,T-s}, \tag{3.97}$$

where π_i is the permutation of $\{1,\ldots,i\}$ which sends j to $i+1-j$, and π_i induces a natural transformation on the tensor product $V^{\otimes i}$. Therefore, we have the following theorem.

Theorem 3.3.3 *Let $X \in \Omega_p(V)$ be a geometric rough path (with running time T). Define $Y \in C_0(\Delta_T, T^{([p])}(V))$ by*

$$Y^i_{s,t} = (-1)^i \pi_i X^i_{T-t,T-s}, \quad i = 1, \ldots, [p] \quad and \quad (s,t) \in \Delta_T .$$

Then Y is a geometric rough path in $T^{([p])}(V)$, and the map $X \to Y$ is continuous in p-variation topology. Here Y is called the time-reversal of the geometric rough path X at time T.

Proof It is clear by definition that the map $X \to Y$ is continuous in p-variation topology. The only thing that we need to show is that Y indeed satisfies the Chen identity, and in fact it is at this stage that we require X to be geometric. Since Y satisfies the Chen identity if X is a smooth rough path, then the continuity allows us to pass to a geometric rough path. ∎

3.4 Comments and notes on Chapter 3

The concept of iterated path integrals goes back, as far as we know, to the fundamental work by E. Picard. The simplest type of iterated path integrals appeared in his famous iterative procedure for solving differential equations and corresponds to approximating the exponential function via the polynomials $\sum_{i=0}^{n} z^i/i!$. In some sense, the work presented here is a generalization of the Picard iteration method, so that differential equations driven by non-smooth paths can be solved. Iterated path integration, as a method for constructing differential forms on (smooth) loop and path spaces over smooth manifolds, was observed by various authors around the 1950s. An extensive study, including many interesting results about the topology and the geometry of path-type space, has been carried out by K. T. Chen in a series of papers, see for example Chen (1954, 1957, 1958, 1961, 1967, 1968, 1971, 1973, 1977). Chen's motivation for studying iterated path integrals was a desire to develop a geometric integration theory and to establish a Hodge–de Rham theorem for path-type spaces, a long-standing question which was recently emphasized again by P. Malliavin (1984) and L. Gross (1998). For early works on the geometric aspects of iterated path integration, see also Whitehead (1947) and Paršin (1958). For details of the algebraic aspect, especially iterated path integrals as Lie elements, see Ree (1958, 1960) and Reutenauer (1993). In fact, Ree (1958) was the first person to discover the so-called 'multiplicative property' of iterated integral sequences in a suitable tensor algebra setting, under the name of shuffle multiplication. Therefore, with historical correctness, Chen's identity should be renamed as Ree's equation. It is the viewpoint that iterated path integrals are Lie elements of some (infinite) Lie groups which has been accepted in Lyons (1998) in order to develop a theory of differential equations driven by rough paths. The same point of view is taken in this monograph in the sense that we regard sequences of iterated path integrals as basic objects like 'generalized differentials', and we leave the important geometric and algebraic significance aside, since our intention is to apply the main techniques to SDEs. However, it is certainly interesting and important to explore the geometric and algebraic meanings of rough paths, but we are not able to develop this here.

Section 3.1. All of the results can be found in Lyons (1998). The basic estimates contained in Theorems 3.1.2 and 3.1.3 were proved in Lyons (1998) by extending the idea and a trick in the important paper of Young (1936). A version for $p < 2$ (Young's estimate) is revisited in Lyons (1994).

Section 3.2. The concept of almost rough paths was introduced in Lyons (1998), under the name of almost multiplicative functionals, as an intermediate step for defining path integrals along rough paths. The basic continuity Theorem 3.2.1 was also proved in Lyons (1998).

Section 3.3. The variation distance has appeared in different situations in mathematics, especially in probability theory. However, it is pointed out in Lyons (1998) that p-variation topology is appropriate to the study of differential equations.

4

BROWNIAN ROUGH PATHS

In this chapter we present several examples of geometric rough paths. As we have seen, to construct a geometric rough path over a continuous path (W_t) is the same question as establishing an iterated path integration theory for this process. One of the interesting cases is when (W_t) is a continuous stochastic process. In this case we want to construct the 'canonical' rough paths associated with the sample paths of the stochastic process (W_t). We consider several stochastic processes which are very often used in the stochastic analysis for which it is desirable to have a (stochastic) integration theory. These stochastic processes include a class of square-integrable processes whose correlation functions satisfy a simple decay condition (see below for detail), Brownian motion, Wiener processes in Banach spaces, fractional Brownian motions with Hurst parameter $h > 1/4$ and square-integrable martingales. Some of these processes are neither Markovian nor Gaussian. Thus we provide a large class of stochastic processes—most of them are not related to semi-martingales—which we may associate with canonical geometric rough paths, and therefore we may apply the integration theory which we will develop in later chapters to them.

4.1 Control variation distances

In this section we establish several technical tools to control p-variation distances between rough paths.

Recall that a multiplicative functional X in $T^{(N)}(V)$, where $X_{s,t} = (1, X^1_{s,t}, \ldots, X^N_{s,t})$, is said to have finite p-variation if

$$\sup_D \sum_l \left| X^i_{t_{l-1},t_l} \right|^{p/i} < +\infty, \quad \forall i = 1, \ldots, N. \tag{4.1}$$

Such a multiplicative functional X is called a geometric rough path if $N \geqslant [p]$ and if there is a sequence of smooth rough paths $X(n)$ (i.e. a sequence of continuous paths $X(n)$ of finite variation together with their iterated path integrals $X(n)^k$ up to degree $[p]$), where $X(n)_{s,t} = (1, X(n)^1_{s,t}, \ldots, X(n)^{[p]}_{s,t})$, such that, according to p-variation distance, $X(n)$ converges to X:

$$\sup_{1 \leqslant i \leqslant [p]} \sup_D \left(\sum_l \left| X(n)^i_{t_{l-1},t_l} - X^i_{t_{l-1},t_l} \right|^{p/i} \right)^{i/p} \to 0,$$

as $n \to \infty$.

We shall establish several basic estimates about p-variation distances which can be applied to any continuous multiplicative functionals.

Let $[S, T]$ be any finite interval. By its dyadic decompositions we mean partitions $\{t_0^n < t_1^n < \cdots < t_{2^n}^n\}$ of $[S, T]$, where

$$t_k^n = \frac{k}{2^n}(T - S) + S, \quad k = 0, \ldots, 2^n, \tag{4.2}$$

and $n \in \mathbb{N}$. The bigger n is, the finer becomes the dyadic partition. The above notations will be used frequently in this chapter, and in the most part of this chapter, it will be the case that $[S, T] = [0, 1]$, though the results can be applied to any bounded intervals. In this case, $t_k^n = k/2^n$ ($k = 0, \ldots, 2^n$) are the dyadic points of the unit interval $[0, 1]$.

Proposition 4.1.1 *Let $X \in C_0(\Delta, T^{(N)}(V))$ (with a fixed running time interval, say $[0, 1]$) be a multiplicative functional. Then for any $1 \leqslant i \leqslant N$, p satisfying $p/i > 1$, and any $\gamma > p/i - 1$, there exists a constant $C_i(p, \gamma)$ depending only on p, γ, and i, such that*

$$\sup_D \sum_l \left| X_{t_{l-1}, t_l}^i \right|^{p/i} \leqslant C_i(p, \gamma) \sum_{n=1}^{\infty} n^{\gamma} \sum_{k=1}^{2^n} \sum_{j=1}^{i} \left| X_{t_{k-1}^n, t_k^n}^j \right|^{p/j}, \tag{4.3}$$

where $\sup_D$ runs over all finite partitions D of $[0, 1]$.

Proof By the Hölder inequality

$$\left| \sum_{n=1}^{\infty} a_n \right|^q \leqslant \left(\sum_{n=1}^{\infty} n^{-\gamma/(q-1)} \right)^{q-1} \sum_{n=1}^{\infty} n^{\gamma} |a_n|^q \tag{4.4}$$

for any sequence of numbers $\{a_n\}$, $q > 1$, and any $\gamma > q - 1$. The last condition guarantees that the series $\sum_{n=1}^{\infty} n^{-\gamma/(q-1)}$ is convergent. This is the reason why we assume that $p/i > 1$. Equation (4.4) is useful only when $\sum_n a_n$ converges to some limit very fast. For simplicity we use $C_0(\gamma, q)$ to denote the constant $\left(\sum_{n=1}^{\infty} n^{-\gamma/(q-1)} \right)^{q-1}$. To prove (4.3) we need the following fact: any sub-interval $[s, t]$ of $[0, 1]$ can be covered by a set of intervals (perhaps infinitely many) of the form $[t_{k-1}^n, t_k^n]$ with the maximum lengths (i.e. for the least possible n). Indeed, we may write any sub-interval $[s, t]$ as a disjoint union of intervals of the form $[t_{k-1}^n, t_k^n]$, such that for each n we need at most two different intervals $[t_{k-1}^n, t_k^n]$. Let us detail the construction as the following. Fix a sub-interval $[s, t] \subseteq [0, 1]$. Choose the least $n_0 \in \mathbb{N}$ and some $k_{n_0} = 0, 1, \ldots$ or $2^{n_0} - 1$, such that $[t_{k_{n_0}}^{n_0}, t_{k_{n_0}+1}^{n_0}]$ is either the sub-interval with maximum length in $[s, t]$ among all intervals with the form $[t_{k-1}^n, t_k^n]$ ($n \in \mathbb{N}$ and $k = 1, \ldots, 2^n$) or exactly

$$\left[t_{k_{n_0}}^{n_0}, t_{k_{n_0}+1}^{n_0} \right] = [s, t]. \tag{4.5}$$

If not the latter case and if $(t_{k_{n_0}+1}^{n_0}, t) \neq \emptyset$, then we may continue our construction. We then choose the least $n_1 > n_0$ and $k_{n_1} = 0, 1, \ldots$ or $2^{n_1} - 1$ such that $t^{n_0}_{k_{n_0}+1} = t^{n_1}_{k_{n_1}}$ and $[t^{n_1}_{k_{n_1}}, t^{n_1}_{k_{n_1}+1}]$ is the maximum sub-interval with the form $[t^n_{k-1}, t^n_k]$ (in the sense of length) of $[t^{n_0}_{k_{n_0}+1}, t]$. By repeating the above procedure we eventually obtain a subsequence:

$$t^{n_0}_{k_{n_0}+1} < t^{n_1}_{k_{n_1}} < t^{n_2}_{k_{n_2}} < \cdots < t^{n_e}_{k_{n_e}} = t \tag{4.6}$$

if $e < \infty$, otherwise it means that $t^{n_e}_{k_{n_e}} \to t$.

The same argument applies to the left end-points, and therefore we have another subsequence

$$t^{n_0}_{k_{n_0}} = t^{n'_0}_{k_{n'_0}} > t^{n'_1}_{k_{n'_1}} > \cdots \to s. \tag{4.7}$$

For simplicity, denote the right end-points $t^{n_i}_{k_{n_i}}$ by s_i, and the left end-points $t^{n'_i}_{k_{n'_i}}$ by s_{-i}. By the construction, for any $n \in \mathbb{N}$, there are at most two intervals $[t^n_{k-1}, t^n_k]$ with the same index n but different k which may appear in the above two sequences. By eqns (4.6) and (4.7),

$$[s, t] = \cup^\infty_{i=-\infty}[s_i, s_{i+1}] \tag{4.8}$$

with each interval $[s_i, s_{i+1}] = [t^n_{k-1}, t^n_k]$ for some n and k.

We are now in a position to prove the domination inequality. First prove the case that $i = 1$. By (4.8) and Chen's identity, we have

$$X^1_{s,t} = \sum^\infty_{i=-\infty} X^1_{s_{i-1}, s_i}. \tag{4.9}$$

Hence,

$$\left| X^1_{s,t} \right|^p = \left| \sum^\infty_{i=-\infty} X^1_{s_{i-1},s_i} \right|^p$$

$$\leqslant C_p \left\{ \left| X^1_{t^{n_0}_{k_{n_0}}, t^{n_0}_{k_{n_0}+1}} \right|^p + \left| \sum_i X^1_{t^{n_i}_{k_{n_i}}, t^{n_i}_{k_{n_i}+1}} \right|^p + \left| \sum_i X^1_{t^{n'_i}_{k_{n'_i}}, t^{n'_i}_{k_{n'_i}+1}} \right|^p \right\}$$

$$\leqslant C_p n_0^\gamma \left| X^1_{t^{n_0}_{k_{n_0}}, t^{n_0}_{k_{n_0}+1}} \right|^p + C_p C_0(\gamma, p) \sum_i i^\gamma \left| X^1_{t^{n_i}_{k_{n_i}}, t^{n_i}_{k_{n_i}+1}} \right|^p$$

$$+ C_p C_0(\gamma, p) \sum_i i^\gamma \left| X^1_{t^{n'_i}_{k_{n'_i}}, t^{n'_i}_{k_{n'_i}+1}} \right|^p, \tag{4.10}$$

where C_p is a positive constant depending only on p. Notice that in (4.10) we always have $i \leqslant n_i$ and $i \leqslant n'_i$. Let $D = \{0 = t_0 < t_1 < \cdots < t_r = 1\}$ be any finite division of $[0, 1]$. Apply (4.10) to each $[t_{l-1}, t_l]$. Since all intervals $[t^n_{k-1}, t^n_k]$

in (4.10) are different for different $[t_{l-1}, t_l]$ by our construction, and therefore, summing up (4.10) over all intervals $[t_{l-1}, t_l]$, we obtain

$$\sum_l \left| X^1_{t_{l-1}, t_l} \right|^p \leqslant C_1(p, \gamma) \sum_{n=1}^{\infty} n^{\gamma} \sum_{k=1}^{2^n} \left| X^1_{t^n_{k-1}, t^n_k} \right|^p. \tag{4.11}$$

Since the right-hand side of (4.11) is independent of the partition D, we have therefore completed the proof for $i = 1$.

We may finish the proof of Proposition 4.1.1 by an induction argument on i. To exhibit the idea, we prove it for $i = 2$. The proof is similar to the case that $i = 1$. In this case we apply Chen's identity to the second component X^2. Consider any finite interval $[s, t]$, and use the same notations as above. By (4.5) we have

$$X^2_{s,t} = \sum_i X^2_{s_{i-1}, s_i} + \sum_{i<j} X^1_{s_{i-1}, s_i} \otimes X^1_{s_{j-1}, s_j} \tag{4.12}$$

and therefore for any $p > 2$,

$$\left| X^2_{s,t} \right|^{p/2} \leqslant 2^{p/2-1} \left| \sum_i X^2_{s_{i-1}, s_i} \right|^{p/2} + 2^{p/2-1} \left| \sum_{i<j} X^1_{s_{i-1}, s_i} \otimes X^1_{s_{j-1}, s_j} \right|^{p/2}$$

$$\leqslant 2^{p/2-1} \left| \sum_i X^2_{s_{i-1}, s_i} \right|^{p/2} + 2^{p/2-1} \left(\sum_i \left| X^1_{s_{i-1}, s_i} \right| \right)^p. \tag{4.13}$$

Apply the above estimate to each $[t_{l-1}, t_l]$ of a finite partition D of $[0, 1]$, where $D = \{0 = t_0 < t_1 < \cdots < t_r = 1\}$, and thus obtain

$$\sum_l \left| X^2_{t_{l-1}, t_l} \right|^{p/2} \leqslant C_2(p, \gamma) \sum_{n=1}^{\infty} n^{\gamma} \sum_{k=1}^{2^n} \left(\left| X^2_{t^n_{k-1}, t^n_k} \right|^{p/2} + \left| X^1_{t^n_{k-1}, t^n_k} \right|^p \right), \tag{4.14}$$

for some constant $C_2(p, \gamma)$ depending only on $p > 2$, $\gamma > p/2 - 1$, which is the required estimate for $i = 2$. ∎

Let $X, Y \in C_0(\Delta, T^{(N)}(V))$ be two multiplicative functionals, and apply the domination control (4.3) to $X^1_{s,t} - Y^1_{s,t}$ ($i = 1$). We therefore get that

$$\sup_D \sum_l \left| X^1_{t_{l-1}, t_l} - Y^1_{t_{l-1}, t_l} \right|^p \leqslant C_1(p, \gamma) \sum_{n=1}^{\infty} n^{\gamma} \sum_{k=1}^{2^n} \left| X^1_{t^n_{k-1}, t^n_k} - Y^1_{t^n_{k-1}, t^n_k} \right|^p, \tag{4.15}$$

where D is any finite partition of $[0, 1]$ and $t^n_k = k/2^n$.

However, the situation for $i \geqslant 2$ is slightly different as the difference $X - Y$ of two multiplicative functionals X, Y is not multiplicative.

Let us consider the case $i = 2$ for simplicity. Therefore let X, Y be two multiplicative functionals in $T^{(2)}(V)$, and let us use the same notations established in the proof of Proposition 4.1.1. Then (4.12) holds both for X and Y. Taking the difference we therefore establish

$$X_{s,t}^2 - Y_{s,t}^2 = \sum_i \left(X_{s_{i-1},s_i}^2 - Y_{s_{i-1},s_i}^2 \right)$$

$$+ \sum_{i<j} \left(X_{s_{i-1},s_i}^1 - Y_{s_{i-1},s_i}^1 \right) \otimes X_{s_{j-1},s_j}^1$$

$$+ \sum_{i<j} Y_{s_{i-1},s_i}^1 \otimes \left(X_{s_{j-1},s_j}^1 - Y_{s_{j-1},s_j}^1 \right). \qquad (4.16)$$

Proposition 4.1.2 *For any $p > 2$, $\gamma > p/2 - 1$, there is a constant $C(p,\gamma) > 0$ depending only on γ and p such that*

$$\sup_D \sum_l \left| X_{t_{l-1},t_l}^2 - Y_{t_{l-1},t_l}^2 \right|^{p/2}$$

$$\leqslant C(p,\gamma) \left(\sum_{n=1}^\infty n^\gamma \sum_{k=1}^{2^n} \left| X_{t_{k-1}^n,t_k^n}^1 - Y_{t_{k-1}^n,t_k^n}^1 \right|^p \right)^{1/2}$$

$$\times \left(\sum_{n=1}^\infty n^\gamma \sum_{k=1}^{2^n} \left| X_{t_{k-1}^n,t_k^n}^1 \right|^p + \left| Y_{t_{k-1}^n,t_k^n}^1 \right|^p \right)^{1/2}$$

$$+ C(p,\gamma) \sum_{n=1}^\infty n^\gamma \sum_{k=1}^{2^n} \left| X_{t_{k-1}^n,t_k^n}^2 - Y_{t_{k-1}^n,t_k^n}^2 \right|^{p/2}. \qquad (4.17)$$

Proof Apply (4.16) to each sub-interval $[t_{l-1}, t_l]$ of a partition D, and use a similar argument as in the proof of Proposition 4.1.1. We may thus control the first term on the right-hand side of (4.16), which gives the second term on the right-hand side of (4.17). Next we estimate the second and the third terms in (4.16). Observe that

$$\left| \sum_{i<j} \left(X_{s_{i-1},s_i}^1 - Y_{s_{i-1},s_i}^1 \right) \otimes X_{s_{j-1},s_j}^1 \right|^{p/2}$$

$$\leqslant \left(\sum_{i<j} \left| X_{s_{i-1},s_i}^1 - Y_{s_{i-1},s_i}^1 \right| \left| X_{s_{j-1},s_j}^1 \right| \right)^{p/2}$$

$$\leqslant \left(\sum_i \left| X_{s_{i-1},s_i}^1 - Y_{s_{i-1},s_i}^1 \right| \right)^{p/2} \left(\sum_j \left| X_{s_{j-1},s_j}^1 \right| \right)^{p/2}$$

$$\leqslant C(p,\gamma) \left(\sum_i i'^\gamma \left| X^1_{t^n_{\bullet\, i'},\, t^n_{\bullet\, i'}} - Y^1_{t^n_{\bullet\, i'},\, t^n_{\bullet\, i'}} \right|^p \right)^{1/2}$$

$$\times \left(\sum_j j'^\gamma \left| X^1_{t^n_{\bullet\, j'},\, t^n_{\bullet\, j'}} \right|^p \right)^{1/2},$$

where the natural numbers $\{n'_i\}$ are the rearrangement of s_i according to the natural order. Summing up over all $[t_{l-1}, t_l]$ in the partition D and then using the Cauchy–Schwartz inequality, we therefore deduce that

$$\sum_l \left| \sum_{i<j} \left(X^1_{s_{i-1},s_i} - Y^1_{s_{i-1},s_i} \right) \otimes X^1_{s_{j-1},s_j} \right|^{p/2}$$

$$\leqslant C(p,\gamma) \left(\sum_{n=1}^{\infty} n^\gamma \sum_{k=1}^{2^n} \left| X^1_{t^n_{k-1},t^n_k} - Y^1_{t^n_{k-1},t^n_k} \right|^p \right)^{1/2}$$

$$\times \left(\sum_{n=1}^{\infty} n^\gamma \sum_{k=1}^{2^n} \left| X^1_{t^n_{k-1},t^n_k} \right|^p + \left| Y^1_{t^n_{k-1},t^n_k} \right|^p \right)^{1/2},$$

which ends the proof. ∎

We may of course, by using the Chen identity, continue the same procedure to the higher-level paths of multiplicative functionals, and therefore we may control the p-variation distance of two multiplicative functionals with any degree. However, the formulae become increasingly complicated. For example, for the third-level paths, it reads as the following:

$$\sup_D \sum_l \left| X^3_{t_{l-1},t_l} - Y^3_{t_{l-1},t_l} \right|^{p/3}$$

$$\leqslant C_1 \sum_{n=1}^{\infty} n^\gamma \sum_{k=1}^{2^n} \left| X^3_{t^n_{k-1},t^n_k} - Y^3_{t^n_{k-1},t^n_k} \right|^{p/3}$$

$$+ C_2 \left(\sum_{n=1}^{\infty} n^\gamma \sum_{k=1}^{2^n} \left| X^1_{t^n_{k-1},t^n_k} - Y^1_{t^n_{k-1},t^n_k} \right|^p \right)^{1/3}$$

$$\times \left(\sum_{n=1}^{\infty} n^\gamma \sum_{k=1}^{2^n} \left| X^2_{t^n_{k-1},t^n_k} \right|^{p/2} + \left| Y^2_{t^n_{k-1},t^n_k} \right|^{p/2} \right)^{2/3}$$

$$+ C_3 \left(\sum_{n=1}^{\infty} n^\gamma \sum_{k=1}^{2^n} \left| X^2_{t^n_{k-1},t^n_k} - Y^2_{t^n_{k-1},t^n_k} \right|^{p/2} \right)^{2/3}$$

$$\times \left(\sum_{n=1}^{\infty} n^\gamma \sum_{k=1}^{2^n} \left| X^1_{t^n_{k-1},t^n_k} \right|^p + \left| Y^1_{t^n_{k-1},t^n_k} \right|^p \right)^{1/3}$$

$$+ C_4 \left(\sum_{n=1}^{\infty} n^{\gamma} \sum_{k=1}^{2^n} \left| X_{t_{k-1}^n,t_k^n}^1 - Y_{t_{k-1}^n,t_k^n}^1 \right|^p \right)^{1/3}$$

$$\times \left(\sum_{n=1}^{\infty} n^{\gamma} \sum_{k=1}^{2^n} \left| X_{t_{k-1}^n,t_k^n}^1 \right|^p + \left| Y_{t_{k-1}^n,t_k^n}^1 \right|^p \right)^{2/3} . \qquad (4.18)$$

Indeed, these are all we need in the remainder of this chapter.

4.2 Dyadic polygonal approximations

The typical situation in applications is the following: we are given a path (W_t) taking in some manifold as a driving force of a dynamical system, and in general (W_t) is a non-smooth path and more often is a stochastic process with very irregular sample paths. Therefore, as in the theory we have developed in the previous chapters, in order to understand such a system better, we need to realize the non-smooth path (W_t) as a geometric rough path. Hence, a natural mathematical question arises: is there any canonical geometric rough path which associates with a non-smooth path? The answer is never unique. As we have seen, if there is a rough path such that its projection is a given path with finite p-variation, then there are infinitely many rough paths with the same projection if $p > 2$. Thus we must come to some specific constructions. In our study, we will choose the so-called dyadic polygonal approximations (see below for details), though other constructions are possible (and generally lead to different geometric rough paths). There is no particular reason that we favor the dyadic polygonal approximations rather than the fact that dyadic polygonal approximations of the Brownian motion paths and its Lévy area are convergent in p-variation distance, for all $2 < p < 3$.

The aim of this section is to give a general construction of dyadic approximations to continuous paths, and prove several technical lemmas. More concrete examples will be given in the following sections. Let $p > 1$ and let V be a separable Banach space with fixed Banach tensor products $V^{\otimes k}$ for $k = 1, \ldots, [p]$.

Let W be a continuous path in V and let $X_{s,t}^1 = W_t - W_s$. For a natural number $m \in \mathbb{N}$, we define a continuous and piecewise-linear path $W(m)$ by

$$W(m)_t = W_{t_{l-1}^m} + 2^m \left(t - t_{l-1}^m \right) \triangle_l^m W, \quad \text{if} \quad t_{l-1}^m \leqslant t \leqslant t_l^m ,$$

for $l = 1, \ldots, 2^m$, where $n \in \mathbb{N}$, $t_k^n = k/2^n$ $(k = 0, 1, \ldots, 2^n)$ are dyadic points, and $\triangle_k^n W = W_{t_k^n} - W_{t_{k-1}^n}$. $W(m)$ is a continuous path such that $W(m)_{t_l^m} = W_{t_l^m}$ and is linear on each interval $[t_{l-1}^m, t_l^m]$. $W(m)$ is called the mth dyadic polygonal approximation to W. The corresponding smooth rough path (of degree k) is denoted by $X(m)$ which is built by taking its iterated path integrals. That is,

$$X(m)_{s,t}^j = \int_{s < t_1 < \cdots < t_j < t} dW(m)_{t_1} \otimes \cdots \otimes dW(m)_{t_j} .$$

By definition,

$$X(m)^j_{s,t} = \lim_{m(D) \to 0} \sum_l \sum_{i=1}^{j-1} X(m)^i_{s,t_{l-1}} \otimes X(m)^{j-i}_{t_{l-1},t_l}.$$

For any $k \in \mathbb{N}$, $X(m)_{s,t} = (1, X(m)^1_{s,t}, X(m)^2_{s,t}, \ldots, X(m)^k_{s,t})$ is a multiplicative functional in $T^{(k)}(V)$ (with any Banach tensor products) with p-finite variation for all $p \geqslant 1$. The main issue is, under what kind of conditions does the sequence (not just a subsequence!) of geometric rough paths $\{X(m)\}_{m \in \mathbb{N}}$ converge to a limit in p-variation distance for some p and $k = [p]$? Since any limit of the sequence $\{X(m)\}_{m \in \mathbb{N}}$ possesses projection (the first-level path) W, then, if the sequence $\{X(m)\}_{m \in \mathbb{N}}$ converges to a geometric rough path X in p-variation topology, we call X the canonical geometric rough path associated to W. In this case we also say that the canonical geometric rough path associated to W exists. Of course we look for the least possible p.

Let us first look at the first-level path $X(m)^1_{s,t}$.

Proposition 4.2.1 *Let (W_t) be a continuous path in a Banach space V (with running time $[0,1]$, as we insist in this part), and $p \geqslant 1$. Define its dyadic approximations $W(m)$ and their associated multiplicative functionals $X(m)$ as above. Then, for all $n \in \mathbb{N}$,*

$$m \to \sum_{k=1}^{2^n} \left| X(m)^1_{t^n_{k-1}, t^n_k} \right|^p$$

is increasing. Therefore,

$$\sup_m \sum_{n=1}^{\infty} n^\gamma \sum_{k=1}^{2^n} \left| X(m)^1_{t^n_{k-1}, t^n_k} \right|^p = \lim_{m \to \infty} \sum_{n=1}^{\infty} n^\gamma \sum_{k=1}^{2^n} \left| X(m)^1_{t^n_{k-1}, t^n_k} \right|^p. \tag{4.19}$$

Proof By definition $X(m)_{t^n_k} = W_{t^n_k}$ for any $n \leqslant m$, and therefore

$$X(m)^1_{t^n_{k-1}, t^n_k} = \triangle^n_k W, \quad k = 1, \ldots, 2^n, \quad \forall n \leqslant m. \tag{4.20}$$

Hence,

$$\sum_{k=1}^{2^n} \left| X(m)^1_{t^n_{k-1}, t^n_k} \right|^p = \sum_{k=1}^{2^n} |\triangle^n_k W|^p.$$

If $n > m$, then we may find a unique integer $0 < l \leqslant 2^m$ such that

$$t^m_{l-1} \leqslant t^n_{k-1} < t^n_k < t^m_l, \tag{4.21}$$

so that

$$X(m)_{t^n_j} = W_{t^m_{l-1}} + 2^m \left(t^n_j - t^m_{l-1} \right) \triangle^m_l W, \quad j = k, k-1. \tag{4.22}$$

Therefore,

$$X(m)^1_{t^n_{k-1}, t^n_k} = 2^{m-n} \triangle^m_l W , \quad \forall n > m . \tag{4.23}$$

Hence, for $n > m$,

$$\sum_{k=1}^{2^n} \left| X(m)^1_{t^n_{k-1}, t^n_k} \right|^p = \sum_{l=1}^{2^m} \sum_{(l-1)/2^m \leqslant t^n_{k-1} < l/2^m} |2^{m-n} \triangle^m_l W|^p$$

$$= \left(\frac{1}{2^n} \right)^p (2^m)^{p-1} \sum_{l=1}^{2^m} |\triangle^m_l W|^p .$$

We claim that the right-hand side is increasing in m, and therefore the conclusion follows. To see this, observe that

$$\triangle^m_l W = \triangle^{m+1}_{2l} W + \triangle^{m+1}_{2l-1} W ,$$

so that

$$(2^m)^{p-1} \sum_{l=1}^{2^m} |\triangle^m_l W|^p = (2^{m+1})^{p-1} \sum_{l=1}^{2^m} \left(\frac{1}{2} \right)^{p-1} |\triangle^{m+1}_{2l} W + \triangle^{m+1}_{2l-1} W|^p$$

$$\leqslant (2^{m+1})^{p-1} \sum_{l=1}^{2^m} \left(|\triangle^{m+1}_{2l} W|^p + |\triangle^{m+1}_{2l-1} W|^p \right)$$

$$= (2^{m+1})^{p-1} \sum_{l=1}^{2^{m+1}} |\triangle^{m+1}_l W|^p .$$

This ends the proof. ∎

We next consider the second-level path $X(m)^2_{s,t}$. By definition,

$$X(m)^2_{s,t} = \int_s^t (W(m)_r - W(m)_s) \otimes dW(m)_r , \quad \forall (s,t) \in \triangle .$$

In order to compute $X(m)^2_{t^n_{k-1}, t^n_k}$, we use Chen's identity. If $n \geqslant m$, then by definition

$$X(m)^N_{t^n_{k-1}, t^n_k} = \frac{1}{N!} 2^{N(m-n)} (\triangle^m_l W)^{\otimes N} . \tag{4.24}$$

For $n < m$,

$$X(m)^2_{t^n_{k-1}, t^n_k} = \sum_{l=2^{m-n}(k-1)+1}^{2^{m-n} k} X(m)^2_{t^m_{l-1}, t^m_l} + X(m)^1_{t^n_{k-1}, t^m_{l-1}} \otimes X(m)^1_{t^m_{l-1}, t^m_l}$$

$$= \sum_{l=2^{m-n}(k-1)+1}^{2^{m-n} k} X(m)^2_{t^m_{l-1}, t^m_l} + X^1_{t^n_{k-1}, t^m_{l-1}} \otimes X^1_{t^m_{l-1}, t^m_l}$$

$$= \sum_{l=2^{m-n}(k-1)+1}^{2^{m-n}k} X(m)^2_{t^m_{l-1},t^m_l}$$

$$+ \sum_{l=2^{m-n}(k-1)+1}^{2^{m-n}k} \sum_{r=1}^{l-1} X^1_{t^m_{r-1},t^m_r} \otimes X^1_{t^m_{l-1},t^m_l}$$

$$= \sum_{l=2^{m-n}(k-1)+1}^{2^{m-n}k} X(m)^2_{t^m_{l-1},t^m_l}$$

$$+ \sum_{l=2^{m-n}(k-1)+1}^{2^{m-n}k} \sum_{r=1}^{l-1} \triangle^m_r W \otimes \triangle^m_l W. \tag{4.25}$$

Using (4.24), we therefore have for $m > n$,

$$X(m)^2_{t^n_{k-1},t^n_k} = \sum_{l=2^{m-n}(k-1)+1}^{2^{m-n}k} \frac{1}{2} \triangle^m_l W \otimes \triangle^m_l W$$

$$+ \sum_{r<l=2^{m-n}(k-1)+1}^{2^{m-n}k} \triangle^m_r W \otimes \triangle^m_l W. \tag{4.26}$$

On the other hand,

$$X(m)^1_{t^n_{k-1},t^n_k} = \sum_{l=2^{m-n}(k-1)+1}^{2^{m-n}k} \triangle^m_l W,$$

so that

$$X(m)^2_{t^n_{k-1},t^n_k} = \frac{1}{2} X(m)^1_{t^n_{k-1},t^n_k} \otimes X(m)^1_{t^n_{k-1},t^n_k}$$

$$+ \frac{1}{2} \sum_{l=2^{m-n}(k-1)+1}^{2^{m-n}k} \sum_{r=1}^{l-1} (\triangle^m_r W \otimes \triangle^m_l W - \triangle^m_l W \otimes \triangle^m_r W)$$

$$= \frac{1}{2} \triangle^n_k W \otimes \triangle^n_k W$$

$$+ \frac{1}{2} \sum_{l=2^{m-n}(k-1)+1}^{2^{m-n}k} \sum_{r=1}^{l-1} (\triangle^m_r W \otimes \triangle^m_l W - \triangle^m_l W \otimes \triangle^m_r W).$$

From this equality, we deduce the following.

Lemma 4.2.1 *If $n < m$, then*

$$X(m+1)^2_{t^n_{k-1}, t^n_k} - X(m)^2_{t^n_{k-1}, t^n_k}$$

$$= \frac{1}{2} \sum_{l=2^{m-n}(k-1)+1}^{2^{m-n}k} \left(\triangle^{m+1}_{2l-1}W \otimes \triangle^{m+1}_{2l}W - \triangle^{m+1}_{2l}W \otimes \triangle^{m+1}_{2l-1}W \right), \quad (4.27)$$

for $k = 1, \ldots, 2^n$.

Finally, let us consider the third-level path $X(m)^3_{s,t}$. The computation is a little complicated, but follows in the same way as for the second-level paths. Let $n < m$. In this case, by Chen's identity, we have

$$X(m)^3_{t^n_{k-1}, t^n_k} = \sum_{l=2^{m-n}(k-1)+1}^{2^{m-n}k} \left(X(m)^3_{t^m_{l-1}, t^m_l} \right.$$

$$\left. + \sum_{j=1}^2 X(m)^{3-j}_{t^n_{k-1}, t^m_{l-1}} \otimes X(m)^j_{t^m_{l-1}, t^m_l} \right)$$

$$= \sum_{l=2^{m-n}(k-1)+1}^{2^{m-n}k} \left(X(m)^3_{t^m_{l-1}, t^m_l} \right.$$

$$+ \sum_{r<l} X(m)^1_{t^m_{r-1}, t^m_r} \otimes X(m)^2_{t^m_{l-1}, t^m_l}$$

$$+ \sum_{r<l} X(m)^2_{t^m_{r-1}, t^m_r} \otimes X(m)^1_{t^m_{l-1}, t^m_l}$$

$$\left. + \sum_{r<u<l} X(m)^1_{t^m_{r-1}, t^m_r} \otimes X(m)^1_{t^m_{u-1}, t^m_u} \otimes X(m)^1_{t^m_{l-1}, t^m_l} \right)$$

$$= \sum_{l=2^{m-n}(k-1)+1}^{2^{m-n}k} \left(\frac{1}{3!} (\triangle^m_l W)^{\otimes 3} \right.$$

$$+ \sum_{r<u<l} \triangle^m_r W \otimes \triangle^m_u W \otimes \triangle^m_l W$$

$$+ \frac{1}{2} \sum_{r<l} \triangle^m_r W \otimes \triangle^m_l W \otimes \triangle^m_l W$$

$$\left. + \frac{1}{2} \sum_{r<l} \triangle^m_r W \otimes \triangle^m_r W \otimes \triangle^m_l W \right),$$

and

$$(\triangle_k^n W)^{\otimes 3} = \left(\sum_{l=2^{m-n}(k-1)+1}^{2^{m-n}k} \triangle_l^m W \right)^{\otimes 3}$$

$$= \sum_{r,u,l=2^{m-n}(k-1)+1}^{2^{m-n}k} \triangle_r^m W \otimes \triangle_u^m W \otimes \triangle_l^m W$$

$$= \sum_{l=2^{m-n}(k-1)+1}^{2^{m-n}k} (\triangle_l^m W)^{\otimes 3}$$

$$+ \sum_{r \neq l} \triangle_r^m W \otimes \triangle_r^m W \otimes \triangle_l^m W$$

$$+ \sum_{r \neq l} \triangle_r^m W \otimes \triangle_l^m W \otimes \triangle_l^m W$$

$$+ \sum_{r \neq l} \triangle_l^m W \otimes \triangle_r^m W \otimes \triangle_l^m W$$

$$+ \sum_{r,u,l \text{ different}} \triangle_r^m W \otimes \triangle_u^m W \otimes \triangle_l^m W .$$

Multiplying the second equality by $1/3!$, and then subtracting from the first equality, we obtain

$$X(m)_{t_{k-1}^n, t_k^n}^3 = \frac{1}{3!}(\triangle_k^n W)^{\otimes 3} + \frac{1}{3}\sum_{r<l} \triangle_r^m W \otimes \triangle_l^m W \otimes \triangle_l^m W$$

$$+ \frac{1}{3}\sum_{r<l} \triangle_r^m W \otimes \triangle_r^m W \otimes \triangle_l^m W$$

$$+ \sum_{r<u<l} \triangle_r^m W \otimes \triangle_u^m W \otimes \triangle_l^m W$$

$$- \frac{1}{3!}\sum_{r>l} \triangle_r^m W \otimes \triangle_r^m W \otimes \triangle_l^m W$$

$$- \frac{1}{3!}\sum_{r>l} \triangle_r^m W \otimes \triangle_l^m W \otimes \triangle_l^m W$$

$$- \frac{1}{3!}\sum_{r \neq l} \triangle_l^m W \otimes \triangle_r^m W \otimes \triangle_l^m W$$

$$- \frac{1}{3!}\sum_{r,u,l \text{ different}} \triangle_r^m W \otimes \triangle_u^m W \otimes \triangle_l^m W , \quad (4.28)$$

where all of the indices r, l, and u run from 1 up to 2^m. Therefore, for any $m > n$, we have

$$X(m+1)^3_{t^n_{k-1}, t^n_k} - X(m)^3_{t^n_{k-1}, t^n_k}$$

$$= \frac{1}{3} \sum_l \triangle^{m+1}_{2l-1} W \otimes \triangle^{m+1}_{2l} W \otimes \triangle^{m+1}_{2l} W$$

$$+ \frac{1}{3} \sum_l \triangle^{m+1}_{2l-1} W \otimes \triangle^{m+1}_{2l-1} W \otimes \triangle^{m+1}_{2l} W$$

$$- \frac{1}{6} \sum_l \triangle^{m+1}_{2l} W \otimes \triangle^{m+1}_{2l} W \otimes \triangle^{m+1}_{2l-1} W$$

$$- \frac{1}{6} \sum_l \triangle^{m+1}_{2l-1} W \otimes \triangle^{m+1}_{2l} W \otimes \triangle^{m+1}_{2l-1} W$$

$$- \frac{1}{6} \sum_l \triangle^{m+1}_{2l} W \otimes \triangle^{m+1}_{2l-1} W \otimes \triangle^{m+1}_{2l-1} W$$

$$- \frac{1}{6} \sum_l \triangle^{m+1}_{2l} W \otimes \triangle^{m+1}_{2l-1} W \otimes \triangle^{m+1}_{2l} W$$

$$+ \frac{1}{2} \sum_{r<l} \left(\triangle^{m+1}_{2r-1} W + \triangle^{m+1}_{2r} W \right) \otimes \triangle^{m+1}_{2l-1} W \otimes \triangle^{m+1}_{2l} W$$

$$- \frac{1}{2} \sum_{r<l} \left(\triangle^{m+1}_{2r-1} W + \triangle^{m+1}_{2r} W \right) \otimes \triangle^{m+1}_{2l} W \otimes \triangle^{m+1}_{2l-1} W$$

$$+ \frac{1}{2} \sum_{r<l} \triangle^{m+1}_{2r-1} W \otimes \triangle^{m+1}_{2r} W \otimes \left(\triangle^{m+1}_{2l-1} W + \triangle^{m+1}_{2l} W \right)$$

$$- \frac{1}{2} \sum_{r<l} \triangle^{m+1}_{2r} W \otimes \triangle^{m+1}_{2r-1} W \otimes \left(\triangle^{m+1}_{2l-1} W + \triangle^{m+1}_{2l} W \right), \quad (4.29)$$

where the last four sums run over the range $2^{m-n}(k-1) + 1 \leqslant r < l \leqslant 2^{m-n}k$, and the first four run over $2^{m-n}(k-1) + 1 \leqslant l \leqslant 2^{m-n}k$.

4.3 Hölder's condition

Let $(\Omega, \mathcal{F}, \mathbb{P})$ be a completed probability space. A stochastic process with values in a separable Banach space V (with running time $[0,1]$) is a family of V-valued random variables $(W_t)_{t \in [0,1]}$. That is, each W_t is a measurable function from Ω into V. For each $\omega \in \Omega$, $t \to W_t(\omega)$ is called a sample path. The process (W_t) is called continuous if for almost all $\omega \in \Omega$, the sample path $t \to W_t(\omega)$ is continuous. Once and for all, we shall suppress the variable ω. We are only interested in the so-called 'almost all' properties. For example, when we say that $(W_t)_{t \in [0,1]}$ has finite p-variation, we mean that for almost all $\omega \in \Omega$, the sample path $(W_t(\omega))$ possesses finite p-variation.

In this section, we work with a *continuous stochastic process* $(W_t)_{t \in [0,1]}$ with values in a separable Banach space V, which satisfies the following *Hölder condition*: there are constants $p > 1$, $h \in (0,1)$ such that $hp > 1$ and C, such that

$$\mathbb{E}|W_t - W_s|^p \leqslant C|t-s|^{hp}, \quad \forall s, t \in [0,1]. \tag{4.30}$$

We continue to use the notation established in the previous section. Thus for each $m \in \mathbb{N}$, $X(m)_{s,t} = (1, X(m)^1_{s,t}, \ldots, X(m)^k_{s,t})$ is the mth dyadic polygonal approximation associated with W. Let $X^1_{s,t} = W_t - W_s$. The Hölder condition implies that $\mathbb{E}|X^1_{s,t}|^p \leqslant C|t-s|^{hp}$. In particular, $\mathbb{E}|X^1_{t^n_{k-1},t^n_k}|^p \leqslant C(1/2^n)^{hp}$. Therefore for any $\gamma > p-1$, there is a constant C (depending only on p and the constant in (4.30)) such that

$$\mathbb{E} \sum_{n=1}^{\infty} n^\gamma \sum_{k=1}^{2^n} |X^1_{t^n_{k-1},t^n_k}|^p \leqslant C \sum_{n=1}^{\infty} n^\gamma \left(\frac{1}{2^n}\right)^{hp-1}. \tag{4.31}$$

By Proposition 4.2.1,

$$\mathbb{E} \sup_D \sum_l |X^1_{t_{l-1},t_l}|^p \leqslant C(p,\gamma) \mathbb{E} \sum_{n=1}^{\infty} n^\gamma \sum_{k=1}^{2^n} |X^1_{t^n_{k-1},t^n_k}|^p$$

$$\leqslant C_1 \sum_{n=1}^{\infty} n^\gamma \left(\frac{1}{2^n}\right)^{hp-1}, \tag{4.32}$$

for constant $C_1 = C(p,\gamma)C$. Since $hp-1 > 0$, the series on the right-hand side of (4.32) is convergent, so that $\sup_D \sum_l |X^1_{t_{l-1},t_l}|^p < \infty$ almost surely. This shows that X^1 has finite p-variation almost surely. In fact we can prove a stronger conclusion.

Indeed, by Proposition 4.2.1,

$$\mathbb{E} \sup_m \sum_{n=1}^{\infty} n^\gamma \sum_{k=1}^{2^n} |X(m)^1_{t^n_{k-1},t^n_k}|^p \leqslant \mathbb{E} \sum_{n=1}^{\infty} n^\gamma \sum_{k=1}^{2^n} |X^1_{t^n_{k-1},t^n_k}|^p$$

$$\leqslant C \sum_{n=1}^{\infty} n^\gamma \left(\frac{1}{2^n}\right)^{hp-1},$$

so that

$$\mathbb{E} \sup_m \sup_D \sum_l |X(m)^1_{t_{l-1},t_l}|^p \leqslant C(p,\gamma) \mathbb{E} \sup_m \sum_{n=1}^{\infty} n^\gamma \sum_{k=1}^{2^n} |X(m)^1_{t^n_{k-1},t^n_k}|^p$$

$$\leqslant C \sum_{n=1}^{\infty} n^\gamma \left(\frac{1}{2^n}\right)^{hp-1} < \infty, \tag{4.33}$$

and therefore, by a scaling argument, we have the following proposition.

Proposition 4.3.1 *Suppose (W_t) is a continuous stochastic process on a completed probability space $(\Omega, \mathcal{F}, \mathbb{P})$ which satisfies Hölder's condition (4.30) for $p > 1$, $h \in (0,1)$ such that $hp > 1$. Then the dyadic polygonal approximations*

$X(m)^1_{s,t}$ *have finite p-variations uniformly in m. Moreover, there is a constant C depending only on p, h, and the constant C in Hölder's condition, such that*

$$\mathbb{E}\sup_{m}\sup_{D_{[s,t]}}\sum_{l}\left|X(m)^1_{t_{l-1},t_l}\right|^p \leqslant C|t-s|^{hp}, \qquad (4.34)$$

for all $[s,t] \subset [0,1]$, *where* $\sup_{D_{[s,t]}}$ *runs over all finite partitions of* $[s,t]$. *In particular,*

$$\sup_{m}\sup_{D}\sum_{l}\left|X(m)^1_{t_{l-1},t_l}\right|^p < \infty$$

almost surely.

Next we want to show that under the Hölder condition, $X(m)^1_{s,t}$ converges to $X^1_{s,t}$ in p-variation distance.

In fact, if $n \leqslant m$ then $X(m)^1_{t^n_{k-1},t^n_k} = X^1_{t^n_{k-1},t^n_k}$, while if $n > m$ then

$$\left|X(m)^1_{t^n_{k-1},t^n_k} - X^1_{t^n_{k-1},t^n_k}\right|^p \leqslant 2^{p-1}\left(\left|X(m)^1_{t^n_{k-1},t^n_k}\right|^p + \left|X^1_{t^n_{k-1},t^n_k}\right|^p\right).$$

Therefore,

$$\mathbb{E}\sum_{n=1}^{\infty} n^{\gamma} \sum_{k=1}^{2^n}\left|X(m)^1_{t^n_{k-1},t^n_k} - X^1_{t^n_{k-1},t^n_k}\right|^p$$

$$= \mathbb{E}\sum_{n=m+1}^{\infty} n^{\gamma} \sum_{k=1}^{2^n}\left|X(m)^1_{t^n_{k-1},t^n_k} - X^1_{t^n_{k-1},t^n_k}\right|^p$$

$$\leqslant C\sum_{n=m+1}^{\infty} n^{\gamma}\left(\frac{1}{2^n}\right)^{hp-1}$$

$$\leqslant C\left(\frac{1}{2^m}\right)^{(hp-1)/2}\sum_{n=m+1}^{\infty} n^{\gamma}\left(\frac{1}{2^n}\right)^{(hp-1)/2}$$

$$\leqslant C\left(\frac{1}{2^m}\right)^{(hp-1)/2},$$

for some constant C depending only on p, h, and the constant which appeared in Hölder's condition. By Proposition 4.1.2,

$$\mathbb{E}\sup_{D}\sum_{l}\left|X(m)^1_{t_{l-1},t_l} - X^1_{t_{l-1},t_l}\right|^p$$

$$\leqslant C(p,\gamma)\mathbb{E}\sum_{n=1}^{\infty} n^{\gamma} \sum_{k=1}^{2^n}\left|X(m)^1_{t^n_{k-1},t^n_k} - X^1_{t^n_{k-1},t^n_k}\right|^p$$

$$\leqslant C\left(\frac{1}{2^m}\right)^{(hp-1)/2}.$$

By the Hölder inequality,

$$\mathbb{E}\sup_{D}\left(\sum_{l}|X(m)^1_{t_{l-1},t_l} - X^1_{t_{l-1},t_l}|^p\right)^{1/p}$$

$$\leqslant \left(\mathbb{E}\sup_{D}\sum_{l}|X(m)^1_{t_{l-1},t_l} - X^1_{t_{l-1},t_l}|^p\right)^{1/p},$$

hence

$$\mathbb{E}\sum_{m=1}^{\infty}\sup_{D}\left(\sum_{l}|X(m)^1_{t_{l-1},t_l} - X^1_{t_{l-1},t_l}|^p\right)^{1/p} \leqslant C\sum_{m=1}^{\infty}\left(\frac{1}{2^m}\right)^{(hp-1)/2p}.$$

By assumption, $hp > 1$, therefore the series on the right-hand side is convergent. Thus we have established the following proposition.

Proposition 4.3.2 *Under the same conditions and notation as in Proposition 4.3.1, we have for any $[s,t] \subset [0,1]$,*

$$\mathbb{E}\sup_{D_{[s,t]}}\sum_{l}|X(m)^1_{t_{l-1},t_l} - X^1_{t_{l-1},t_l}|^p \leqslant C\left(\frac{1}{2^m}\right)^{(hp-1)/2}|t-s|^{hp}, \qquad (4.35)$$

for some constant C depending only on p, h, and the constant in Hölder's condition. Therefore,

$$\sum_{m=1}^{\infty}\sup_{D}\left(\sum_{l}|X(m)^1_{t_{l-1},t_l} - X^1_{t_{l-1},t_l}|^p\right)^{1/p} < \infty$$

almost surely. In particular, $(X(m)^1_{s,t})$ converges to $(X^1_{s,t})$ in p-variation distance almost surely.

As an application of estimate (4.35), we show the following corollary.

Corollary 4.3.1 *Suppose a continuous process $(W_t)_{t\geqslant 0}$ satisfies Hölder's condition (4.30) for some $ph > 1$, and $W(m)$ is the mth dyadic approximation of W, namely*

$$W(m) = W_{t^m_{l-1}} + 2^m(t - t^m_{l-1})\triangle^m_l W, \quad \text{if} \quad t^m_{l-1} \leqslant t < t^m_l,$$

for $l = 1, \ldots, 2^m$. Then there is a constant C depending only on p, h, and the constant in Hölder's condition (4.30) such that

$$\mathbb{E}\sup_{t\in[0,1]}|W_t - W(m)_t| \leqslant C\left(\frac{1}{2^m}\right)^{(hp-1)/2p}, \qquad (4.36)$$

for all $m \in \mathbb{N}$.

Proof The claim follows from

$$\sup_{t\in[0,1]} |W_t - W(m)_t| \leqslant \sup_D \left(\sum_l |X(m)^1_{t_{l-1},t_l} - X^1_{t_{l-1},t_l}|^p\right)^{1/p}$$

and (4.35). ∎

Under Hölder's condition (4.30), we also have the following partial result for the second component $X(m)^2_{s,t}$ of the dyadic approximation.

Proposition 4.3.3 *Suppose (W_t) is a continuous stochastic process on a completed probability space $(\Omega, \mathcal{F}, \mathbb{P})$ which satisfies Hölder's condition (4.30) with $ph > 1$ (and $p \geqslant 2$, otherwise we don't need to consider the second-level paths). Then for $n > m$,*

$$\sum_{k=1}^{2^n} \mathbb{E}\left|X(m+1)^2_{t^n_{k-1},t^n_k} - X(m)^2_{t^n_{k-1},t^n_k}\right|^{p/2} \leqslant C\left(\frac{1}{2^{n+m}}\right)^{(ph-1)/2}, \qquad (4.37)$$

where C is a constant depending only on p, h, and the constant in (4.30).

Proof If $n > m$, then by (4.24),

$$\sum_{k=1}^{2^n} \mathbb{E}\left|X(m+1)^2_{t^n_{k-1},t^n_k} - X(m)^2_{t^n_{k-1},t^n_k}\right|^{p/2}$$

$$= \sum_{l=1}^{2^{m+1}} \sum_{t^{m+1}_{l-1} \leqslant t^n_{k-1} < t^{m+1}_l} \mathbb{E}\Big|2^{2(m-n)+1}\triangle^{m+1}_l W \otimes \triangle^{m+1}_l W$$

$$- \frac{1}{2}2^{2(m-n)}\triangle^m_l W \otimes \triangle^m_l W\Big|^{p/2}$$

$$\leqslant C\left(\frac{2^m}{2^n}\right)^p \sum_{l=1}^{2^{m+1}} \sum_{t^{m+1}_{l-1} \leqslant t^n_{k-1} < t^{m+1}_l} \left(\frac{1}{2^m}\right)^{ph}$$

$$= C\left(\frac{2^m}{2^n}\right)^{p-ph}\left(\frac{1}{2^n}\right)^{ph-1}. \qquad (4.38)$$

Equation (4.37) follows immediately. ∎

However, estimating the left-hand side of (4.37) is more delicate in the case when $n \leqslant m$, and should be treated separately for different kinds of stochastic processes.

4.4 Processes with long-time memory

In this section we deal with the second-level paths $X(m)^2_{s,t}$. Our aim is to show that, under a certain condition, $X(m)^2_{s,t}$ converges to some limit. We need some further conditions together with the Hölder condition on the original stochastic process $W_t = (w_t^1, \ldots, w_t^d)$. By Proposition 4.3.3, we only have to control the following term:

$$\mathbb{E}\left| X(m+1)^2_{t^n_{k-1}, t^n_k} - X(m)^2_{t^n_{k-1}, t^n_k} \right|^{p/2},$$

when $n < m$. Notice that

$$X(m+1)^2_{t^n_{k-1}, t^n_k} - X(m)^2_{t^n_{k-1}, t^n_k}$$

$$= \frac{1}{2} \sum_{l=2^{m-n}(k-1)+1}^{2^{m-n}k} \left(\triangle^{m+1}_{2l-1}W \otimes \triangle^{m+1}_{2l}W - \triangle^{m+1}_{2l}W \otimes \triangle^{m+1}_{2l-1}W \right).$$

Therefore, we need to estimate the mixed moments of the process W. In order to explain the conditions that we will introduce to handle the second-level paths, we consider a simple, but a very important, class, namely the fractional Brownian motions.

Recall that a real, centered, and continuous Gaussian process $(b_t)_{t \geq 0}$ is called a standard one-dimensional Brownian motion if $b_0 = 0$ and its covariance function $R(s,t) \equiv \mathbb{E}(b_t b_s) = s \wedge t$. A d-dimensional Brownian motion $B_t = (b_t^1, \ldots, b_t^d)$ comprises d independent copies of a one-dimensional standard Brownian motion. Clearly, for any $p > 0$,

$$\mathbb{E}|B_t - B_s|^p = C_{p,d}|t - s|^{p/2},$$

and therefore the Hölder condition is satisfied for any $p > 0$ when $h = 1/2$. However, the condition that $hp > 1$ forces $p > 2$, so that $[p] = 2$ if we restrict $p < 3$ as well. Thus any geometric rough path associated with the Brownian motion must have degree greater than 2. Therefore, we need to know the second-level paths.

The increments of the Brownian motion B_t are independent, i.e. for any $0 < t_1 < t_2 < \cdots < t_k$, the random variables $B_{t_1}, B_{t_2} - B_{t_1}, \ldots, B_{t_k} - B_{t_{k-1}}$ are mutually independent. Therefore, we may think that the Brownian motion has little memory with time t.

There is a class of Gaussian processes called fractional Brownian motions which share most of the properties of the Brownian motion, except independence of increments. A real, centered Gaussian process (b_t) with initial zero is called a fractional Brownian motion with Hurst parameter h $(0 < h < 1)$ if it possesses a covariance function

$$R(s,t) \equiv \mathbb{E}(b_t b_s) = \frac{1}{2}\left(|t|^{2h} + |s|^{2h} - |t-s|^{2h}\right), \tag{4.39}$$

for all s, t. Therefore, $\mathbb{E}|b_t - b_s|^2 = |t-s|^{2h}$, so that

$$\mathbb{E}|b_t - b_s|^p = C_p|t-s|^{hp}, \quad \forall s, t \in \mathbb{R}. \tag{4.40}$$

A d-dimensional fractional Brownian motion with Hurst parameter h is the $\mathbb{R}^d$-valued stochastic process $B_t = (b_t^1, \ldots, b_t^d)$ of d independent copies of the fractional Brownian motion (b_t) with the same Hurst parameter h. Again we have

$$\mathbb{E}|B_t - B_s|^p = C_p |t - s|^{hp}, \quad \forall s, t \in \mathbb{R}. \tag{4.41}$$

Let us consider the correlation of the increments of the fractional Brownian motion (b_t) with Hurst parameter h. Set $t > s$, $\tau > 0$, and $\varepsilon = (t - s)/\tau$. Then we have

$$\mathbb{E}(b_t - b_s)(b_{t+\tau} - b_{s+\tau}) = \frac{1}{2} \left(|t - s + \tau|^{2h} + |t - s - \tau|^{2h} - 2\tau^{2h} \right)$$
$$= \frac{\tau^{2h}}{2} \left(|1 + \varepsilon|^{2h} + |1 - \varepsilon|^{2h} - 2 \right),$$

which doesn't vanish if $h \neq 1/2$, $t \neq s$, and $\tau \geqslant 0$. That is, a fractional Brownian motion with Hurst parameter $h \neq 1/2$ possesses long-time memory.

Using Taylor's formula at $\varepsilon = 0$, we can show that there is a constant C_h depending on h such that

$$\left| |1 + \varepsilon|^{2h} + |1 - \varepsilon|^{2h} - 2 \right| \leqslant C_h \varepsilon^2, \quad \forall 0 \leqslant \varepsilon \leqslant 1,$$

and therefore

$$|\mathbb{E}(b_t - b_s)(b_{t+\tau} - b_{s+\tau})| \leqslant C\tau^{2h} \left| \frac{t - s}{\tau} \right|^2, \tag{4.42}$$

when $|(t - s)/\tau| \leqslant 1$. The inequality (4.42) indicates that the increments of the fractional Brownian motion (b_t) are asymptotically independent. Since we shall show that (4.42) allows us to control the second-level paths (and the third-level paths as well if the process is Gaussian), we introduce the following definition.

Definition 4.4.1 *We say that a real-valued, continuous stochastic process $(w_t)_{t \in [0,1]}$ on a completed probability space $(\Omega, \mathcal{F}, \mathbb{P})$ has (h, p)-long-time memory for some $h \in (0, 1)$, $p > 1$, if there is a constant C such that (w_t) satisfies Hölder's condition, namely*

$$\mathbb{E}|w_t - w_s|^p \leqslant C|t - s|^{hp}, \quad \forall s, t \in [0, 1], \tag{4.43}$$

and, for all $1 \geqslant t > s \geqslant 0$, $\tau > 0$ such that $(t - s)/\tau \leqslant 1$, we have

$$|\mathbb{E}(w_t - w_s)(w_{t+\tau} - w_{s+\tau})| \leqslant C\tau^{2h} \left| \frac{t - s}{\tau} \right|^2. \tag{4.44}$$

Remark 4.4.1 For a Gaussian process (w_t) we may replace (4.43) by the following equivalent condition:

$$\mathbb{E}|w_t - w_s|^2 \leqslant C|t - s|^{2h}, \quad \forall s, t \in [0, 1]. \tag{4.45}$$

By (4.44), for any $|l - r| \geqslant 1$ we have

$$\left|\mathbb{E}(\triangle_{2l}^{m+1}w)(\triangle_{2r}^{m+1}w)\right| \leqslant C\left(\frac{1}{2^{m+1}}\right)^{2h}\frac{1}{|l - r|^{2-2h}}. \qquad (4.46)$$

Indeed, it is this decay condition which will be used in handling the second-level paths.

Remark 4.4.2 Regarding the fractional Brownian motion (b_t) with Hurst parameter h, we can get a better estimate than (4.46). Let $[s, t]$, $[t, u] \subset [0, 1]$ be two sub-intervals with the same length, namely $t - s = u - t$. Let $\varepsilon = (t - s)/\tau$. Then,

$$\mathbb{E}(b_u - b_t)(b_{u+\tau} - b_{t+\tau}) = \frac{\tau^{2h}}{2}\left(|1 + \varepsilon|^{2h} + |1 - \varepsilon|^{2h} - 2\right),$$

$$\mathbb{E}(b_t - b_s)(b_{u+\tau} - b_{t+\tau}) = \frac{\tau^{2h}}{2}\left(|1 + 2\varepsilon|^{2h} + 1 - 2|1 + \varepsilon|^{2h}\right),$$

and

$$\mathbb{E}(b_u - b_t)(b_{t+\tau} - b_{s+\tau}) = \frac{\tau^{2h}}{2}\left(1 + |1 - 2\varepsilon|^{2h} - 2|1 - \varepsilon|^{2h}\right).$$

Hence, we have

$$\begin{aligned}
&\left(\mathbb{E}(b_t - b_s)(b_{t+\tau} - b_{s+\tau})\right)\left(\mathbb{E}(b_u - b_t)(b_{u+\tau} - b_{t+\tau})\right) \\
&- \left(\mathbb{E}(b_t - b_s)(b_{u+\tau} - b_{t+\tau})\right)\left(\mathbb{E}(b_u - b_t)(b_{t+\tau} - b_{s+\tau})\right) \\
&\qquad = \frac{\tau^{4h}}{4}\left\{\left(|1 + \varepsilon|^{2h} + |1 - \varepsilon|^{2h} - 2\right)^2 - \left(|1 + 2\varepsilon|^{2h} + 1 - 2|1 + \varepsilon|^{2h}\right)\right. \\
&\qquad\qquad\qquad \left. \times \left(1 + |1 - 2\varepsilon|^{2h} - 2|1 - \varepsilon|^{2h}\right)\right\}.
\end{aligned}$$
$$(4.47)$$

Using Taylor's formula at $\varepsilon = 0$, we may conclude that

$$\begin{aligned}
&\left|\left(|1 + \varepsilon|^{2h} + |1 - \varepsilon|^{2h} - 2\right)^2 - \left(|1 + 2\varepsilon|^{2h} + 1 - 2|1 + \varepsilon|^{2h}\right)\right. \\
&\qquad\qquad \left. \times \left(1 + |1 - 2\varepsilon|^{2h} - 2|1 - \varepsilon|^{2h}\right)\right| \leqslant C_h \varepsilon^5,
\end{aligned}$$

for all $0 \leqslant \varepsilon \leqslant 1$. Therefore,

$$\begin{aligned}
&\left(\mathbb{E}(b_t - b_s)(b_{t+\tau} - b_{s+\tau})\right)\left(\mathbb{E}(b_u - b_t)(b_{u+\tau} - b_{t+\tau})\right) \\
&- \left(\mathbb{E}(b_t - b_s)(b_{u+\tau} - b_{t+\tau})\right)\left(\mathbb{E}(b_u - b_t)(b_{t+\tau} - b_{s+\tau})\right) \\
&\qquad\qquad\qquad\qquad\qquad \leqslant C\tau^{4h}\left|\frac{t - s}{\tau}\right|^5,
\end{aligned}$$

for any $s < t < u$ such that $t - s = u - t$ and $|(t - s)/\tau| \leqslant 1$. The last estimate is enough to treat the second-level paths, see (4.49) below.

Let $W_t = (w_t^1, \ldots, w_t^d)$ be d independent real continuous processes with (h, p)-long-time memory. We use the notation established in the previous sections for its dyadic polygonal approximations, and we consider the second-level path $X(m)_{s,t}^2$ of the dyadic polygonal approximation to W.

Proposition 4.4.1 *Let $2 < p < 4$, $3/4 > h > 1/4$ such that $hp > 1$, and let $W_t = (w_t^1, \ldots, w_t^d)$ be d independent real continuous processes with (h, p)-long-time memory. Then there exists a constant C, depending only on p and h, such that, for any n, m,*

$$\mathbb{E} \left| X(m+1)_{t_{k-1}^n, t_k^n}^2 - X(m)_{t_{k-1}^n, t_k^n}^2 \right|^{p/2} \leqslant C \left(2^{m-n} \right)^{p/4} \left(\frac{1}{2^m} \right)^{hp}. \tag{4.48}$$

Proof For $m \leqslant n$, by Proposition 4.3.3 we have

$$\mathbb{E} \left| X(m+1)_{t_{k-1}^n, t_k^n}^2 - X(m)_{t_{k-1}^n, t_k^n}^2 \right|^{p/2} \leqslant C \left(\frac{1}{2^{n+m}} \right)^{(hp-1)/2}.$$

Let us consider the case $m > n$. As we have mentioned before, we need to estimate the following term:

$$\mathbb{E} \left| \sum_{l=2^{m-n}(k-1)+1}^{2^{m-n}k} \triangle_{2l-1}^{m+1} W \otimes \triangle_{2l}^{m+1} W - \triangle_{2l}^{m+1} W \otimes \triangle_{2l-1}^{m+1} W \right|^2,$$

where we use the Hilbert–Schmidt norm on the finite-dimensional space $(\mathbb{R}^d)^{\otimes 2}$. Using independence of $(w_t^1, \ldots, w_t^d)$, we have

$$\mathbb{E} \left| \sum_{l=2^{m-n}(k-1)+1}^{2^{m-n}k} \triangle_{2l-1}^{m+1} W \otimes \triangle_{2l}^{m+1} W - \triangle_{2l}^{m+1} W \otimes \triangle_{2l-1}^{m+1} W \right|^2$$

$$= \sum_{i \neq j} \mathbb{E} \left(\sum_l \triangle_{2l-1}^{m+1} w^i \triangle_{2l}^{m+1} w^j - \triangle_{2l}^{m+1} w^i \triangle_{2l-1}^{m+1} w^j \right)^2$$

$$= \sum_{i \neq j} \mathbb{E} \sum_{l,r} (\triangle_{2l-1}^{m+1} w^i \triangle_{2l}^{m+1} w^j - \triangle_{2l}^{m+1} w^i \triangle_{2l-1}^{m+1} w^j)$$

$$\times (\triangle_{2r-1}^{m+1} w^i \triangle_{2r}^{m+1} w^j - \triangle_{2r}^{m+1} w^i \triangle_{2r-1}^{m+1} w^j)$$

$$= 2 \sum_{i \neq j} \sum_{l,r} \mathbb{E}(\triangle_{2r-1}^{m+1} w^i \triangle_{2l-1}^{m+1} w^i) \mathbb{E}(\triangle_{2r}^{m+1} w^j \triangle_{2l}^{m+1} w^j)$$

$$- 2 \sum_{i \neq j} \sum_{l,r} \mathbb{E}(\triangle_{2r-1}^{m+1} w^i \triangle_{2l}^{m+1} w^i) \mathbb{E}(\triangle_{2r}^{m+1} w^j \triangle_{2l-1}^{m+1} w^j). \tag{4.49}$$

Apply (4.46) to each term on the right-hand side of (4.49). Then we obtain

$$\mathbb{E}\left|\sum_{l=2^{m-n}(k-1)+1}^{2^{m-n}k} \triangle_{2l-1}^{m+1}W \otimes \triangle_{2l}^{m+1}W - \triangle_{2l}^{m+1}W \otimes \triangle_{2l-1}^{m+1}W\right|^2$$

$$\leqslant Cd^2 2^{m-n}\left(\frac{1}{2^m}\right)^{4h} + 4d^2C^2 \sum_{l=2}^{2^{m-n}}\left(\frac{1}{2^m}\right)^{4h}\sum_{r=1}^{l-1}\frac{1}{(l-r)^{4-4h}}$$

$$\leqslant C(d,h)2^{m-n}\left(\frac{1}{2^m}\right)^{4h},$$

and therefore (4.48) follows from the Hölder inequality as $p < 4$. ∎

If we take into account the terms with the same nature, the above estimate is useful only if $hp > p/4$, i.e. only if $h > 1/4$.

Corollary 4.4.1 *Under the same assumptions as in Proposition 4.4.1, we have*

$$\sup_m \sum_n n^\gamma \sum_{k=1}^{2^n}\left|X(m)_{t_{k-1}^n,t_k^n}^2\right|^{p/2} < \infty.$$

Theorem 4.4.1 *Let $W_t = (w_t^1,\ldots,w_t^d)$ be d independent stochastic processes with (h,p)-long-time memory for some $0 < h < 3/4$ such that $ph > 1$. Then there is a unique function X^i on $\triangle$ which takes values in $(\mathbb{R}^d)^{\otimes i}$ ($i = 1, 2$) such that*

$$\sum_{i=1}^2 \sup_D \left(\sum_l \left|X(m)_{t_{l-1},t_l}^i - X_{t_{l-1},t_l}^i\right|^{p/i}\right)^{i/p} \to 0$$

both almost surely and in $L^1(\Omega, \mathcal{F}, \mathbb{P})$, as $m \to \infty$.

In fact, the case $i = 1$ was proved in Proposition 4.3.2. By Proposition 4.1.2 and Proposition 4.4.1, we have

$$\mathbb{E}\sup_D \sum_l \left|\left(X(m+1)_{t_{l-1},t_l}^2 - X(m)_{t_{l-1},t_l}^2\right)\right|^{p/2}$$

$$\leqslant C\sum_{n=1}^\infty n^\gamma \left(\frac{1}{2^{n+m}}\right)^{(hp-1)/2} + C\left(\sum_{n=m+1}^\infty n^\gamma \left(\frac{1}{2^n}\right)^{(hp-1)}\right)^{1/2}$$

$$\leqslant C\left(\frac{1}{2^m}\right)^{(hp-1)/2}.$$

Therefore, $(X(m)^2)_{m\in\mathbb{N}}$ is a Cauchy sequence in p-variation distance, and the conclusion follows.

In particular, we have the following theorem.

Theorem 4.4.2 *Let W be a d-dimensional fractional Brownian motion with Hurst parameter h, $1/3 < h \leqslant 1/2$. Then, for any $2 < p < 3$ such that $hp > 1$, the*

sequence of its dyadic polygonal approximations $X(m) = (1, X(m)_{s,t}^1, X(m)_{s,t}^2)$
converges both almost surely and in $L^1(\Omega, \mathcal{F}, \mathbb{P})$ *to a unique geometric rough path*
$X_{s,t} = (1, X_{s,t}^1, X_{s,t}^2)$ *according to the p-variation distance, and* $X_{s,t}^1 = W_t - W_s$.
Therefore, the canonical geometric rough path associated with such a fractional
Brownian motion exists.

This theorem covers the case in which W is the Brownian motion on $\mathbb{R}^d$
$(h = 1/2)$.

We will call the geometric rough path $X_{s,t} = (1, X_{s,t}^1, X_{s,t}^2)$, constructed in
Theorem 4.4.2, the canonical rough path associated with the fractional Brownian
motion with Hurst parameter $h > 1/3$.

Theorem 4.4.3 *Let* $X_{s,t} = (1, X_{s,t}^1, X_{s,t}^2)$ *be the canonical geometric rough path*
associated with the d-dimensional standard Brownian motion W. *Then, for any*
$s < t$, $X_{s,t}^1 = W_t - W_s$ *and*

$$X_{s,t}^2 = \int_{s<t_1<t_2<t} \circ \, dW_{t_1} \otimes \circ \, dW_{t_2}$$

almost surely, where $\circ \, dW_t$ *means the Stratonovich integral.*

Proof The conclusion is almost obvious. In fact, for any n and $1 \leqslant k \leqslant 2^n$,

$$X_{t_{k-1}^n, t_k^n}^2 = \int_{t_{k-1}^n<t_1<t_2<t_k^n} \circ \, dW_{t_1} \otimes \circ \, dW_{t_2}$$

$$= \frac{1}{2}\triangle_k^n W \otimes \triangle_k^n W + \frac{1}{2}\int_{t_{k-1}^n}^{t_k^n} (W_s - W_{t_{k-1}^n}) \otimes dW_s$$

$$- \frac{1}{2}\int_{t_{k-1}^n}^{t_k^n} dW_s \otimes (W_s - W_{t_{k-1}^n}),$$

where the two integrals on the right-hand side are in Itô's sense. By definition,
for $m > n$,

$$X(m)_{t_{k-1}^n, t_k^n}^2 = \frac{1}{2}\triangle_k^n W \otimes \triangle_k^n W$$

$$+ \frac{1}{2}\sum_{l=2^{m-n}(k-1)+1}^{2^{m-n}k} (W_{t_{l-1}^n} - W_{t_{k-1}^n}) \otimes \triangle_l^m W$$

$$- \frac{1}{2}\sum_{l=2^{m-n}(k-1)+1}^{2^{m-n}k} \triangle_l^m W \otimes (W_{t_{l-1}^n} - W_{t_{k-1}^n}).$$

However,

$$\sum_{l=2^{m-n}(k-1)+1}^{2^{m-n}k} (W_{t_{l-1}^n} - W_{t_{k-1}^n}) \otimes \triangle_l^m W$$

converges to $\int_{t_{k-1}^n}^{t_k^n} (W_s - W_{t_{k-1}^n}) \otimes dW_s$, and

$$\sum_{l=2^{m-n}(k-1)+1}^{2^{m-n}k} \triangle_l^m W \otimes (W_{t_{l-1}^n} - W_{t_{k-1}^n})$$

to $\int_{t_{k-1}^n}^{t_k^n} dW_s \otimes (W_s - W_{t_{k-1}^n})$ in probability as $m \to \infty$. Since $X_{t_{k-1}^n, t_k^n}^2$ is the p-variation limit of $X(m)_{t_{k-1}^n, t_k^n}^2$ almost surely, the conclusion therefore follows. ∎

Since any square-integrable martingale satisfies (4.44) (with $C = 0$ and any h), we therefore have the following theorem.

Theorem 4.4.4 *Let $W_t = (w_t^1, \ldots, w_t^d)$ be d independent square-integrable martingales on a completed probability space $(\Omega, \mathcal{F}, \mathbb{P})$. Suppose, for some $3 > p > 2$, $hp > 1$, $C > 0$,*

$$\mathbb{E}|w_t^i - w_s^i|^p \leqslant C|t-s|^{ph},$$

for all $t, s \in [0,1]$ and $i = 1, \ldots, d$. Then its dyadic polygonal approximation $X(m)_{s,t} = (1, X(m)_{s,t}^1, X(m)_{s,t}^2)$ converges both almost surely and in $L^1(\Omega, \mathcal{F}, \mathbb{P})$ to a geometric rough path $X_{s,t} = (1, X_{s,t}^1, X_{s,t}^2)$ according to the p-variation distance, and moreover $X_{s,t} = W_t - W_s$.

4.5 Gaussian processes

In the previous sections, we have shown that if $W_t = (w_t^1, \ldots, w_t^d)$ are d independent processes with (h, p)-long-time memory, for some $2 < p < 4$, $1/4 < h \leqslant 1/2$ such that $hp > 1$ (we remind the reader that the smaller h, the rougher is the path W), then the first two level paths of its dyadic polygonal approximation $X(m)_{s,t}^1$ and $X(m)_{s,t}^2$ converge to unique limits $X_{s,t}^1$ and $X_{s,t}^2$, respectively, as $m \to \infty$. However, if $h \leqslant 1/3$, then the condition that $hp > 1$ forces $[p] \geqslant 3$, and therefore $(1, X_{s,t}^1, X_{s,t}^2)$ is not yet a geometric rough path. Thus we have to consider the higher-level paths. It is the case for the example of fractional Brownian motions with Hurst parameter $h \leqslant 1/3$.

Recall that a real-valued stochastic process (w_t) on a completed probability space $(\Omega, \mathcal{F}, \mathbb{P})$ is called a (centered) Gaussian process if, for any (non-trivial) linear combination, $\sum_i c_i w_{t_i}$ possesses a Gaussian distribution $\mathcal{N}(0, \sigma^2)$, where

$$\sigma^2 = \mathbb{E}\left|\sum_i c_i w_{t_i}\right|^2.$$

The law of such a real Gaussian process is determined by its covariance function $R(s,t) \equiv \mathbb{E}(w_t w_s)$. In particular,

$$\mathbb{E}|w_t - w_s|^2 = R(t,t) + R(s,s) - 2R(s,t). \tag{4.50}$$

Let $\sigma^2 = \mathbb{E}|w_t - w_s|^2$. Then by definition, $w_t - w_s$ possesses a distribution $\mathcal{N}(0, \sigma^2)$, and therefore, for any $p > 0$,

$$\mathbb{E}|w_t - w_s|^p = \frac{1}{\sqrt{2\pi}\sigma} \int_{\mathbb{R}} |x|^p e^{-|x|^2/2\sigma^2} \, dx$$
$$= C_p \left(\mathbb{E}|w_t - w_s|^2 \right)^{p/2}, \tag{4.51}$$

where C_p is a universal constant depending only on $p > 0$.

Take d independent Gaussian processes $W_t = (w_t^1, \ldots, w_t^d)$, and assume that each (w_t^i) is a continuous, centered Gaussian process with covariance function $R_i(s, t)$, and that each (w_t^i) has (h, p)-long-time memory. In terms of covariance functions,

$$R_i(t, t) + R_i(s, s) - 2R_i(s, t) \leqslant C|t - s|^{2h}, \quad \forall s, t \in [0, 1] \tag{4.52}$$

and

$$\left| R_i(t, t + \tau) + R_i(s, s + \tau) - R_i(s, t + \tau) - R_i(t, s + \tau) \right| \leqslant C\tau^{2h} \left| \frac{t - s}{\tau} \right|^2, \tag{4.53}$$

for all $|(t - s)/\tau| \leqslant 1$.

The following lemma follows from the Wick formula for Gaussian random variables.

Lemma 4.5.1 *Let G_1, G_2, G_3, and G_4 be four centered, real Gaussian variables. Set $\sigma_{ij} = \mathbb{E}(G_i G_j)$. Then,*

$$\mathbb{E}(G_1 G_2 G_3 G_4) = \sigma_{14}\sigma_{23} + \sigma_{24}\sigma_{13} + \sigma_{34}\sigma_{12} \tag{4.54}$$

and

$$\begin{aligned}
\mathbb{E}\left((G_1 G_2)^2 G_3 G_4 \right) = {} & 2\sigma_{22}\sigma_{13}\sigma_{14} + 4\sigma_{12}\sigma_{23}\sigma_{14} \\
& + 4\sigma_{12}\sigma_{13}\sigma_{24} + 2\sigma_{12}^2\sigma_{34} \\
& + 2\sigma_{11}\sigma_{23}\sigma_{24} + \sigma_{11}\sigma_{22}\sigma_{34}.
\end{aligned} \tag{4.55}$$

Corollary 4.5.1 *If (w_t) is a Gaussian process with (h, p)-long-time memory, then, for any $l \neq r$,*

$$\left| \mathbb{E}\left((\triangle_{2r}^{m+1} w)^2 (\triangle_{2l}^{m+1} w)^2 (\triangle_{2r-1}^{m+1} w)(\triangle_{2l-1}^{m+1} w) \right) \right| \leqslant C_h \left(\frac{1}{2^m} \right)^{6h} \frac{1}{|l - r|^{2 - 2h}}, \tag{4.56}$$

where C_h is a constant depending only on h.

Proof Let $G_1 = \triangle_{2r}^{m+1}w$, $G_2 = \triangle_{2l}^{m+1}w$, $G_3 = \triangle_{2r-1}^{m+1}w$, and $G_4 = \triangle_{2l-1}^{m+1}w$. Using (4.52) and (4.53), we have

$$|\sigma_{ij}| = |\mathbb{E}G_i G_j| \leqslant C\left(\frac{1}{2^m}\right)^{2h}$$

and

$$|\sigma_{14}| = \left|\mathbb{E}\left((\triangle_{2r}^{m+1}w)(\triangle_{2l-1}^{m+1}w)\right)\right| \leqslant C\left(\frac{1}{2^m}\right)^{2h}\frac{1}{|l-r|^{2-2h}}.$$

Similarly, $|\sigma_{23}|$, $|\sigma_{12}|$, $|\sigma_{34}|$ are bounded above by $C\,(1/2^m)^{2h}\,(1/|l-r|^{2-2h})$. Therefore, by (4.55), we establish (4.56). ∎

Actually, estimate (4.56) is the only place where we need to assume that W is Gaussian.

In the remainder of this section, we assume that $W_t = (w_t^1, \ldots, w_t^d)$ are d independent, centered Gaussian processes with (h, p)-long-time memory. $X(m)_{s,t} = (1, X(m)_{s,t}^1, X(m)_{s,t}^2, X(m)_{s,t}^3)$ are the geometric rough paths associated with the dyadic polygonal approximations of W. We next consider the third-level paths $X(m)_{s,t}^3$.

Proposition 4.5.1 *If $h > 1/4$, $p < 4$ such that $hp > 1$, then, for any n, m,*

$$\sum_{k=1}^{2^n} \mathbb{E}\left|X(m+1)_{t_{k-1}^n, t_k^n}^3 - X(m)_{t_{k-1}^n, t_k^n}^3\right|^{p/3} \leqslant C\left(\frac{1}{2^{m+n}}\right)^{(hp-1)/2}, \qquad (4.57)$$

where C is a finite constant depending only on p, h, and the dimension d.

Proof If $m \leqslant n$, then, by the formula for $X(m)_{t_{k-1}^n, t_k^n}^3$, one can easily verify by using only (4.52) that

$$\sum_{k=1}^{2^n} \mathbb{E}\left|X(m+1)_{t_{k-1}^n, t_k^n}^3 - X(m)_{t_{k-1}^n, t_k^n}^3\right|^{p/3} \leqslant C\left(\frac{1}{2^{m+n}}\right)^{(hp-1)/2}.$$

For $m > n$, we use the formula (4.29). First estimate the following term:

$$\sum_{r<l} \left(\triangle_{2r-1}^{m+1}W + \triangle_{2r}^{m+1}W\right) \otimes \triangle_{2l-1}^{m+1}W \otimes \triangle_{2l}^{m+1}W$$

$$- \sum_{r<l} \left(\triangle_{2r-1}^{m+1}W + \triangle_{2r}^{m+1}W\right) \otimes \triangle_{2l}^{m+1}W \otimes \triangle_{2l-1}^{m+1}W.$$

We want to bound

$$\mathbb{E}\left|\sum_{r<l} \left(\triangle_{2r-1}^{m+1}W + \triangle_{2r}^{m+1}W\right) \otimes \left(\triangle_{2l-1}^{m+1}W \otimes \triangle_{2l}^{m+1}W - \triangle_{2l}^{m+1}W \otimes \triangle_{2l-1}^{m+1}W\right)\right|^2,$$

where l runs from $2+2^{m-n}(k-1)$ up to $2^{m-n}k$. To simplify the notation, denote by I_1 the integrand inside the expectation $\mathbb{E}$, and let $A_l^i = w_{t_{2l-2}^{m+1}}^i - w_{t_{k-1}^n}^i$. Then,

$$
\begin{aligned}
I_1 &= \sum_{i\neq j,u}\left(\sum_l A_l^u(\triangle_{2l-1}^{m+1}w^i\triangle_{2l}^{m+1}w^j - \triangle_{2l}^{m+1}w^i\triangle_{2l-1}^{m+1}w^j)\right)^2 \\
&= \sum_{i\neq j,u}\sum_l (A_l^u)^2\,(\triangle_{2l-1}^{m+1}w^i\triangle_{2l}^{m+1}w^j - \triangle_{2l}^{m+1}w^i\triangle_{2l-1}^{m+1}w^j)^2 \\
&\quad + 2\sum_{i\neq j,u}\sum_{r<l}\{A_r^u A_l^u\triangle_{2l-1}^{m+1}w^i\triangle_{2r-1}^{m+1}w^i\triangle_{2l}^{m+1}w^j\triangle_{2r}^{m+1}w^j \\
&\qquad\qquad - A_r^u A_l^u\triangle_{2l}^{m+1}w^i\triangle_{2r-1}^{m+1}w^j\triangle_{2l}^{m+1}w^i\triangle_{2r}^{m+1}w^j \\
&\qquad\qquad - A_r^u A_l^u\triangle_{2l-1}^{m+1}w^i\triangle_{2l}^{m+1}w^j\triangle_{2r}^{m+1}w^i\triangle_{2r-1}^{m+1}w^j \\
&\qquad\qquad + A_r^u A_l^u\triangle_{2l}^{m+1}w^i\triangle_{2l-1}^{m+1}w^j\triangle_{2r}^{m+1}w^i\triangle_{2r-1}^{m+1}w^j\}.
\end{aligned}
$$

For $i\neq j$, $l\neq r$, we have by (4.53) that

$$
\left|\mathbb{E}(\triangle_{2l}^{m+1}w^j\triangle_{2r}^{m+1}w^j)\right| \leqslant C\left(\frac{1}{2^m}\right)^{2h}\frac{1}{|l-r|^{2-2h}}. \tag{4.58}
$$

If $u\neq i$, then

$$
\begin{aligned}
\left|\mathbb{E}(A_r^u A_l^u\triangle_{2l-1}^{m+1}w^i\triangle_{2r-1}^{m+1}w^i)\right| &= \left|\mathbb{E}(A_r^u A_l^u)\right|\left|\mathbb{E}(\triangle_{2l-1}^{m+1}w^i\triangle_{2r-1}^{m+1}w^i)\right| \\
&\leqslant C\left(\frac{1}{2^{m+n}}\right)^{2h}\frac{1}{|l-r|^{2-2h}}, \tag{4.59}
\end{aligned}
$$

and for $u=i$, using (4.54) we have

$$
\left|\mathbb{E}(A_l^i A_r^i\triangle_{2l-1}^{m+1}w^i\triangle_{2r-1}^{m+1}w^i)\right| \leqslant C\left(\frac{1}{2^{m+n}}\right)^{2h}. \tag{4.60}
$$

Using (4.58)–(4.60), for $i\neq j$, we thus deduce that

$$
\begin{aligned}
&\left|\mathbb{E}(A_r^u A_l^u\triangle_{2l-1}^{m+1}w^i\triangle_{2r-1}^{m+1}w^i\triangle_{2l}^{m+1}w^j\triangle_{2r}^{m+1}w^j)\right| \\
&= \left|\mathbb{E}(A_r^u A_l^u\triangle_{2l-1}^{m+1}w^i\triangle_{2r-1}^{m+1}w^i)\right|\left|\mathbb{E}(\triangle_{2l}^{m+1}w^j\triangle_{2r}^{m+1}w^j)\right| \\
&\leqslant C\left(2^{m-n}\right)^{2h}\left(\frac{1}{2^m}\right)^{6h}\frac{1}{|l-r|^{2-2h}}. \tag{4.61}
\end{aligned}
$$

With estimate (4.61) we therefore can estimate I_1. It turns out that

$$
|\mathbb{E}I_1| \leqslant C\left(2^{m-n}\right)^{1+2h}\left(\frac{1}{2^m}\right)^{6h}, \tag{4.62}
$$

for some constant C depending only on d and h.

Similarly, we may estimate the other two summations which appeared in the right-hand side of (4.29) with the index $r < l$, and eventually we get the same upper bound as above (with a different constant C).

Next we estimate the remaining terms in (4.29). By symmetry, consider for example the following term:

$$\mathbb{E}\sum_l \left| \triangle_{2l-1}^{m+1}W \otimes \triangle_{2l}^{m+1}W \otimes \triangle_{2l}^{m+1}W \right|^2$$

$$= \mathbb{E}\sum_{i,j,k}\sum_l \left(\triangle_{2l-1}^{m+1}w^i\right)^2 \left(\triangle_{2l}^{m+1}w^j\right)^2 \left(\triangle_{2l}^{m+1}w^k\right)^2$$

$$+ \sum_{\substack{i,j,k \\ i\neq k \text{ or } j\neq k \text{ or } i\neq j}} \sum_{r\neq l} \mathbb{E}\Big\{ \left(\triangle_{2r-1}^{m+1}w^i \triangle_{2l-1}^{m+1}w^i\right) \left(\triangle_{2r}^{m+1}w^j \triangle_{2l}^{m+1}w^j\right)$$

$$\times \left(\triangle_{2r}^{m+1}w^k \triangle_{2l}^{m+1}w^k\right)\Big\}$$

$$+ \sum_i \sum_{r\neq l} \mathbb{E}\left(\triangle_{2l}^{m+1}w^i\right)^2 \left(\triangle_{2r}^{m+1}w^i\right)^2 \left(\triangle_{2r-1}^{m+1}w^i \triangle_{2l-1}^{m+1}w^i\right).$$

The first term can be estimated by (4.52), and the second term can be treated by (4.53), at least to one of its factors. To estimate the last term, we use (4.56). Thus we finally establish the following upper bound:

$$\mathbb{E}\sum_l \left| \triangle_{2l-1}^{m+1}W \otimes \triangle_{2l}^{m+1}W \otimes \triangle_{2l}^{m+1}W \right|^2 \leqslant C 2^{m-n} \left(\frac{1}{2^m}\right)^{6h}.$$

Putting all of the above estimates together, we obtain

$$\mathbb{E}\left| X(m+1)_{t_{k-1}^n,t_k^n}^3 - X(m)_{t_{k-1}^n,t_k^n}^3 \right|^2 \leqslant C \left(2^{m-n}\right)^{1+2h} \left(\frac{1}{2^m}\right)^{6h},$$

for any $m > n$, $1 \leqslant k \leqslant 2^n$, and for some constant C depending only on p, h. Therefore, by Hölder's inequality, for $p < 4$,

$$\mathbb{E}\left| X(m+1)_{t_{k-1}^n,t_k^n}^3 - X(m)_{t_{k-1}^n,t_k^n}^3 \right|^{p/3} \leqslant C \left(2^{m-n}\right)^{(1+2h)p/6} \left(\frac{1}{2^m}\right)^{hp}$$

$$\leqslant C \left(\frac{1}{2^m}\right)^{(hp-1)/2},$$

as $ph > 1$ and $p < 4$, where C is a constant depending only on p, h, d. ∎

Theorem 4.5.1 *Let $W_t = (w_t^1,\ldots,w_t^d)$ be d independent centered Gaussian processes with (h,p)-long-time memory in the sense that*

$$R_i(t,t) + R_i(s,s) - 2R_i(s,t) \leqslant C|t-s|^{2h}, \quad \forall s,t \in [0,1]$$

and

$$|R_i(t, t+\tau) + R_i(s, s+\tau) - R_i(s, t+\tau) - R_i(t, s+\tau)| \leqslant C\tau^{2h} \left|\frac{t-s}{\tau}\right|^2,$$

for any $|(t-s)/\tau| \leqslant 1$, *where* $R_i(s,t)$ *is the covariance function of* w^i. *Let* $X(m)_{s,t} = (1, X(m)^1_{s,t}, X(m)^2_{s,t}, X(m)^3_{s,t})$ *be the geometric rough paths associated with the dyadic polygonal approximations of* W. *If* $p < 4$ *and* $h > 1/4$ *such that* $hp > 1$, *then there is a unique function* X^j *on* Δ *which takes values in* $(\mathbb{R}^d)^{\otimes j}$, *such that*

$$\sup_D \left(\sum_l \left| X(m)^j_{t_{l-1}, t_l} - X^j_{t_{l-1}, t_l} \right|^{p/j} \right)^{j/p} \to 0$$

both almost surely and in $L^1(\Omega, \mathcal{F}, \mathbb{P})$ *as* $m \to \infty$, *for* $j = 1, 2, 3$.

Proof We only need to prove the claims for $j = 3$. Indeed, by (4.57),

$$\mathbb{E}\sup_D \sum_l \left| X(m+1)^3_{t_{l-1}, t_l} - X(m)^3_{t_{l-1}, t_l} \right|^{p/3}$$

$$\leqslant C_1 \sum_{n=1}^{\infty} n^\gamma \left(\frac{1}{2^{m+n}}\right)^{(hp-1)/2} + C_2 \left(\sum_{n=1}^{\infty} n^\gamma \left(\frac{1}{2^{m+n}}\right)^{(hp-1)/2} \right)^{2/3}$$

$$+ C_3 \left(\sum_{n=m+1}^{\infty} n^\gamma \left(\frac{1}{2^n}\right)^{(hp-1)} \right)^{1/3}$$

$$\leqslant C_4 \left(\frac{1}{2^m}\right)^{(hp-1)/4}.$$

Therefore,

$$\mathbb{E}\sup_D \left(\sum_l \left| X(m+1)^3_{t_{l-1}, t_l} - X(m)^3_{t_{l-1}, t_l} \right|^{p/3} \right)^{3/p} \leqslant C_5 \left(\frac{1}{2^m}\right)^{3(hp-1)/4p},$$

which implies that $X(m)^3$ converges to some function X^3 in p-variation distance as $m \to \infty$. Being the limit of smooth rough paths, $(1, X^1_{s,t}, X^2_{s,t}, X^3_{s,t})$ is, by definition, a geometric rough path. We have thus proved the theorem. ∎

Corollary 4.5.2 *Let* $W_t = (w^1_t, \ldots, w^d_t)$ *be a d-dimensional fractional Brownian motion with Hurst parameter* $h > 1/4$. *Then its dyadic polygonal approximation* $(1, X(m)^1_{s,t}, X(m)^2_{s,t}, X(m)^3_{s,t})$ *converges to the unique geometric rough path* $(1, X^1_{s,t}, X^2_{s,t}, X^3_{s,t})$ *in p-variation distance both almost surely and in* $L^1(\Omega, \mathcal{F}, \mathbb{P})$, *for any* $p < 4$ *such that* $ph > 1$.

One would ask what can happen if $h \leqslant 1/4$? Recall that the parameter h appears in (4.43) and (4.44). In particular, from (4.43), it seems that the smaller is h, the rougher is the stochastic process (w_t), and the more 'iterated path integrals' we need. In fact, the two conditions that $h \leqslant 1/4$ and $hp > 1$ imply that $p > 4$, and therefore, if there is anything, we at least hope to have the second-level paths. However, if $h \leqslant 1/4$, the approach via dyadic polygonal approximations is hopeless even for the second-level path.

Consider the two-dimensional fractional Brownian motion $W_t = (w_t^1, w_t^2)$ with Hurst parameter $h \leqslant 1/4$. The process W_t possesses finite p-variation only if $p > 4$ (so that the condition $hp > 1$ is valid). Consider the dyadic polygonal approximation $X(m)_{s,t}^2$. Since, for all $n < m$, we have

$$
\sup_D \sum_l \left| X(m+1)_{t_{l-1},t_l}^2 - X(m)_{t_{l-1},t_l}^2 \right|^{p/2}
$$

$$
\geqslant \sum_{k=1}^{2^n} \left| X(m+1)_{t_{k-1}^n,t_k^n}^2 - X(m)_{t_{k-1}^n,t_k^n}^2 \right|^{p/2}
$$

$$
= \frac{1}{2} \sum_{k=1}^{2^n} \left| \sum_{l=2^{m-n}(k-1)+1}^{2^{m-n}k} \triangle_{2l-1}^{m+1}W \otimes \triangle_{2l}^{m+1}W - \triangle_{2l}^{m+1}W \otimes \triangle_{2l-1}^{m+1}W \right|^{p/2},
$$

and therefore

$$
\mathbb{E}\sup_D \sum_l \left| X(m+1)_{t_{l-1},t_l}^2 - X(m)_{t_{l-1},t_l}^2 \right|^{p/2}
$$

$$
\geqslant \frac{1}{2} \sum_{k=1}^{2^n} \mathbb{E} \left| \sum_{l=2^{m-n}(k-1)+1}^{2^{m-n}k} \triangle_{2l-1}^{m+1}W \otimes \triangle_{2l}^{m+1}W - \triangle_{2l}^{m+1}W \otimes \triangle_{2l-1}^{m+1}W \right|^{p/2}.
$$

However,

$$
\mathbb{E} \left| \sum_{l=2^{m-n}(k-1)+1}^{2^{m-n}k} \triangle_{2l-1}^{m+1}W \otimes \triangle_{2l}^{m+1}W - \triangle_{2l}^{m+1}W \otimes \triangle_{2l-1}^{m+1}W \right|^{p/2}
$$

$$
\geqslant \left(\mathbb{E} \left| \sum_{l=2^{m-n}(k-1)+1}^{2^{m-n}k} \triangle_{2l-1}^{m+1}W \otimes \triangle_{2l}^{m+1}W - \triangle_{2l}^{m+1}W \otimes \triangle_{2l-1}^{m+1}W \right|^2 \right)^{p/4}.
$$

Let us compute

$$
H_k \equiv \mathbb{E} \left| \sum_{l=2^{m-n}(k-1)+1}^{2^{m-n}k} \triangle_{2l-1}^{m+1}W \otimes \triangle_{2l}^{m+1}W - \triangle_{2l}^{m+1}W \otimes \triangle_{2l-1}^{m+1}W \right|^2.
$$

It turns out that

$$
H_k = 4 \sum_{l,r} \mathbb{E}(\triangle_{2r-1}^{m+1}w^1 \triangle_{2l-1}^{m+1}w^1)\mathbb{E}(\triangle_{2r}^{m+1}w^2 \triangle_{2l}^{m+1}w^2)
$$

$$
- 4 \sum_{l,r} \mathbb{E}(\triangle_{2r}^{m+1}w^1 \triangle_{2l-1}^{m+1}w^1)\mathbb{E}(\triangle_{2r-1}^{m+1}w^2 \triangle_{2l}^{m+1}w^2)
$$

$$= \left[4 - (2^{2h} - 1)^2\right] 2^{m-n} \left(\frac{1}{2^{m+1}}\right)^{4h}$$

$$+ 2 \sum_{l > r} \left(\frac{1}{2^{m+1}}\right)^{4h} \left(|2(l-r)+1|^{2h} + |2(l-r)-1|^{2h} - 2|2(l-r)|^{2h}\right)^2$$

$$- 2 \sum_{l > r} \left(\frac{1}{2^{m+1}}\right)^{4h} \left(|2(l-r)|^{2h} + |2(l-r)-2|^{2h} - 2|2(l-r)-1|^{2h}\right)$$

$$\times \left(|2(l-r)+2|^{2h} + |2(l-r)|^{2h} - 2|2(l-r)+1|^{2h}\right).$$

We may conclude that, for $h < 1/2$,

$$H_k \geqslant C 2^{m-n} \left(\frac{1}{2^{m+1}}\right)^{4h}.$$

In particular, if $h \leqslant 1/4$, then $H_k \geqslant C\left(1/2^n\right)$. Thus, for any $n < m$,

$$\mathbb{E} \sup_D \sum_l \left| X(m+1)^2_{t_{l-1},t_l} - X(m)^2_{t_{l-1},t_l} \right|^{p/2} \geqslant C \left(\frac{1}{2^n}\right)^{(p-4)/4},$$

which cannot go to zero. This implies that $(X(m)^j)_{j \geqslant 2}$ is not a Cauchy sequence according to the p-variation distance in $L^1(\Omega, \mathcal{F}, \mathbb{P})$, for any $p \geqslant 4$. However, we don't know from the above argument if it is Cauchy almost surely.

4.6 Wiener processes in Banach spaces

The reader may notice that all of the estimates which we have done before depend on the dimension of the space V. It turns out that it is a quite delicate issue to carry over our arguments to infinite-dimensional spaces. One of the difficulties is that there are many different choices of infinite-dimensional tensor products, and choosing an appropriate one becomes an important task. In this section we consider Wiener processes with values in a separable Banach space. Since any Wiener process possesses finite p-variation for any $p > 2$, the main issue is therefore to construct the second-level path, also called the Lévy area process.

4.6.1 *Gaussian analysis*

Let us collect several facts about Gaussian processes with a general index set. Thus we temporarily move away from our standard notation and we use t to denote an element in a general index set.

Gaussian processes

Let $(\Omega, \mathcal{F}, \mathbb{P})$ be a completed probability space. A family $\{X_t : t \in \boldsymbol{T}\}$ of real-valued random variables is called a centered Gaussian process (with index set $\boldsymbol{T}$) if any finite subset $\{X_{t_1}, \ldots, X_{t_k}\}$ (for $t_1, \ldots, t_k \in \boldsymbol{T}$) has k-dimensional normal distribution with expectation vector zero and variance matrix (a_{ij}) (of course, $a_{ij} = \mathbb{E}(X_{t_i} X_{t_j}))$. This definition is equivalent to saying that any finite, non-trivial linear combination $\sum_{i=1}^{k} c_i X_{t_i}$ (where k is an integer, c_i are real numbers, and $t_i \in \boldsymbol{T}$) possesses a one-dimensional Gaussian distribution with mean zero. The law of a centered Gaussian process $\{X_t : t \in \boldsymbol{T}\}$ is thus completely determined by its covariance function as follows:

$$R(s,t) = \mathbb{E}(X_s X_t), \quad \forall s, t \in \boldsymbol{T}.$$

Conversely, given a positive-definite function R on an index set $\boldsymbol{T}$, there is a centered Gaussian process on $\boldsymbol{T}$ with R as its covariance function.

Let B be a separable Banach space endowed with the Borel σ-field of the weak topology, namely the smallest σ-algebra on B such that all bounded linear functionals are measurable. A B-valued random variable G on a probability space $(\Omega, \mathcal{F}, \mathbb{P})$ is called a B-valued, centered Gaussian variable if the family $\{l(G) : l \in B^*\}$ is a (real-valued) centered Gaussian process. We may extend the definition of real Gaussian processes to that of Banach space-valued processes. Thus a family of B-valued random variables $\{X_t : t \in \boldsymbol{T}\}$ is called a B-valued, centered Gaussian process if any non-trivial, finite linear combination $\sum_i c_i X_{t_i}$ (where $c_i \in \mathbb{R}$ and t_i belong to the index set $\boldsymbol{T}$) is a centered, B-valued Gaussian variable.

Let $\boldsymbol{X} = \{X_t : t \in \boldsymbol{T}\}$ be a centered B-valued Gaussian process. Then the L^2-distance induced by $\boldsymbol{X}$ is given by

$$r(s,t) = \sqrt{\mathbb{E}|X_s - X_t|_B^2}, \quad \forall s, t \in \boldsymbol{T},$$

which is actually a quasi-distance $r(s,t)$ on the index set $\boldsymbol{T}$. The Gaussian process $\boldsymbol{X} = \{X_t : t \in \boldsymbol{T}\}$ is said to be continuous on $\boldsymbol{T}$ if $\lim_{r(s,t) \to 0} X_s = X_t$ almost surely. It turns out that the continuity of $\boldsymbol{X}$ is equivalent to the boundedness of $\boldsymbol{X} = \{X_t : t \in \boldsymbol{T}\}$, namely

$$\sup_{F \subset \boldsymbol{T}} \mathbb{E} \sup_{t \in F} |X_t|_B < \infty,$$

where the supremum is taken over all finite subsets $F \subset \boldsymbol{T}$.

If d is a quasi-metric on $\boldsymbol{T}$, then $B_t(\varepsilon) = \{s \in \boldsymbol{T} : d(s,t) < \varepsilon\}$ denotes the d-ball of center t and radius ε. The metric entropy (also called d-entropy), denoted by $N(\boldsymbol{T}, d, \varepsilon)$, is the minimal number of balls with radius ε for the metric d that are necessary to cover $\boldsymbol{T}$.

Theorem 4.6.1 (R. M. Dudley, 1967) *If $(X_t)_{t \in T}$ is a centered B-valued Gaussian process, and if d is a quasi-metric on $\boldsymbol{T}$ such that*

$$\sqrt{\mathbb{E}|X_s - X_t|_B^2} \leqslant d(s,t), \quad \forall s, t \in \boldsymbol{T},$$

then there is a universal constant C, where

$$\mathbb{E} \sup_{t \in F} |X_t|_B \leqslant C \int_0^\infty \sqrt{\log N(\boldsymbol{T}, d, \varepsilon)} \, d\varepsilon,$$

for all finite subsets $F \subset \boldsymbol{T}$.

Moreover, under some additional structure condition on T, the above integrability of the metric entropy is also a necessary condition for a *real-valued*, centered Gaussian process $(X_t)_{t \in T}$ to be continuous.

Theorem 4.6.2 (X. Fernique, 1975) *If T is a compact group, or a compact subset of the Euclidean space $\mathbb{R}^d$, and $(X_t)_{t \in T}$ is a centered, real, and stationary Gaussian process, then $\sup_{F \subset T} \mathbb{E} \sup_{t \in F} X_t < \infty$ if and only if*

$$\int_0^\infty \sqrt{\log N(T, r, \varepsilon)} \, d\varepsilon < \infty,$$

where r is the induced quasi-metric of $(X_t)_{t \in T}$.

For a general index set T, the integrability of the metric entropy is not a necessary condition for a Gaussian process to be bounded. A necessary and sufficient condition has been found by M. Talagrand via the majorizing measures, for details see Talagrand (1987), Adler (1990), Ledoux and Talagrand (1991), and Ledoux (1996).

Theorem 4.6.3 *There is a universal constant $C > 0$ such that*

(i) *(Fernique, 1975) If $(X_t)_{t \in T}$ is centered, B-valued Gaussian process, then, for any probability m on T,*

$$\mathbb{E} \sup_{t \in T} |X_t|_B \leqslant C \sup_{t \in T} \int_0^\infty \sqrt{-\log m(B(t, \varepsilon))} \, d\varepsilon,$$

where $B(t, \varepsilon)$ denotes the metric ball at t with radius ε with respect to the L^2-metric $r(s, t) = \sqrt{\mathbb{E}|X_t - X_s|_B^2}$.

(ii) *(Talagrand, 1987) If $(X_t)_{t \in T}$ is a real-valued and centered Gaussian process, then there is a probability m on T such that*

$$\mathbb{E} \sup_{t \in T} |X_t| \geqslant C^{-1} \sup_{t \in T} \int_0^\infty \sqrt{-\log m(B(t, \varepsilon))} \, d\varepsilon.$$

In fact we will only need the following corollary.

Corollary 4.6.1 *Let $(X_t)_{t \in T}$ be a centered, B-valued Gaussian process and let $(Y_t)_{t \in T}$ be a centered, real-valued Gaussian process such that*

$$\mathbb{E}|X_t - X_s|_B^2 \leqslant \mathbb{E}|Y_t - Y_s|^2, \quad \forall s, t \in T.$$

Then

$$\mathbb{E} \sup_{t \in T} |X_t|_B \leqslant C^2 \mathbb{E} \sup_{t \in T} |Y_t|,$$

where C is the universal constant defined in Theorem 4.6.3.

The following Borel inequality is one of the main tools used in proving the above theorems, and is itself one of the important results in the theory of Gaussian processes.

Theorem 4.6.4 (Borel inequality) *Let $(X_t)_{t \in F}$ be a centered Gaussian process on a finite index set F. Set $M = \mathbb{E} \sup_{t \in F} X_t$ and $\sigma^2 = \sup_{t \in F} \mathbb{E} X_t^2$. Then, for all $\varepsilon > 0$,*

$$\mathbb{P} \left\{ \sup_{t \in F} X_t \geq M + \varepsilon \right\} \leq \exp \left(-\frac{\varepsilon^2}{2\sigma^2} \right).$$

Abstract Wiener spaces

The concept of an abstract Wiener space was introduced by L. Gross in the 1960s. An abstract Wiener space consists of three items, namely a separable Banach space B, a Hilbert space H which is a vector subspace of B such that the inclusion $i : H \hookrightarrow B$ is bounded, and a Gaussian measure μ supported in B with a mean of zero and covariance given by the Hilbert space H. The σ-field $\mathcal{B}(B)$ on B is the Borel σ-field generated by its weak topology, and μ is the unique probability on $(B, \mathcal{B}(B))$ such that, under μ, every continuous linear functional ξ on B has a one-dimensional normal distribution $\mathcal{N}(0, |\xi|_H^2)$. Therefore, μ is characterized by

$$\int_B \exp \left(i \langle x, \xi \rangle \right) \mu(\mathrm{d}x) = \exp \left(-\frac{1}{2} |\xi|_H^2 \right), \quad \forall \xi \in B^*. \tag{4.63}$$

We will call the triple (B, H, μ) an abstract Wiener space. A basic fact about abstract Wiener spaces is that the embedding $i : H \hookrightarrow B$ is actually compact, and so is $i : B^* \hookrightarrow H$.

Given an abstract Wiener space (B, H, μ) and any $\xi \in B^*$, we define $X_\xi(x) = \langle x, \xi \rangle \equiv \xi(x)$ for $x \in B$. Then $\{X_\xi : \xi \in B^*\}$ is a centered Gaussian process under μ, with covariance function $\mathbb{E}(X_\xi X_\eta) = \langle \xi, \eta \rangle_H$, the inner product of H. Moreover, X_ξ is continuous on the unit ball B_1^*, and therefore

$$a \equiv \mathbb{E} \sup_{\xi \in B_1^*} X_\xi = \int_B |x|_B \mu(\mathrm{d}x) < \infty. \tag{4.64}$$

The Borel inequality then takes the following form:

$$\mu \left\{ x \in B : |x|_B \geq a + \varepsilon \right\} \leq \exp \left(-\frac{\varepsilon^2}{2\sigma^2} \right), \tag{4.65}$$

for all $\varepsilon > 0$, where $\sigma^2 = \sup_{t \in B_1^*} \mathbb{E} X_t^2$. As a consequence we have the following theorem.

Theorem 4.6.5 (Landau and Shepp, 1970; Fernique, 1975) *Let (B, H, μ) be an abstract Wiener space and let $\sigma^2 = \sup_{t \in B_1^*} \mathbb{E} X_t^2$. Then,*

$$\int_B \exp \left(\alpha |x|_B^2 \right) \mu(\mathrm{d}x) < \infty \quad \text{if and only if} \quad \alpha < \frac{1}{2\sigma^2}.$$

In particular, for any $p \geq 0$, $\int_B |x|_B^p \mu(\mathrm{d}x) < \infty$.

Let $G : \Omega \to B$ be a random variable on a completed probability space $(\Omega, \mathcal{F}, \mathbb{P})$, taking values in B. Then we say that G possesses a distribution μ, where (B, H, μ) is an abstract Wiener space, if, for any finite many-continuous linear functionals $\xi_1, \ldots, \xi_k$, the real random variables $(\xi_1(G), \ldots, \xi_k(G))$ have a centered normal distribution with variance matrix $(\langle \xi_i, \xi_j \rangle_H)$.

The basic example of an abstract Wiener space is the so-called classical Wiener space.

The standard d-dimensional Brownian motion induces a unique probability measure ν (called the Wiener measure) on the space $\boldsymbol{W}_0^d \equiv C_0([0,1], \mathbb{R}^d)$ of all $\mathbb{R}^d$-valued continuous paths with running time $[0,1]$ and initial point zero. The Cameron–Martin space $\boldsymbol{H}_0^1$ is the vector subspace of all paths h in $\boldsymbol{W}_0^d$ with square-integrable first derivatives, that is

$$\boldsymbol{H}_0^1 = \{ h \in \boldsymbol{W}_0^d : \dot{h} \in L^2[0,1] \} \, .$$

$\boldsymbol{W}_0^d$ is a Banach space with the maximum norm, and $\boldsymbol{H}_0^1$ is a Hilbert space under the scaler product norm $|h|_{\boldsymbol{H}_0^1}^2 = \int_0^1 |\dot{h}_s|^2 \, \mathrm{d}s$. The value ν is the unique probability on $\boldsymbol{W}_0^d$ such that

$$\int_{\boldsymbol{W}_0^d} \exp \left(\int_0^1 \dot{h}_s \, \mathrm{d}x_s \right) \nu(\mathrm{d}x) = \exp \left(-\frac{1}{2} |h|_{\boldsymbol{H}_0^1}^2 \right), \quad \forall h \in \boldsymbol{H}_0^1 \, ,$$

where the integral $\int_0^1 \dot{h}_s \, \mathrm{d}x_s$ can be identified as the Itô integral. Therefore, $(\boldsymbol{W}_0^d, \boldsymbol{H}_0^1, \nu)$ is an abstract Wiener space, called the classical Wiener space. The probability ν, of course, is the distribution of the Brownian motion in $\mathbb{R}^d$.

Wiener processes

Let (B, H, μ) be an abstract Wiener space. A continuous stochastic process $(W_t)_{t \geqslant 0}$ on a completed probability space $(\Omega, \mathcal{F}, \mathbb{P})$, with values in the Banach space B, is called a Wiener process with marginal law μ if $W_0 = 0$, for any $0 \leqslant t_1 < t_2 < \cdots < t_k$, the random variables $W_{t_1}, W_{t_2} - W_{t_1}, \ldots, W_{t_k} - W_{t_{k-1}}$ are mutually independent, and finally, for all $t > s$, $W_t - W_s$ possesses the distribution $\mathcal{N}(0, (t-s)| \cdot |_H^2)$, i.e. $(t-s)^{-1/2} (W_t - W_s)$ is a B-valued Gaussian variable with distribution μ. Therefore, for any $t > s, p > 0$,

$$\mathbb{E} \left| (t-s)^{-1/2} (W_t - W_s) \right|_B^p = \int_B |x|_B^p \mu(\mathrm{d}x) < \infty \, ,$$

so that

$$\mathbb{E} |W_t - W_s|_B^p = C_p |t - s|^{p/2} \, , \quad \forall t, s \geqslant 0 \, , \tag{4.66}$$

where $C_p = \int_B |x|_B^p \mu(\mathrm{d}x)$ is a constant.

Theorem 4.6.6 (Gross, 1970) *Let (B, H, μ) be an abstract Wiener space. Then there is a B-valued Wiener process with marginal law μ.*

Proof Define an index set $T = \mathbb{R}_+ \times B^*$, and a covariance function $R(\tilde{s}, \tilde{t}) = (s \wedge t)\langle \xi, \eta \rangle_H$ if $\tilde{s} = (s, \xi)$, $\tilde{t} = (t, \eta) \in \mathbb{R}_+ \times B^*$. Then one can check that R is positive definite. Hence, there is a centered Gaussian process $(X_{\tilde{t}})_{\tilde{t} \in T}$ on some completed probability space $(\Omega, \mathcal{F}, \mathbb{R})$ with covariance function R. For any $t \geqslant 0$, set $Y_t(\xi) = X_{(t,\xi)}$. Then $\xi \to Y_t(\xi)$ is linear almost surely. Thus, $t^{-1/2}Y_t$ is a random variable taking values in the algebraic dual of B^* with distribution μ, and therefore $Y_t \in B$ almost surely as $\mu(B) = 1$. Moreover,

$$\mathbb{E}|Y_t - Y_s|_B^p = C_p|t - s|^{p/2}, \quad \forall t, s \geqslant 0,$$

for any $p > 0$, so that, by the Kolmogorov–Čentsov theorem, there is a continuous version $(W_t)_{t \geqslant 0}$ of $(Y_t)_{t \geqslant 0}$, which is a Wiener process with marginal distribution μ.   ∎

4.6.2   *Wiener processes as geometric rough paths*

Now we come to the question of constructing a canonical geometric rough path which is associated with a Banach space-valued Wiener process.

Let (B, H, μ) be an abstract Wiener space, and let (W_t) be a B-valued Wiener process on a completed probability space $(\Omega, \mathcal{F}, \mathbb{P})$ with marginal distribution μ. We need a Banach tensor product $B \otimes B$ to build a truncated tensor algebra $T^{(2)}(B)$ in which a canonical geometric rough path takes its values.

As in the first two sections, we use $X(m)_{s,t} = (1, X(m)_{s,t}^1, X(m)_{s,t}^2)$ to denote the dyadic polygonal approximations. By (4.66), all of the estimates in the first two sections of this chapter hold for any Wiener process. Thus we need only to control the second-level path $X(m)_{s,t}^2$. Let $2 < p < 3$. By Proposition 4.3.3 and Lemma 4.2.1, we have to estimate

$$X(m+1)_{t_{k-1}^n, t_k^n}^2 - X(m)_{t_{k-1}^n, t_k^n}^2$$

$$= \frac{1}{2} \sum_{l=2^{m-n}(k-1)+1}^{2^{m-n}k} (\triangle_{2l-1}^{m+1}W \otimes \triangle_{2l}^{m+1}W - \triangle_{2l}^{m+1}W \otimes \triangle_{2l-1}^{m+1}W),$$

$$\text{(4.67)}$$

for $m \geqslant n$. Let us isolate a condition that allows us to control the above expression.

Definition 4.6.1 *We say a Banach tensor product $B \otimes B$ is exact with respect to the Gaussian measure μ if there are constants C and $0 \leqslant \alpha < 1$ such that, for any sequence $\{G_l, \hat{G}_l : l = 1, \ldots, N\}$ of independent, B-valued random variables with the common distribution μ, we have*

$$\mathbb{E}\left|\sum_{l=1}^{N} G_l \otimes \hat{G}_l\right|_{B \otimes B} \leqslant CN^\alpha. \quad \text{(4.68)}$$

Remark 4.6.1 Let us consider the real variable case. If $\{\xi_i\}$ is a sequence of real, independent random variables with identical distributions such that $\mathbb{E}\xi_i = 0$ and $\mathbb{E}\xi_i^2 = 1$, then, for all N,

$$\mathbb{E}\left|\sum_{i=1}^{N} \xi_i\right| \leqslant \sqrt{N}.$$

Hence, the exact condition (4.68) is set to catch cancellation due to independence.

Suppose $B \otimes B$ is exact. Then,

$$\left|\sum_{l=1}^{N} G_l \otimes \hat{G}_l\right|_{B \otimes B}^2 = \sup_{\varphi \in (B \otimes B)_1^*} \left(\sum_{l=1}^{N} \varphi(G_l \otimes \hat{G}_l)\right)^2$$

$$\leqslant \sum_{l=1}^{N} \sup_{\varphi \in (B \otimes B)_1^*} \varphi(G_l \otimes \hat{G}_l)^2$$

$$+ 2 \sup_{\varphi \in (B \otimes B)_1^*} \sum_{k<l}^{N} \varphi(G_l \otimes \hat{G}_l)\varphi(G_k \otimes \hat{G}_k)$$

$$\leqslant \sum_{l=1}^{N} |G_l|^2|\hat{G}_l|^2 + 2\sum_{l=2}^{N} |G_l||\hat{G}_l|\left|\sum_{k=1}^{l-1} G_k \otimes \hat{G}_k\right|.$$

Therefore, by independence, we have

$$\mathbb{E}\left|\sum_{l=1}^{N} G_l \otimes \hat{G}_l\right|_{B \otimes B}^2 \leqslant N\left(\mathbb{E}|G_1|^2\right)^2 + 2N^\alpha N(\mathbb{E}|G_1|)^2$$

$$\leqslant CN^{1+\alpha}, \qquad (4.69)$$

for some finite constant $C > 0$.

Let $B \otimes B$ be a Banach tensor product which is exact with respect to μ. Since, for each $k \leqslant 2^n$,

$$\left\{\sqrt{2^{m+1}}\triangle_{2l-1}^{m+1}W, \quad \sqrt{2^{m+1}}\triangle_{2l}^{m+1}W, \quad l = 2^{m-n}(k-1)+1,\ldots,2^{m-n}k\right\}$$

are independent with the same distribution μ, then, by (4.67) and (4.68), and the Hölder inequality (as $p/2 > 1$), there is a constant C such that, for any $m \geqslant n$,

$$\mathbb{E}\sum_{k=1}^{2^n} \left|X(m+1)_{t_{k-1}^n,t_k^n}^2 - X(m)_{t_{k-1}^n,t_k^n}^2\right|_{B \otimes B}^{p/2}$$

$$\leqslant C\left(\frac{1}{2^m}\right)^{(1-\alpha)(p-2)/4}\left(\frac{1}{2^n}\right)^{(1+\alpha)(p-2)/4}. \qquad (4.70)$$

With the above estimate, the estimate (4.37) for $m < n$, and together with the estimates in the first two sections of this chapter for any continuous process under Hölder's condition, we may deduce, by using a quite similar (if not identical) argument as in the d-dimensional case, the following key estimate. For any natural numbers $m \geqslant n$,

$$\mathbb{E} \sup_{D} \sum_{l} \left| X(m+1)^2_{t_{l-1},t_l} - X(m)^2_{t_{l-1},t_l} \right|^{p/2} \leqslant C \left(\frac{1}{2^m} \right)^{(1-\alpha)(p-2)/4}, \qquad (4.71)$$

for some constant C depending only on p and the exponent α in the exact condition. By the Hölder inequality, we may thus conclude that

$$\mathbb{E} \left(\sup_{D} \sum_{l} \left| X(m+1)^2_{t_{l-1},t_l} - X(m)^2_{t_{l-1},t_l} \right|^{p/2} \right)^{2/p} \leqslant C \left(\frac{1}{2^m} \right)^{((1-\alpha) \vee \frac{1}{2})(p-2)/2p},$$

for all m, and therefore, almost surely, $(X(m)^2_{s,t})$ converges to a path $(X^2_{s,t})$ in the Banach tensor product $B \otimes B$ in $p/2$-variation distance. It is routine to check that almost surely $(1, X^1_{s,t}, X^2_{s,t})$ (with $X^1_{s,t} = W_t - W_s$) are indeed geometric rough paths in $T^{([p])}(B)$ (via an exact tensor product $B \otimes B$). We summarize the above discussion as the following theorem.

Theorem 4.6.7 *Let (B, H, μ) be an abstract Wiener space, let $B \otimes B$ be an exact Banach tensor product with respect to the Wiener measure μ, and let $(W_t)_{t \geqslant 0}$ be a B-valued Wiener process with marginal law μ. Then, for any $2 < p < 3$, the dyadic polygonal approximation $(1, X(m)^1_{s,t}, X(m)^2_{s,t})$ converges almost surely and in $L^1(\Omega, \mathcal{F}, \mathbb{P})$ to a unique geometric rough path $(1, X^1_{s,t}, X^2_{s,t})$ according to the p-variation distance. Moreover, $X^1_{s,t} = W_t - W_s$.*

Injective tensor products

The proof of the following proposition is due to M. Ledoux (private communication).

Proposition 4.6.1 *Let $\{G_l, \hat{G}_l\}$ be an independent identically-distributed (i.i.d.) sequence with common distribution μ. Then, under the injective tensor norm $|\cdot|_{B \otimes B}$,*

$$\sup_{N} \mathbb{E} \left| \frac{1}{\sqrt{N}} \sum_{l=1}^{N} G_l \otimes \hat{G}_l \right|_{B \otimes B} \leqslant C \mathbb{E} |G_1|^2_B,$$

for some constant C, which implies that the injective tensor space is exact.

Proof For simplicity, we move away from our standard notation. We will denote the unit ball B^*_1 by $\boldsymbol{T}$, and use s, t, etc. to denote an element in B^*_1. Then, by definition of the injective norm, we may write

$$\left| \frac{1}{\sqrt{N}} \sum_{l=1}^{N} G_l \otimes \hat{G}_l \right|_{B \otimes B} = \sup_{s,t \in \boldsymbol{T}} \frac{1}{\sqrt{N}} \sum_{l=1}^{N} G_l(t) \hat{G}_l(s)$$

$$= \sup_{t \in \boldsymbol{T}} \left| \frac{1}{\sqrt{N}} \sum_{l=1}^{N} G_l(t) \hat{G}_l \right|_{B}.$$

By homogeneity, we may assume without loss of generality that $\mathbb{E}|G_1|_B^2 = 1$. Conditionally on $\{G_1, \ldots, G_N\}$ (and denoting the conditional expectation by $\mathbb{E}_1$), consider the following centered, B-valued Gaussian process:

$$\hat{Z}_t = \frac{1}{\sqrt{N}} \sum_{l=1}^{N} G_l(t)\hat{G}_l, \quad t \in \boldsymbol{T}.$$

Since $\{\hat{G}_1, \ldots, \hat{G}_N\}$ are i.i.d. Gaussian variables with common distribution μ, so that the Gaussian process $(1/\sqrt{N}) \sum_{l=1}^{N} G_l(t)\hat{G}_l$ possesses the same distribution as $\sqrt{N^{-1} \sum_{l=1}^{N} G_l(t)^2}\hat{G}_1$, and therefore

$$\mathbb{E}\big|\hat{Z}_t - \hat{Z}_s\big|^2 = \frac{1}{N} \sum_{l=1}^{N} |G_l(t) - G_l(s)|^2 .$$

Let $\boldsymbol{g} = \{g_1, \ldots, g_N\}$ be real i.i.d. with the standard normal distribution $\mathcal{N}(0,1)$, each of which is independent of all G_l and $\hat{G}_l$, and set $H_t = N^{-1/2} \sum_{l=1}^{N} g_l G_l(t)$. Then H_t is a real Gaussian process with L^2 distance

$$N^{-1} \sum_{l=1}^{N} |G_l(t) - G_l(s)|^2 .$$

Therefore,

$$\mathbb{E}_1 \big|\hat{Z}_t - \hat{Z}_s\big|^2 = \mathbb{E}_{\boldsymbol{g}} |H_t - H_s|^2 .$$

($\mathbb{E}_{\boldsymbol{g}}$ denotes the conditional expectation for $\boldsymbol{g} = \{g_1, \ldots, g_N\}$.) By Corollary 4.6.1,

$$\mathbb{E}_1 \sup_{t \in \boldsymbol{T}}\big|\hat{Z}_t\big|_B \leqslant C^2 \mathbb{E}_{\boldsymbol{g}} \sup_{t \in \boldsymbol{T}} |H_t| ,$$

which yields that

$$
\begin{aligned}
\mathbb{E}\left|\frac{1}{\sqrt{N}} \sum_{l=1}^{N} G_l \otimes \hat{G}_l\right|_{B \otimes B} &= \mathbb{E}\mathbb{E}_1 \sup_{t \in \boldsymbol{T}}\big|\hat{Z}_t\big|_B \\
&\leqslant C^2 \mathbb{E}\mathbb{E}_{\boldsymbol{g}} \sup_{t \in \boldsymbol{T}} |H_t| \\
&= C^2 \mathbb{E}_{\boldsymbol{g}} \mathbb{E} \sup_{t \in \boldsymbol{T}} |H_t| \\
&= C^2 \mathbb{E}_{\boldsymbol{g}} \mathbb{E}\left|\frac{1}{\sqrt{N}} \sum_{l=1}^{N} g_l G_l\right|_B .
\end{aligned}
$$

Now, conditional on $\boldsymbol{g} = \{g_1, \ldots, g_N\}$, under $\mathbb{E}$, the centered, B-valued Gaussian variable $N^{-1/2} \sum_{l=1}^{N} g_l G_l$ has the same distribution as $\sqrt{N^{-1} \sum_{l=1}^{N} g_l^2}\,G_1$. Therefore,

$$\mathbb{E}\left|\frac{1}{\sqrt{N}}\sum_{l=1}^{N}g_{l}G_{l}\right|_{B}=\sqrt{N^{-1}\sum_{l=1}^{N}g_{l}^{2}\,\mathbb{E}\,|G_{1}|_{B}}\,,$$

so that

$$\mathbb{E}\left|\frac{1}{\sqrt{N}}\sum_{l=1}^{N}G_{l}\otimes\hat{G}_{l}\right|_{B\otimes B}\leqslant C^{2}\mathbb{E}\,|G_{1}|_{B}\,\mathbb{E}_{g}\sqrt{N^{-1}\sum_{l=1}^{N}g_{l}^{2}}$$

$$\leqslant C^{2}\mathbb{E}\,|G_{1}|_{B}\,.$$

This ends the proof.   ■

Finite-dimensional projections

Let (B,H,μ) be an abstract Wiener space, and let $B\otimes B$ be a Banach tensor product. In this subsection, we are going to describe an idea of how to control (up to a right growth order in N) the key term

$$\mathbb{E}\left|\sum_{l=1}^{N}G_{l}\otimes\hat{G}_{l}\right|_{B\otimes B},$$

by projecting the Wiener measure (the common distribution of G_{l}) to finite-dimensional subspaces. The idea is to split each Gaussian variable G_{l} (and $\hat{G}_{l}$) into two parts. One part is a finite-dimensional projection which we should be able to handle easily, and for which we can use the cancellation due to independence of the centered random variables G_{l}, $\hat{G}_{l}$. The remaining part should tend to zero at a fast rate, which can be determined by the nature of the Wiener measure μ.

We choose a sequence $\{\xi_{n}\}$ in the dual space B^{*} (actually in H is enough) with uniformly bounded B-norms, where, as usual, we identify an element in B^{*} as a vector in H, and therefore in B, via the canonical embedding. In particular, the previous requirement is satisfied if $\{\xi_{n}\}$ is a bounded sequence in H. Let G be a B-valued Gaussian variable with distribution μ, and define, for any natural number d, a random variable $G(d)$ by $G(d)=\sum_{i=1}^{d}\langle G,\xi_{i}\rangle\xi_{i}$, where $\langle\cdot,\cdot\rangle$ denotes the pairing between B and B^{*}. Applying the above notation to independent copies $\{G_{l},\hat{G}_{l}:l=1,\ldots,N\}$ of G, we obtain

$$\left|\sum_{l=1}^{N}G_{l}(d)\otimes\hat{G}_{l}(d)\right|_{B\otimes B}=\left|\sum_{i,j}^{d}\left(\sum_{l=1}^{N}\langle G_{l},\xi_{i}\rangle\langle\hat{G}_{l},\xi_{j}\rangle\right)\xi_{i}\otimes\xi_{j}\right|_{B\otimes B}$$

$$\leqslant\sum_{i,j}^{d}\left|\sum_{l=1}^{N}\langle G_{l},\xi_{i}\rangle\langle\hat{G}_{l},\xi_{j}\rangle\right||\xi_{i}|_{B}\,|\xi_{j}|_{B}$$

$$\leqslant C\sum_{i,j}^{d}\left|\sum_{l=1}^{N}\langle G_{l},\xi_{i}\rangle\langle\hat{G}_{l},\xi_{j}\rangle\right|,$$

as the sequence $\{\xi_n\}$ is bounded in B. Therefore

$$\mathbb{E}\left|\sum_{l=1}^{N} G_l(d) \otimes \hat{G}_l(d)\right|_{B \otimes B} \leqslant C\sqrt{N}\left(\sum_{i}^{d} \sqrt{\mathbb{E}\langle G, \xi_i\rangle^2}\right)^2,$$

and the right-hand side can be bounded by $\sqrt{N}d^2$ and $\mathbb{E}\langle e, \xi_i\rangle^2$ (e is a Gaussian variable with law μ). Define

$$\varepsilon(d) = \mathbb{E}|G - G(d)|$$

$$= \int_B \left|x - \sum_{i=1}^{d}\langle x, \xi_i\rangle\xi_i\right|_B \mu(\mathrm{d}x).$$

This quantity depends not only on d but also on $\{\xi_n\}$. Let $X^l = G_l - G_l(d)$, $\hat{X}^l = \hat{G}_l - \hat{G}_l(d)$, and let $S = \sum_{l=1}^{N} G_l \otimes \hat{G}_l$. Then

$$S = \sum_{l=1}^{N} G_l(d) \otimes \hat{G}_l(d) + \sum_{l=1}^{N} X^l \otimes \hat{G}_l + \sum_{l=1}^{N} G_l \otimes \hat{X}^l - \sum_{l=1}^{N} X^l \otimes \hat{X}^l,$$

and therefore

$$\mathbb{E}|S| \leqslant \mathbb{E}\left|\sum_{l=1}^{N} G_l(d) \otimes \hat{G}_l(d)\right|_{B \otimes B} + 2N\mathbb{E}|G_l \otimes \hat{X}^l|_{B \otimes B} + N\mathbb{E}|X^l \otimes \hat{X}^l|_{B \otimes B}$$

$$\leqslant \mathbb{E}\left|\sum_{l=1}^{N} G_l(d) \otimes \hat{G}_l(d)\right|_{B \otimes B} + N\varepsilon(d)^2 + 2N\varepsilon(d). \tag{4.72}$$

Both the first term and $\varepsilon(d)$ on the right-hand side can be estimated via the knowledge of the Wiener measure μ.

Suppose that B is the classical Wiener space $W_0^1 \equiv C_0([0,1];\mathbb{R})$ of real-valued continuous functions on the unit interval $[0,1]$, and H is the Cameron–Martin space H_0^1. Let μ be the Wiener measure ν, namely the law of the standard Brownian motion, or more generally the law of a continuous, real Gaussian process. The only fact which we will use is that the embedding $H_0^1 \hookrightarrow W_0^1$ is bounded (actually compact, due to the existence of the Wiener measure ν).

Suppose that μ is the law of a real, centered, continuous Gaussian process. Let us, in this case, estimate $\varepsilon(d)$ and $\left|\sum_{l=1}^{N} G_l(d) \otimes \hat{G}_l(d)\right|_{B \otimes B}$. Then μ is a Gaussian measure on W_0^1, but it may have a different Martin space H and therefore a different abstract Wiener space from the classical one.

We choose the Haar basis (see Bass, 1988), so that $\gamma_{0,0}(t)$ is the constant function 1 on $[0,1]$. For any $m \geqslant 0$ and $j = 1,\ldots,2^m$,

$$\gamma_{j,m+1}(t) = \begin{cases} 2^{m/2}, & \text{if } (2j-2)/2^{m+1} \leqslant t < (2j-1)/2^{m+1}, \\ -2^{m/2}, & \text{if } (2j-1)/2^{m+1} \leqslant t < 2j/2^{m+1}, \\ 0, & \text{otherwise}. \end{cases}$$

Then $\{\gamma_{j,m} : 1 \leqslant j \leqslant 2^{m-1}, m \in \mathbb{N}\}$ is an orthonormal base of the L^2 space $L^2[0,1]$. Set $\eta_{j,m}(t) = \int_0^t \gamma_{j,m}(s)\,\mathrm{d}s$, so that $\{\eta_{j,m} : m = 0,1,2,\ldots\}$ is an orthonormal base of the Cameron–Martin space $\boldsymbol{H}_0^1$. Let us rearrange the basis $\{\eta_{j,m}\}$ to be $\{\xi_k : k = 1,2,\ldots\}$, first according to m, and then on j. We want to estimate

$$\varepsilon(d) = \int_{\boldsymbol{W}_0^1} \sup_{0 \leqslant t \leqslant 1} \left| x_t - \sum_{k=1}^d \langle x, \xi_k \rangle \xi_k(t) \right| \mu(\mathrm{d}x). \qquad (4.73)$$

The observation is that $\sum_{m=0}^{n-1} \sum_{j=1}^{2^m} \langle x, \eta_{j,m+1} \rangle \eta_{j,m+1}(t)$ is the nth dyadic polygonal approximation

$$x^n(t) = x_{t_{l-1}^n} + 2^n(t - t_{l-1}^n)\triangle_l^n x, \quad \text{if} \quad t_{l-1}^n \leqslant t < t_l^n,$$

for $l = 1,\ldots,2^n$.

In fact, for any continuous function x_t on $[0,1]$ with initial point zero, any m, and $j = 1,\ldots,2^m$,

$$\begin{aligned}
\langle x, \eta_{j,m+1} \rangle &= \int_0^1 \dot{\eta}_{j,m+1}(s)\,\mathrm{d}x(s) \\
&= 2^{m/2} \int_{(2j-2)/2^{m+1}}^{(2j-1)/2^{m+1}} \mathrm{d}x(s) - 2^{m/2} \int_{(2j-1)/2^{m+1}}^{2j/2^{m+1}} \mathrm{d}x(s) \\
&= 2^{m/2} \left(\triangle_{2j-1}^{m+1} x - \triangle_{2j}^{m+1} x \right),
\end{aligned}$$

and therefore

$$\langle x, \eta_{j,m+1} \rangle = \langle x^n, \eta_{j,m+1} \rangle, \quad \text{if} \quad m+1 \leqslant n,$$

and, if $m+1 > n$, then $\langle x^n, \eta_{j,m+1} \rangle = 0$. Clearly $\langle x, \eta_{0,0} \rangle = \langle x^n, \eta_{0,0} \rangle$. Therefore, as functions in $L^2[0,1]$,

$$x^n = \sum_{m=0}^{n-1} \sum_{j=1}^{2^m} \langle x, \eta_{j,m+1} \rangle \eta_{j,m+1}(t),$$

with $\langle x, \eta_{j,m+1} \rangle = 2^{m/2} \left(\triangle_{2j-1}^{m+1} x - \triangle_{2j}^{m+1} x \right)$. The latter can be identified as Itô's integral if (x_t) is a standard Brownian motion (i.e. μ is the classical Wiener measure ν). However, both functions are continuous, so that they are equal everywhere.

Now let us take independent copies of the real Gaussian process G with distribution μ, namely $G_l, \hat{G}_l$, for $l = 1,\ldots,N$. By (4.72), we obtain

$$\mathbb{E} \left| \sum_{l=1}^N G_l^n \otimes \hat{G}_l^n \right|_{B \otimes B}$$

$$\leqslant C\sqrt{N} \left(\sum_{m=0}^{n-1} \sum_{j=1}^{2^m} \sqrt{\mathbb{E}\langle G_l, \xi_{j,m+1} \rangle^2} \right)^2$$

$$= C\sqrt{N}\left(\sum_{m=0}^{n-1}\sum_{j=1}^{2^m}\sqrt{2^m\mathbb{E}\left(\triangle_{2j-1}^{m+1}G - \triangle_{2j}^{m+1}G\right)^2}\right)^2$$

$$\leqslant C\sqrt{N}\left(\sum_{m=0}^{n-1}\sum_{j=1}^{2^m}2^{m/2}\sqrt{\mathbb{E}\left(\triangle_{2j-1}^{m+1}G\right)^2 + \mathbb{E}\left(\triangle_{2j}^{m+1}G\right)^2}\right)^2. \tag{4.74}$$

The last term in the above inequality can be estimated by the covariance function of the real Gaussian process G which is determined by μ. In fact,

$$\varepsilon(2^n) = \int_{\boldsymbol{W}_0^1}\left|x - \sum_{m=0}^{n-1}\sum_{j=1}^{2^m}\langle x, \xi_{j,m+1}\rangle \xi_{j,m+1}(t)\right|\mu(\mathrm{d}x)$$

$$= \mathbb{E}^{\mu}\sup_{0\leqslant t\leqslant 1}|x_t - x(n)_t|, \tag{4.75}$$

which can be controlled via Hölder's condition, as we did in the second section of this chapter, see (4.36).

Real-valued Gaussian processes

Let us assume that the Gaussian measure μ satisfies the Hölder condition

$$\mathbb{E}^{\mu}|x_t - x_s|^p \leqslant C_p|t - s|^{hp}, \quad \forall s, t \geqslant 0,$$

for some constants $h \leqslant 1/2$ and C_p (for some p, and therefore for any p). Then,

$$\mathbb{E}^{\mu}|x_t - x_s|^2 \leqslant C_p|t - s|^{2h}.$$

By (4.74), we get that

$$\mathbb{E}\left|\sum_{l=1}^{N}G_l(n)\otimes\hat{G}_l(n)\right|_{B\otimes B} \leqslant C\sqrt{N}\left(\sum_{m=0}^{n-1}\sum_{j=1}^{2^m}2^{m/2}\left(\frac{1}{2^m}\right)^h\right)^2$$

$$\leqslant C(h)\sqrt{N}\left(2^{3/2-h}\right)^{2n}.$$

By (4.75) and (4.36),

$$\varepsilon(2^n) = \mathbb{E}^{\mu}\sup_{0\leqslant t\leqslant 1}|x(t) - x_n(t)|$$

$$\leqslant C(p,h)\left(\frac{1}{2^n}\right)^{(hp-1)/2p},$$

for any $hp > 1$. Now choose n such that

$$\sqrt{N}\left(2^{3/2-h}\right)^{2n} \leqslant N\left(\frac{1}{2^n}\right)^{(hp-1)/2p}.$$

For example, set

$$(2^n)^{(3-2h)+(hp-1)/2p} = \sqrt{N} \,.$$

Then, by (4.72), we obtain

$$\mathbb{E} \left| \sum_{l=1}^{N} G_l \otimes \hat{G}_l \right| \leqslant C(p, h) N^{1-\beta} \,,$$

where

$$\beta = \frac{p}{hp - 1 + p(6 - 4h)} > 0 \,.$$

Thus we have proved the following theorem.

Theorem 4.6.8 *Let $B = W_0^1$ be the space of continuous functions on $[0, 1]$, and let μ be the law of a real centered Gaussian process which satisfies the Hölder condition*

$$\mathbb{E}^\mu |x_t - x_s|^p \leqslant C_p |t - s|^{hp} \,,$$

for some $p > 0$ and $h \leqslant 1/2$ such that $hp > 1$. Then any Banach tensor product $B \otimes B$ is exact with respect to the Gaussian measure μ.

Examples of Gaussian measures on W_0^1 which satisfy the Hölder condition include the measure induced by the standard Brownian motion, the law of a fractional Brownian motion with fractional exponent $h \in (0, 1/2]$, etc.

Brownian sheet measure

Let us point out that the above idea can be applied to some other situations. In this part, we treat the Gaussian measure induced by the Brownian sheet. For simplicity we only consider Brownian motion on the square $[0, 1]^2$, which is a real centered Gaussian process $\{x_t : t \in [0, 1]^2\}$ on some probability space with covariance function

$$\mathbb{E}(x_t x_s) = (s_1 \wedge t_1)(s_2 \wedge t_2) \,,$$

if $t = (t_1, t_2)$ and $s = (s_1, s_2)$. The notation with bold letters will apply to two parameters in what follows. The Brownian sheet $\{x_t : t \in [0, 1]^2\}$ induces a Gaussian measure μ on the Banach space B of continuous functions on $[0, 1]^2$ with initial point zero (with the supremum norm). The Cameron–Martin space H of the Gaussian measure μ is

$$H = \left\{ h : h(s, t) = \int_0^t \int_0^s f(s, t) \, ds \, dt, \quad \text{for} \quad f \in L^2[0, 1]^2 \right\} \,,$$

with the natural inner product.

Theorem 4.6.9 *Using the above notation, any Banach tensor product $B \otimes B$ is exact with respect to the Brownian sheet measure.*

The proof uses the same technique of finite-dimensional approximation, but this time applied to the Brownian sheet. The Haar orthonormal basis of $L^2[0,1]^2$ comprises functions of the form

$$\varphi_{j,m}(t) = \gamma_{j_1,m_1}(t_1)\gamma_{j_2,m_2}(t_2), \quad \text{if} \quad 1 \leqslant j_i \leqslant 2^{m_i-1},$$

or $j_i = 0$ if $m_i = 0$. Therefore, $\{\phi_{j,m}\}$ is an orthonormal basis for H, where

$$\phi_{j,m}(t) = \int_0^{t_1} \int_0^{t_2} \varphi_{j,m}(t)\, dt.$$

Then, again, we rearrange $\{\phi_{j,m}\}$ to be $\{\xi_k : k = 1, 2, \ldots\}$ according to the lexicographic order. Since μ is a Brownian sheet, then, for any independent copies x_t^l, $\hat{x}_t^l$ of the Brownian sheet,

$$\mathbb{E}\left| \sum_{l=1}^{N} x(d)^l \otimes \hat{x}(d)^l \right| \leqslant Cd^2,$$

and (see Goodman and Kuelbs, 1991)

$$\varepsilon(d) = \mathbb{E}^\mu |x - x(d)|_B \leqslant C \left(\frac{(\log d)^4}{d} \right)^{1/2}$$

$$\leqslant C_p \left(\frac{1}{d} \right)^{1/p},$$

for any $p > 2$, where

$$x(d) = \sum_{i=1}^{d} \langle x, \xi_i \rangle \xi_i.$$

The above two estimates are enough to conclude that $B \otimes B$ is exact with respect to the Brownian sheet measure.

An approach via metric entropy

Let (B, H, μ) be an abstract Wiener space. Recall that for any tensor norm $|\cdot|_{B\otimes B}$ we have

$$|x|_{B\otimes B} = \sup_{\substack{\varphi \in (B\otimes B)^* \\ \|\varphi\| \leqslant 1}} \sum_i \varphi(\xi_i)\eta_i,$$

if $x = \sum_i \xi_i \otimes \eta_i$, where we identify $\varphi \in (B \otimes B)^*$ as an element in $L(B, B^*)$ by $\varphi(\xi)\eta = \varphi(\xi \otimes \eta)$.

For any $\varphi \in L(B, B^*)$, we identify it as a bilinear functional on $B \times B$, and therefore a linear functional on $B \otimes_a B$. Let $\{G_l, \hat{G}_l : l = 1, \ldots, N\}$ be independent B-valued Gaussian random variables with identical distribution μ. For any $\varphi \in L(B, B^*)$, define

$$
Z_\varphi = \frac{1}{\sqrt{N}} \sum_{l=1}^{N} \varphi(G_l, \hat{G}_l).
$$

Notice that, by using the independence, conditional on $\{G_l : l = 1, \ldots, N\}$, $\varphi(G_l)\hat{G}_l$ possesses a central normal distribution $\mathcal{N}(0, |\varphi(G_l)|_H^2)$, and therefore

$$
\mathbb{E}\exp\left(\mathrm{it}Z_\varphi\right) = \prod_l \mathbb{E}e^{\mathrm{it}\varphi(G_l, \hat{G}_l)/\sqrt{N}}.
$$

Consider $\mathbb{E}\exp\left(\mathrm{it}\varphi(G, \hat{G})\right)$, where G, $\hat{G}$ are independent B-valued Gaussian random variables with distribution μ. Write $G = \sum_i g_l^i \xi_i$ and $\hat{G} = \sum_j \hat{g}_l^j \xi_j$, where $\{\xi_i\}$ is an orthonormal basis of H. Then all of the g_l^i, $\hat{g}_l^j$ are independent with a normal distribution $\mathcal{N}(0, 1)$, and therefore

$$
\mathbb{E}\exp\left[\mathrm{it}\varphi(G, \hat{G})\right] = \prod_{i,j} \exp\left(\mathrm{it}\varphi_{ij} g_l^i \hat{g}_l^j\right)
$$

$$
= \prod_{i,j} \left(\frac{1}{1 + \varphi_{ij}^2 t^2}\right)^{1/2},
$$

where $\varphi_{ij} = \varphi(\xi_i, \xi_j)$. By independence we get that

$$
\mathbb{E}\exp\left(\mathrm{it}Z_\varphi\right) = \prod_{i,j} \left(\frac{1}{1 + \varphi_{ij}^2 t^2/N}\right)^{N/2},
$$

which is analytic in t, and therefore, for small $a > 0$,

$$
\mathbb{E}\exp\left(aZ_\varphi\right) = \prod_{i,j} \left(\frac{1}{1 - \varphi_{ij}^2 a^2/N}\right)^{N/2}.
$$

Choose $a > 0$ such that $\sum_{i,j} \varphi_{ij}^2 a^2 = 1/2$, then we have

$$
\log \mathbb{E}\exp\left(aZ_\varphi\right) = \sum_{i,j} -\frac{N}{2} \log\left(1 - \frac{\varphi_{ij}^2 a^2}{N}\right)
$$

$$
\leqslant C,
$$

for some universal constant C. Therefore

$$
\mathbb{E}\exp\left(\sqrt{\frac{1}{2 \sum_{i,j} \varphi_{ij}^2}} Z_\varphi\right) \leqslant C,
$$

and, for all $r > 0$,

$$\mathbb{P}\{Z_\varphi > r\} \leqslant C \exp\left(-\frac{r}{\sqrt{2\sum_{i,j}\varphi_{ij}^2}}\right).$$

Let

$$r(\varphi, \psi) = \sqrt{\int_B |\varphi(x) - \psi(x)|_H^2 \mu(\mathrm{d}x)}. \tag{4.76}$$

Then,

$$\mathbb{E}|Z_\varphi - Z_\psi|^2 = r(\varphi, \psi)^2, \quad \forall \varphi, \psi \in L(B, B^*). \tag{4.77}$$

Since $\sqrt{\sum_{i,j}\varphi_{ij}^2} = \int_B |\varphi(x)|_H^2 \mu(\mathrm{d}x)$, which we denote by $||\varphi||_H$, the exponential decay can be rewritten as

$$\mathbb{P}\{Z_\varphi > r\} \leqslant C_1 \exp\left(-\frac{C_2 r}{||\varphi||_H}\right), \quad \forall r > 0. \tag{4.78}$$

Now we are in a position to prove the following theorem.

Theorem 4.6.10 *Using the above notation, for any subset $F \subset L(B, B^*)$,*

$$\mathbb{E}\sup_{\varphi \in F} \frac{1}{\sqrt{N}} \sum_{l=1}^N \varphi(G_l, \hat{G}_l) \leqslant C \int_0^\infty \log N(F, r, \varepsilon) \, \mathrm{d}\varepsilon, \tag{4.79}$$

for some universal finite constant C.

Proof The left-hand side of (4.79) can be written as $\mathbb{E}\sup_{\varphi \in F} Z_\varphi$. The proof is standard, and is based on the exponential decay (4.77) and (4.78), although the process $\{Z_\varphi : \varphi \in F\}$ is no longer Gaussian. For the simplicity of the notation, we will denote a point in F by a lower-case letter t, s, etc.

Choose $q > 1$, and n_0 to be the largest integer such that $N(F, r, q^{-n}) = 1$. For integer $n \geqslant n_0$, we construct a partition A_n of F with cardinality $N(n)$ by covering F with balls of diameter less than $2q^{-n}$. In each $A \in A_n$, we choose a point, and the collection thus obtained is denoted by B_n. For any point $t \in F$, and any $n \geqslant n_0$, let $s_n(t)$ be the point in B_n such that $r(t, s_n(t)) \leqslant 2q^{-n}$. Therefore, we have, for any $t \in F$,

$$Z_t = \sum_{n>n_0} \left(Z_{s_n(t)} - Z_{s_{n-1}(t)}\right), \quad \text{with} \quad s_{n_0-1}(t) = 0.$$

By assumption,

$$r(s_n(t), s_{n-1}(t)) \leqslant 2(q+1)q^{-n}.$$

Let c_n be a sequence of non-negative numbers to be chosen later. By splitting the related integrals, we then have

$$\mathbb{E}\sup_{t \in F} Z_t \leqslant \sum c_n + \mathbb{E}\sup_{t \in F} \sum_{n>n_0} |Z_{s_n(t)} - Z_{s_{n-1}(t)}| 1_{\{|Z_{s_n(t)} - Z_{s_{n-1}(t)}| > c_n\}}$$

$$\leqslant \sum c_n + \sum_{n>n_0} \sum_{(t,s)\in \boldsymbol{H}_n} \mathbb{E}\,|\,Z_s - Z_t|\,1_{\{|Z_{t-s}|>c_n\}}$$

$$\leqslant \sum c_n + \sum_{n>n_0} \mathrm{Card}\,(\boldsymbol{H}_n)\,2(q+1)q^{-n}\exp\left(-\frac{C_2 c_n}{2(q+1)q^{-n}}\right),$$

where $\boldsymbol{H}_n = \{(s,t) \in \boldsymbol{B}_n \times \boldsymbol{B}_{n-1}\}$. By definition, $\mathrm{Card}\,(\boldsymbol{H}_n) \leqslant N(n)^2$. Thus, by choosing c_n such that

$$\frac{C_2 c_n}{2(q+1)q^{-n}} = \log N(n)^2\,,$$

we obtain

$$\mathbb{E}\sup_{t\in \boldsymbol{F}} \boldsymbol{Z}_t \leqslant \frac{4}{C_2}\sum_n (q+1)q^{-n}\log N(n) + \sum_{n>n_0} 2(q+1)q^{-n}$$

$$\leqslant C_3 \int_0^\infty \log N(\boldsymbol{F},d,\varepsilon)\,\mathrm{d}\varepsilon\,. \qquad\blacksquare$$

Remark 4.6.2 The best possible bound we would expect is

$$\int_0^\infty \sqrt{\log N(\boldsymbol{F},d,\varepsilon)}\,\mathrm{d}\varepsilon\,,$$

but, at the time of writing this book, we have no idea of how to recover the root.

4.7   Comments and notes on Chapter 4

The fact that knowing the sample paths of the Brownian motion and its Lévy area process is enough to solve all stochastic differential equations, as initially conjectured by H. Föllmer, is the main motivation of the rough path approach, which was achieved in Lyons (1998). The construction of the Lévy area of the Brownian motion was, of course, given by P. Lévy in his famous Lévy (1948), independently of Itô's theory of stochastic integration. The approach presented in this chapter uses Lévy's original idea, namely the dyadic polygonal approximations.

Section 4.1. The idea of controlling the p-variations of continuous paths via the variations through dyadic partitions originates from Lévy's work, and the results in this section have been established in Bass *et al.* (1998), Hambly and Lyons (1998), and Ledoux *et al.* (2000) in one or another form.

Section 4.2. All results are taken from Bass *et al.* (1998) and Ledoux *et al.* (2000).

Section 4.3. The Lévy area of the two-dimensional Brownian motion was constructed by P. Lévy. The fact that almost all Brownian motion paths together with their Lévy area processes are geometric rough paths was observed in Sipilainen (1993) using Lévy's martingale convergence theorem.

Section 4.4. In Coutin and Qian (2000, 2002), it is shown how to turn fractional Brownian motions with Hurst parameter $h > 1/4$ into geometric paths (with level three). Other examples of geometric rough paths with level three can be found in Bass *et al.* (1998). In Capitaine and Donati-Martin (2001), the authors construct a Lévy area process for the free Brownian motion.

Section 4.5. All of the results are taken from Ledoux *et al.* (2000), and therein the reader may find some more general results than we have proved here.

5

PATH INTEGRATION ALONG ROUGH PATHS

In this chapter we are going to define path integrals of one-forms along rough paths. We first treat rough paths with roughness $p < 3$. We have shown in Chapter 4 that many stochastic processes fall into this class. In this particular case we have a complete theory in which one may integrate one-forms (under a certain number of regular conditions) against these rough paths. We show that the operation of integrating a one-form is continuous with respect to p-variation topology, and it is also an extension of the classical path integration. Moreover, for one-forms which are sufficiently smooth, the integration we have defined is Lipschitz continuous (see below for a precise meaning). We extend the integration theory to all geometric rough paths with any roughness p as a continuous operator if the one-form is sufficiently regular (see below), but leave it open for non-geometric rough paths. By a reason which we will explain later, the integration theory of geometric rough paths can be regarded as an extension of the Stratonovich integration for Brownian motion, while the integration theory for non-geometric rough paths corresponds to Itô's integration theory. We should keep in mind that generally one cannot expect to have an integration theory for defining integrals such as $\int Y_s \, dX_s$ for rough paths X and Y, as in the theory of semi-martingales. However, the path integration theory of one-forms along rough paths which we are going to present here applies to much rougher paths, goes beyond the semi-martingale setting, and is enough to run differential equations.

5.1 Lipschitz one-forms

In this section, we introduce a class of one-forms (on Banach spaces, and valued in Banach spaces as well) which may be integrated against rough paths. Recall that we aim to define path integrals of the following type: $\int \alpha(X) \, dX$ for a rough path X in $T^{([p])}(V)$, where $\alpha : V \to \boldsymbol{L}(V, W)$. Such an α is a function which sends elements of V linearly to W-valued one-forms on V. A special case is where $\alpha = df$ is the Fréchet differential of a W-valued smooth function f on V. We call such a function α a (W-valued) one-form, which actually takes vectors $v \in V$ linearly to W-valued one-forms $\alpha(\cdot)v$ on V.

Any permutation π of $\{1, 2, \dots, k\}$ naturally induces a linear transformation, still denoted by π, on $V^{\otimes_a k}$ such that

$$\pi(v_1 \otimes \cdots \otimes v_k) = v_{\pi(1)} \otimes \cdots \otimes v_{\pi(k)} . \tag{5.1}$$

The tensor norm $|\cdot|_k$ on $V^{\otimes_a k}$ will be assumed to satisfy the following invariant condition, for any permutation π of $\{1, \dots, k\}$:

$$|\pi(v)|_k = |v|_k\,, \quad \forall v \in V^{\otimes_a k}\,. \tag{5.2}$$

The above condition is equivalent to the following weaker one, for any permutation π of $\{1, \dots, k\}$:

$$|\pi(v)|_k \leqslant |v|_k\,, \quad \forall v \in V^{\otimes_a k}\,. \tag{5.3}$$

Definition 5.1.1 *Let $p \geqslant 1$ and $p < \gamma \leqslant [p] + 1$. Let $\alpha : V \to L(V, W)$, where V and W are two Banach spaces, and let $V^{\otimes 2}, \dots, V^{\otimes [p]}$ and $W^{\otimes 2}, \dots, W^{\otimes [p]}$ be their Banach tensor spaces up to degree $[p]$. We say that the system $(\alpha, V^{\otimes j}, W^{\otimes j} : 1 \leqslant j \leqslant [p])$ is admissible if*

(i) α is a $\mathrm{Lip}\,(\gamma)$ one-form (with respect to p) in the sense that, for $j = 1, \dots, [p]$, there exist functions $\alpha^j : V \to L(V^{\otimes j}, W)$ and $R_j : V \times V \to L(V^{\otimes j}, W)$ such that $\alpha^1 = \alpha$, and, for any Lipschitz path X in V, we have (recall that $X_{s,t}^0 = 1$)

$$\alpha^j(X_t) = \sum_{i=0}^{[p]-j} \alpha^{i+j}(X_s)(X_{s,t}^i) + R_j(X_s, X_t)\,, \tag{5.4}$$

$$\int_s^t \alpha^{j+1}(\mathrm{d}X_u) = \alpha^j(X_t) - \alpha^j(X_s)\,, \tag{5.5}$$

for all $t > s$, and

$$\alpha^1(\xi) \leqslant M(1 + |\xi|)\,, \quad |\alpha^{j+1}(\xi)| \leqslant M\,, \quad \forall \xi \in V\,, \tag{5.6}$$

$$|R_j(\xi, \eta)| \leqslant M|\xi - \eta|^{\gamma-j}\,, \quad \forall \xi, \eta \in V\,, \tag{5.7}$$

for $j = 1, \dots, [\gamma]$. In this case, M is called a Lipschitz constant M of α.

(ii) For all $\boldsymbol{j} = (j_1, \dots, j_k)$ (integers $j_i \geqslant 0$) such that $|\boldsymbol{j}| = \sum_{i=1}^k j_i \leqslant [p]$, the linear operator

$$\alpha^{j_1}(\xi) \otimes \cdots \otimes \alpha^{j_k}(\xi)$$

from $V^{\otimes|\boldsymbol{j}|}$ to $W^{\otimes|\boldsymbol{j}|}$ is bounded (with bound M), where

$$\alpha^{j_1}(\xi) \otimes \cdots \otimes \alpha^{j_k}(\xi)\left(\sum_i v_i^{j_1} \otimes \cdots \otimes v_i^{j_k}\right)$$
$$= \sum_i \left(\alpha^{j_1}(\xi)(v_i^{j_1})\right) \otimes \cdots \otimes \left(\alpha^{j_k}(\xi)(v_i^{j_k})\right)\,,$$

for all $v_i^{j_l} \in V^{\otimes j_l}$.

We emphasize that the above definition not only depends on the Banach spaces V and W, but also on all Banach tensor products up to degree $[p]$ as well.

In what follows, we always assume that $p < \gamma \leqslant [p] + 1$.

Fréchet derivatives and projective tensor products

Given a smooth one-form $\alpha : V \to L(V, W)$, possible candidates for functions α^{j+1} in the definition of the Lip (γ)-property are Fréchet derivatives of α. It is easy to see that, if V, W are two Euclidean spaces, and if α possesses bounded $([p] + 1)$th derivatives, then α, together with any tensor products, is a Lip (γ) one-form, and therefore forms an admissible system. However, if V is infinite-dimensional, this cannot be assumed since there are restrictions on the tensor norms, even for smooth one-forms.

Suppose $\alpha : V \to L(V, W)$ possesses all kth continuous Fréchet derivatives $d^k \alpha$ up to degree $[p] + 1$. The kth Fréchet derivative $d^k \alpha$ is, by definition (for the basic properties of Fréchet derivatives, see the appendix for this chapter), a bounded multilinear functional on V valued in $L(V, W)$. That is,

$$d^k \alpha : V \to L(V \times \cdots \times V, L(V, W))$$

such that

$$|d^{k-1}\alpha(v + h) - d^{k-1}\alpha(v) - d^k\alpha(v)h|_{L(V \times \cdots \times V, L(V,W))} = o(h)$$

as $h \to 0$ in V, where $V \times \cdots \times V$ is the product Banach space of k copies of V. Define

$$\alpha^{k+1} : V \to L(V^{\otimes_a(k+1)}, W)$$

by

$$\alpha^{k+1}(v)(\boldsymbol{v}) = \sum_i d^k\alpha(v)(v_{k+1}^i, \ldots, v_2^i)v_1^i \,,$$

for all $\boldsymbol{v} = \sum_i v_{k+1}^i \otimes \cdots \otimes v_2^i \otimes v_1^i$ in $V^{\otimes_a(k+1)}$. With this convention, $\alpha^{k+1}(v)(v_{k+1} \otimes \cdots \otimes v_2 \otimes v_1)$ is symmetric in the first (from the left) k elements v_i, but, in general, not in all vectors $v_1, \ldots, v_k, v_{k+1}$. Clearly,

$$|\alpha^{k+1}(v)(\boldsymbol{v})|_W \leqslant |d^k\alpha(v)| \left(\sum_i |v_{k+1}^i| \cdots |v_2^i||v_1^i| \right),$$

and therefore

$$|\alpha^{k+1}(\xi)(\boldsymbol{v})|_W \leqslant |d^k\alpha(\xi)||\boldsymbol{v}|_\vee \,,$$

where

$$|\boldsymbol{v}|_\vee = \inf \left\{ \sum_i |v_{k+1}^i| \cdots |v_2^i||v_1^i| : \boldsymbol{v} = \sum_i v_{k+1}^i \otimes \cdots \otimes v_2^i \otimes v_1^i \right\}$$

is exactly the projective tensor norm on $V^{\otimes_a(k+1)}$. Together with Taylor's theorem, we have the following lemma.

Lemma 5.1.1 *If* $\alpha : V \to L(V, W)$ *possesses all bounded, continuous Fréchet derivatives up to degree* $[p] + 1$, *and if* $V^{\otimes k}$ *are projective tensor products (*$k = 1, \ldots, [p]$*), then* α *is a Lip (γ) one-form (with respect to any p such that $p < \gamma \leqslant [p] + 1$).*

On the other hand, any multilinear bounded operator on $V^{\otimes j_1} \times \cdots \times V^{\otimes j_k}$ into a Banach space E can be lifted uniquely to a linear bounded operator on the projective tensor product $V^{\otimes j} \otimes_{\mathrm{proj}} \cdots \otimes_{\mathrm{proj}} V^{\otimes j_k}$. Therefore we obtain the following proposition.

Proposition 5.1.1 *Let p and γ be such that $p < \gamma \leqslant [p] + 1$ and let $\alpha : V \to L(V, W)$ possess all bounded, continuous Fréchet derivatives up to degree $[p] + 1$. Then α, together with the projective tensor products $V^{\otimes j}$, $W^{\otimes j}$, $1 \leqslant j \leqslant [p]$, is admissible.*

To understand the constraint condition (5.5), let us look at the case $j = 1$. If α is Fréchet differentiable, $\mathrm{d}\alpha$ is a map from the Banach space V into the Banach space $L(V, L(V, W))$ which, by definition, for any $v \in V$, $\mathrm{d}\alpha(v) \in L(V, L(V, W))$ such that

$$\alpha(v + h) - \alpha(v) = \mathrm{d}\alpha(v)h + o(|h|), \quad \text{as} \quad h \to 0.$$

We regard it as a bounded bilinear map from $V \times V$ into W, namely

$$(\eta, \zeta) \to (\mathrm{d}\alpha(\xi)\eta)\,\zeta, \tag{5.8}$$

and suppose that it can be lifted to the tensor product $\alpha^2 : V \to L(V^{\otimes 2}, W)$ (for example, $V^{\otimes 2}$ is the projective tensor product).

Suppose now (X_t) is a smooth path in V, then by definition

$$\alpha(X_{t+s}) - \alpha(X_s) = \mathrm{d}\alpha(X_s)(X_{t+s} - X_s) + o(|X_{t+s} - X_s|),$$

and therefore

$$\frac{\mathrm{d}\alpha(X_t)}{\mathrm{d}t} = \mathrm{d}\alpha(X_t)(\dot{X}_t).$$

Hence

$$\int_s^t \frac{\mathrm{d}\alpha(X_r)}{\mathrm{d}r}\,\mathrm{d}r = \int_s^t \mathrm{d}\alpha(X_r)(\dot{X}_r)\,\mathrm{d}r$$

$$= \int_s^t \alpha^2(X_r)(\mathrm{d}X_r),$$

so that

$$\int_s^t \alpha^2(X_r)(\mathrm{d}X_r) = \alpha^1(X_t) - \alpha^1(X_s).$$

Let us end this section with a remark.

Remark 5.1.1 This remark makes a crucial difference between the case $[p] = 2$ and the case when $[p] \geqslant 3$. The basic observation is that if α is a Lip (γ) one-form for $2 < \gamma \leqslant 3$, then, for any two points $x, y \in V$,

$$\alpha^1(x)v = \alpha^1(y)v + \alpha^2(x)((x - y) \otimes v) + R_1(x, y)v, \tag{5.9}$$

$$\alpha^2(x) = \alpha^2(y) + R_2(x, y), \tag{5.10}$$

and no iterated integrals are needed to join x and y.

One-forms induced by vector fields

We will see that the projective tensor norms imposed on the algebraic tensor product $V^{\otimes_a k}$ are too big when we discuss differential equations. However, on one hand, one cannot do better since, in order for *all* smooth one-forms (or even only all linear one-forms) to be Lip(γ), then the tensor norms on $V^{\otimes_a k}$ $(k \leqslant [p])$ must be equivalent to the projective norms. On the other hand, for some one-forms, one is really able to relax this condition. This point becomes crucial in the next chapter when we consider differential equations.

The one-forms which we will meet when studying differential equations possess a special feature which allows us to choose reasonable tensor norms.

Let $f : W \to L(V, W)$ be a Fréchet differentiable function. We will call such a function a vector field, since f can be regarded as a map which sends elements of V linearly to vector fields on W. Define a one-form

$$\hat{f} : V \oplus W \to L(V \oplus W, V \oplus W)$$

by

$$\hat{f}(v, w)(v_1, w_1) = (v_1, f(w)v_1), \quad \forall (v, w), (v_1, w_1) \in V \oplus W, \qquad (5.11)$$

where $V \oplus W$ is the direct sum of two Banach spaces endowed with the direct sum norm

$$|(v, w)| = |v| + |w|.$$

Again, $\hat{f}$ is Fréchet differentiable (with the same smoothness as f). The algebraic tensor product $(V \oplus W)^{\otimes_a k}$ possesses a natural direct sum decomposition, namely

$$(V \oplus W)^{\otimes_a k} = V^{\otimes_a k} \oplus (V^{\otimes_a(k-1)} \otimes_a W) \oplus \cdots \oplus (W^{\otimes_a(k-1)} \otimes V) \oplus W^{\otimes_a k}.$$

On each component $V^{\otimes_a(k-i)} \otimes_a W^{\otimes_a i}$ (and $W^{\otimes_a(k-i)} \otimes_a V^{\otimes_a i}$), we endow it with the projective tensor norm, namely if $\boldsymbol{v} \in (V \oplus W)^{\otimes_a k}$ then

$$|\boldsymbol{v}|_\vee = \inf\left\{\sum_j |v_1^j| \cdots |v_{k-i}^j||w_1^j| \cdots |w_i^j| : \right.$$

$$\left. \text{if } \boldsymbol{v} = \sum_j v_1^j \otimes \cdots \otimes v_{k-i}^j \otimes w_1^j \otimes \cdots \otimes w_i^j \right\},$$

and also the direct sum norm on $(V \oplus W)^{\otimes_a k}$, namely

$$|\boldsymbol{v}|_\mathrm{m} = \sum_{i=0}^{k} |\pi_{V^{\otimes(k-i)} \otimes W^{\otimes i}}(\boldsymbol{v})|_\vee.$$

Then $|\boldsymbol{v}|_\mathrm{m}$ is a tensor norm on $(V \oplus W)^{\otimes_a k}$. Here, and in what follows, we use π_V to denote the natural projection of $V \oplus W$ onto V.

Define

$$\hat{f}^{k+1}(v,w)(\boldsymbol{v}) = \sum_j \mathrm{d}^k \hat{f}(v,w)\big((v_{k+1}^j, w_{k+1}^j), \dots, (v_2^j, w_2^j)\big)(v_1^j, w_1^j)\,,$$

for all $\boldsymbol{v} = \sum_j (v_{k+1}^j, w_{k+1}^j) \otimes \cdots \otimes (v_2^j, w_2^j) \otimes (v_1^j, w_1^j)$ in $(V \oplus W)^{\otimes_a k}$. Since $\mathrm{d}^k \alpha$ is multilinear, $\hat{f}^{k+1}$ is a linear operator from $(V \oplus W)^{\otimes_a k}$ into $V \oplus W$. It turns out that

$$\hat{f}^{k+1}(v,w)(\boldsymbol{v}) = \left(0, \sum_j \mathrm{d}^k f(w)(w_{k+1}^j, \dots, w_2^j)v_1^j\right).$$

Therefore,

$$|\hat{f}^{k+1}(v,w)(\boldsymbol{v})| \leqslant \sum_j |\mathrm{d}^k f(w)||w_{k+1}^j| \cdots |w_2^j||v_1^j|$$

and

$$\begin{aligned}
|\hat{f}^{k+1}(v,w)(\boldsymbol{v})| &\leqslant |\mathrm{d}^k f(w)||\pi_{W^{\otimes k} \otimes V}(\boldsymbol{v})|_\vee \\
&\leqslant |\mathrm{d}^k f(w)||\boldsymbol{v}|_\mathrm{m}\,.
\end{aligned}$$

We summarize the above discussion in the following proposition.

Proposition 5.1.2 *Let $f : W \to \boldsymbol{L}(V, W)$ possess bounded Fréchet derivatives up to degree $[p] + 1$. Let $\hat{f} : V \oplus W \to \boldsymbol{L}(V \oplus W, V \oplus W)$ be the induced one-form defined by (5.11). Endow $(V \oplus W)^{\otimes_a k}$ with the tensor norm $|\cdot|_\mathrm{m}$. Then $\hat{f}$ is Lip (γ) with respect to these Banach tensor products, and $\hat{f}$, together with $(V \oplus W)^{\otimes k}$ (the completion of $(V \oplus W)^{\otimes_a k}$ under $|\cdot|_\mathrm{m}$), for $k = 1, \dots, [p]$, is an admissible system (the target space coincides with the initial space).*

Remark 5.1.2 In particular, we require that $V^{\otimes j}$ and $W^{\otimes j}$ are projective tensor products.

Finite-dimensional vector fields

We are also interested in finite systems of differential equations driven by a (possibly infinite-dimensional) noise, namely

$$\mathrm{d}Y_t = f(Y_t)\,\mathrm{d}X_t\,,$$

where X is a geometric rough path in $T^{([p])}(V)$, a solution Y should be a rough path in a finite-dimensional space $W = \mathbb{R}^K$, and $f : W \to \boldsymbol{L}(V, W)$. Since W is finite-dimensional, it is reasonable to call such a function f a finite-dimensional vector field. We may write $f = (f^j)_{j \leqslant K}$ and each $f^j : W \to V^*$. In this case, we may release the restriction on the Banach tensor product $V^{\otimes k}$. All finite tensor products of W are equivalent.

The induced one-form $\hat{f}$ (defined by (5.11)) lives in the direct Banach space $V \oplus W$, with one component W being a finite-dimensional space. If we assume that f is Fréchet differentiable with bounded derivatives up to degree $[p] + 1$, then so is $\hat{f}$. For simplicity, we consider the case $[p] = 2$ only, although a similar result must hold in general.

First we consider the tensor product $\hat{f}(v, w) \otimes \hat{f}(v, w)$, for $(v, w) \in V \oplus W$, which is extended uniquely to a linear operator on the algebraic tensor product $(V \oplus W)^{\otimes_a 2}$. Namely, for any finite sum $\boldsymbol{v} = \sum_i (\xi_i^1, \eta_i^1) \otimes (\xi_i^2, \eta_i^2)$ in $(V \oplus W)^{\otimes_a 2}$,

$$\hat{f}(v, w) \otimes \hat{f}(v, w)(\boldsymbol{v})$$

$$= \sum_i (\xi_i^1, f(w)\xi_i^1) \otimes (\xi_i^2, f(w)\xi_i^2)$$

$$= \left(\sum_i \xi_i^1 \otimes \xi_i^2, \sum_i f(w)\xi_i^1 \otimes \xi_i^2, \sum_i \xi_i^1 \otimes f(w)\xi_i^2, \sum_i f(w)\xi_i^1 \otimes f(w)\xi_i^2 \right).$$

Therefore,

$$\left| \sum_i f(w)\xi_i^1 \otimes \xi_i^2 \right| = \left| \sum_{i,j} f^j(w)(\xi_i^1) \otimes \xi_i^2 \right|_{W \otimes V}$$

$$\leqslant \sum_j \left| \sum_i f^j(w)(\xi_i^1) \otimes \xi_i^2 \right|_{\mathbb{R} \otimes V}$$

$$= \sum_j \left| \sum_i f^j(w)(\xi_i^1)\xi_i^2 \right|_V$$

$$\leqslant K|f(w)| \sup_{s \in V^*, |s| \leqslant 1} \left| \sum_i s(\xi_i^1)\xi_i^2 \right|_V$$

$$\leqslant K|f(w)| \left| \sum_i \xi_i^1 \otimes \xi_i^2 \right|_{\text{inj}},$$

where $|\cdot|_{\text{inj}}$ denotes the injective norm. Similarly, we may verify that

$$\left| \sum_i f(w)\xi_i^1 \otimes f(w)\xi_i^2 \right| \leqslant K^2|f(w)|^2 \left| \sum_i \xi_i^1 \otimes \xi_i^2 \right|_{\text{inj}}.$$

We endow $(V \oplus W)^{\otimes_a 2}$ with the following norm:

$$|\boldsymbol{v}| = |\pi_{V^{\otimes 2}}(\boldsymbol{v})|_{V^{\otimes 2}} + |\pi_{V \otimes W}(\boldsymbol{v})|_V + |\pi_{W \otimes V}(\boldsymbol{v})|_V + |\pi_{W^{\otimes 2}}(\boldsymbol{v})|_V, \qquad (5.12)$$

where the last three terms on the right-hand side of (5.12) are projective norms. The above argument shows that

$$|\hat{f}(v, w) \otimes \hat{f}(v, w)(\boldsymbol{v})| \leqslant \left(1 + 2K|f(w)| + K^2|f(w)|^2 \right) |\pi_{V^{\otimes 2}}(\boldsymbol{v})|_{\text{inj}}$$

$$\leqslant C\left(1 + 2K|f(w)| + K^2|f(w)|^2 \right) |\pi_{V^{\otimes 2}}(\boldsymbol{v})|_{V^{\otimes 2}}$$

$$\leqslant C\left(1 + 2K|f(w)| + K^2|f(w)|^2 \right) |\boldsymbol{v}|,$$

for some constant C and for all $\boldsymbol{v} \in (V \oplus W)^{\otimes_a 2}$.

Next consider the Fréchet derivative $\hat{f}^2$. By definition,

$$\hat{f}^2(v,w)(\boldsymbol{v}) = \left(0, \sum_i f^1(w)(\xi_i^1)\eta_i^2\right),$$

for any $\sum_i(\xi_i^1, \eta_i^1) \otimes (\xi_i^2, \eta_i^2)$, so that

$$\begin{aligned}
|\hat{f}^2(v,w)(\boldsymbol{v})|_{V\oplus W} &= \left|\sum_i f^1(w)(\xi_i^1)\eta_i^2\right| \\
&\leqslant |f^1(w)| \sum_i |\xi_i^1||\eta_i^2|,
\end{aligned}$$

and therefore

$$\begin{aligned}
|\hat{f}^2(v,w)(\boldsymbol{v})|_{V\oplus W} &\leqslant |f^1(w)||\pi_{V\otimes W}(\boldsymbol{v})| \\
&\leqslant |f^1(w)||\boldsymbol{v}|.
\end{aligned}$$

We summarize the above discussion in the following proposition.

Proposition 5.1.3 *Let W be a Banach space of finite dimension, and let V be any Banach space together with its injective tensor products $V^{\otimes_{\mathrm{inj}}2}$. Let $f : W \to L(V,W)$ be a map (called a vector field). Let $W^{\otimes 2}$ be the projective tensor product, and let $(V \oplus W)^{\otimes 2}$ be the Banach tensor product defined as in (5.12). Then $\hat{f}$ defined by (5.11), together with $(V \oplus W)^{\otimes i}$ ($i = 1,2$), is admissible for any $2 < \gamma \leqslant 3$ ([p] = 2).*

5.2 Integration theory: degree two

In this section we restrict ourselves to rough paths with roughness p, where $2 \leqslant p < 3$. In this case, we will establish a complete integration theory. As usual, we use $\Omega_p(V)$ to denote the space of all rough paths with roughness p in a separable Banach space V together with a given Banach tensor product $V^{\otimes 2}$. We have shown in Chapter 4 that a continuous stochastic process with (h,p)-long-time memory such that $hp > 1$ (for example, a Brownian motion, or a fractional Brownian motion with Hurst parameter $h > 1/3$) can be regarded canonically as a geometric rough path (or, more precisely, a family of geometric rough paths) with roughness p, for some $2 < p < 3$.

Before going into the details, we should emphasize that the whole theory depends not only on given Banach spaces, but also on their tensor products, although it will be irrelevant if these Banach spaces are finite-dimensional.

Let $\alpha : V \to L(V,W)$ be a one-form. We want to define path integrals of one-forms against rough paths. In fact, we will define an integration operator $\int \alpha : \Omega_p(V) \to \Omega_p(W)$, for a given one-form α. To this end, we define an almost rough path whose associated rough path should be our path integral. It is very easy to formally write down what the almost rough path should be by using

Taylor's expansion. Let $X \in \Omega_p(V)$ and let $X_{s,t} = (1, X_{s,t}^1, X_{s,t}^2)$. The almost rough path which defines the path integral $\int \alpha(X) \, dX$ is $Y \in C_0(\Delta, T^{(2)}(W))$, where $Y_{s,t} = (1, Y_{s,t}^1, Y_{s,t}^2)$ and

$$Y_{s,t}^1 = \alpha^1(X_s)(X_{s,t}^1) + \alpha^2(X_s)(X_{s,t}^2), \tag{5.13}$$

$$Y_{s,t}^2 = \alpha^1(X_s) \otimes \alpha^1(X_s)(X_{s,t}^2), \tag{5.14}$$

for all $(s,t) \in \Delta$. In order for Y to be indeed an almost rough path, we need to add some conditions on the one-form α and the tensor product $V^{\otimes 2}$.

We assume that α is a Lip (γ) one-form, for $2 < \gamma < 3$. Thus, there are continuous maps $\alpha^2 : V \to L(V^{\otimes 2}, W)$, $R_1 : V \times V \to L(V, W)$, and $R_2 : V \times V \to L(V^{\otimes 2}, W)$ such that (with $\alpha^1 = \alpha$)

$$\alpha^1(x)v = \alpha^1(y)v + \alpha^2(y)((x-y) \otimes v) + R_1(y, x)v, \tag{5.15}$$

$$\alpha^2(x)\boldsymbol{v} = \alpha^2(y)\boldsymbol{v} + R_2(y, x)\boldsymbol{v}, \quad \text{for} \quad v \in V, \quad \boldsymbol{v} \in V^{\otimes 2}, \tag{5.16}$$

and

$$|\alpha^i(x)| \leqslant M(1 + |x|), \quad \forall x \in V, \quad i = 1, 2, \tag{5.17}$$

$$|R_1(y, x)|_{L(V,W)} \leqslant M|x - y|^{\gamma - 1}, \quad \forall x, y \in V, \tag{5.18}$$

$$|R_2(y, x)|_{L(V^{\otimes 2}, W)} \leqslant M|x - y|^{\gamma - 2}, \quad \forall x, y \in V. \tag{5.19}$$

In this case, M is called a Lipschitz constant of α. Notice that the restriction on the tensor norm on $V^{\otimes_a 2}$ enters the picture via α^2 and (5.19), but not on the Banach tensor product $W^{\otimes 2}$, which will be brought in now.

Recall $\alpha^1(v) \otimes \alpha^1(v) : V^{\otimes_a 2} \to W^{\otimes_a 2}$ is defined by

$$\alpha^1(v) \otimes \alpha^1(v)(\boldsymbol{v}) = \sum_i \alpha^1(v)v_i \otimes \alpha^1(v)\hat{v}_i,$$

for $\boldsymbol{v} = \sum_i v_i \otimes \hat{v}_i \in V^{\otimes_a 2}$. Assume that the tensor norm on $W^{\otimes_a 2}$ is chosen so that

$$|\alpha^1(v) \otimes \alpha^1(v)(\boldsymbol{v})|_{W^{\otimes 2}} \leqslant (C_1 + C_2|\alpha^1(v)|^2)|\boldsymbol{v}|_{V^{\otimes 2}}, \tag{5.20}$$

for all $v \in V$ and $\boldsymbol{v} \in V^{\otimes_a 2}$, where C_1, C_2 are two non-negative constants.

Condition 5.2.1 In this section we will assume that $(\alpha, V^{\otimes j}, W^{\otimes j} : j = 1, 2)$ is admissible with respect to $p < \gamma \leqslant [p] + 1$, where p and γ are fixed constants such that $[p] = 2$.

Theorem 5.2.1 (Under Condition 5.2.1) *If $X \in \Omega_p(V)$, then $Y \in C_0(\Delta, T^{(2)}(W))$ defined by (5.13), (5.14) is an almost rough path with roughness p in $T^{(2)}(W)$.*

Proof It is clear that $Y \in C_{0,p}(\Delta, T^{(2)}(W))$. We only need to show that Y is almost multiplicative. By definition, for all $(s,t), (t,u) \in \Delta$,

$$Y_{s,t} \otimes Y_{t,u} = (1, Y_{s,t}^1 + Y_{t,u}^1, Y_{s,t}^2 + Y_{t,u}^2 + Y_{s,t}^1 \otimes Y_{t,u}^1),$$
$$Y_{s,u} = (1, Y_{s,u}^1, Y_{s,u}^2), \tag{5.21}$$

and

$$Y_{s,t}^1 + Y_{t,u}^1 = \alpha^1(X_s)(X_{s,t}^1) + \alpha^2(X_s)(X_{s,t}^2) + \alpha^1(X_t)(X_{t,u}^1) + \alpha^2(X_t)(X_{t,u}^2),$$

We move X_t to X_s via Taylor's formula as follows:

$$\alpha^1(X_t) = \alpha^1(X_s) + \alpha^2(X_s)(X_{s,t}^1) + R_1(X_s, X_t),$$
$$\alpha^2(X_t) = \alpha^2(X_s) + R_2(X_s, X_t).$$

Then we have

$$\alpha^1(X_t)(X_{t,u}^1) = \alpha^1(X_s)(X_{t,u}^1) + \alpha^2(X_s)(X_{s,t}^1 \otimes X_{t,u}^1) + R_1(X_s, X_t)(X_{t,u}^1),$$

and

$$\alpha^2(X_t)(X_{t,u}^2) = \alpha^2(X_s)(X_{t,u}^2) + R_2(X_s, X_t)(X_{t,u}^2).$$

Therefore,

$$\begin{aligned}
Y_{s,t}^1 + Y_{t,u}^1 &= \alpha^1(X_s)(X_{s,t}^1 + X_{t,u}^1) \\
&\quad + \alpha^2(X_s)(X_{s,t}^2 + X_{t,u}^2 + X_{s,t}^1 \otimes X_{t,u}^1) \\
&\quad + R_1(X_s, X_t)(X_{t,u}^1) + R_2(X_s, X_t)(X_{t,u}^2) \\
&= \alpha^1(X_s)(X_{s,u}^1) + \alpha^2(X_s)(X_{s,u}^2) \\
&\quad + R_1(X_s, X_t)(X_{t,u}^1) + R_2(X_s, X_t)(X_{t,u}^2) \\
&= Y_{s,u}^1 + R_1(X_s, X_t)(X_{t,u}^1) + R_2(X_s, X_t)(X_{t,u}^2). \tag{5.22}
\end{aligned}$$

Now suppose that X is controlled by a control ω, namely

$$|X_{s,t}^i| \leqslant \omega(s,t)^{i/p}, \quad \forall i = 1, 2, \quad (s,t) \in \Delta.$$

Then, by (5.22), we have

$$\begin{aligned}
|Y_{s,t}^1 + Y_{t,u}^1 - Y_{s,u}^1| &\leqslant |R_1(X_s, X_t)| \, |X_{t,u}^1| + |R_2(X_s, X_t)| \, |X_{t,u}^2| \\
&\leqslant M|X_{s,t}^1|^{\gamma-1} \omega(t,u)^{1/p} + M|X_{s,t}^1|^{\gamma-2} \omega(t,u)^{2/p} \\
&\leqslant M\omega(s,t)^{(\gamma-1)/p} \omega(t,u)^{1/p} + M\omega(s,t)^{(\gamma-2)/p} \omega(t,u)^{2/p} \\
&\leqslant 2M\omega(s,u)^{\gamma/p}. \tag{5.23}
\end{aligned}$$

Next consider the term $(Y_{s,t} \otimes Y_{t,u})^2$. Again, by definition,

$$Y_{s,t}^2 + Y_{t,u}^2 = \alpha^1(X_s) \otimes \alpha^1(X_s)(X_{s,t}^2) + \alpha^1(X_t) \otimes \alpha^1(X_t)(X_{t,u}^2).$$

Now let us move X_t to X_s. Since, in this case, the term $\alpha^2(X_s)(X_{s,t}^1 \otimes \cdot)$ is not significant, and therefore for simplicity we set

$$\bar{R}_1(X_s, X_t) = \alpha^2(X_s)(X_{s,t}^1 \otimes \cdot) + R_1(X_s, X_t).$$

Then
$$\left|\bar{R}_1(X_s, X_t)\right| \leqslant M|X_{s,t}^1|^{1/p} + M|X_{s,t}^1|^{\gamma-1},$$

and

$$
\begin{aligned}
Y_{s,t}^2 + Y_{t,u}^2 &= \alpha^1(X_s) \otimes \alpha^1(X_s)(X_{s,t}^2) \\
&\quad + \left(\alpha^1(X_s) + \bar{R}_1(X_s, X_t)\right) \otimes \left(\alpha^1(X_s) + \bar{R}_1(X_s, X_t)\right)(X_{t,u}^2) \\
&= \alpha^1(X_s) \otimes \alpha^1(X_s)(X_{s,t}^2 + X_{t,u}^2) \\
&\quad + \bar{R}_1(X_s, X_t) \otimes \bar{R}_1(X_s, X_t)(X_{t,u}^2) \\
&\quad + \alpha^1(X_s) \otimes \bar{R}_1(X_s, X_t)(X_{t,u}^2) \\
&\quad + \bar{R}_1(X_s, X_t) \otimes \alpha^1(X_s)(X_{t,u}^2) \\
&\equiv \alpha^1(X_s) \otimes \alpha^1(X_s)(X_{s,t}^2 + X_{t,u}^2) + R_3.
\end{aligned}
\tag{5.24}
$$

The other term can be written as

$$
\begin{aligned}
Y_{s,t}^1 \otimes Y_{t,u}^1 &= \left(\alpha^1(X_s)(X_{s,t}^1) + \alpha^2(X_s)(X_{s,t}^2)\right) \\
&\quad \otimes \left(\alpha^1(X_t)(X_{t,u}^1) + \alpha^2(X_t)(X_{t,u}^2)\right) \\
&= \left(\alpha^1(X_s)(X_{s,t}^1) + \alpha^2(X_s)(X_{s,t}^2)\right) \\
&\quad \otimes \big(\alpha^1(X_s)(X_{t,u}^1) + \alpha^2(X_s)(X_{s,t}^1 \otimes X_{t,u}^1) + R_1(X_s, X_t)X_{t,u}^1 \\
&\qquad\qquad + \alpha^2(X_s)(X_{t,u}^2) + R_2(X_s, X_t)(X_{t,u}^2)\big) \\
&= \alpha^1(X_s) \otimes \alpha^1(X_s)\left(X_{s,t}^1 \otimes X_{t,u}^1\right) \\
&\quad + \alpha^1(X_s)(X_{s,t}^1) \otimes \left(\alpha^2(X_s)(X_{s,t}^1 \otimes X_{t,u}^1)\right) \\
&\quad + \alpha^1(X_s)(X_{s,t}^1) \otimes R_1(X_s, X_t)X_{t,u}^1 \\
&\quad + \alpha^1(X_s)(X_{s,t}^1) \otimes \alpha^2(X_s)(X_{t,u}^2) \\
&\quad + \alpha^1(X_s)(X_{s,t}^1) \otimes R_2(X_s, X_t)(X_{t,u}^2) \\
&\quad + \alpha^2(X_s)(X_{s,t}^2) \otimes \alpha^1(X_s)(X_{t,u}^1) \\
&\quad + \alpha^2(X_s)(X_{s,t}^2) \otimes \alpha^2(X_s)(X_{s,t}^1 \otimes X_{t,u}^1) \\
&\quad + \alpha^2(X_s)(X_{s,t}^2) \otimes R_1(X_s, X_t)X_{t,u}^1 \\
&\quad + \alpha^2(X_s)(X_{s,t}^2) \otimes \alpha^2(X_s)(X_{t,u}^2) \\
&\quad + \alpha^2(X_s)(X_{s,t}^2) \otimes R_2(X_s, X_t)(X_{t,u}^2) \\
&\equiv \alpha^1(X_s) \otimes \alpha^1(X_s)\left(X_{s,t}^1 \otimes X_{t,u}^1\right) + R_4.
\end{aligned}
\tag{5.25}
$$

Putting (5.24) and (5.25) together we have

$$
\begin{aligned}
Y_{s,t}^2 + Y_{t,u}^2 &+ Y_{s,t}^1 \otimes Y_{t,u}^1 \\
&= \alpha^1(X_s) \otimes \alpha^1(X_s) \left(X_{s,t}^2 + X_{t,u}^2 + X_{s,t}^1 \otimes X_{t,u}^1 \right) + R_3 + R_4 \\
&= \alpha^1(X_s) \otimes \alpha^1(X_s)(X_{s,u}^2) + R_3 + R_4 \\
&= Y_{s,u}^2 + R_3 + R_4 \,.
\end{aligned}
$$

Therefore

$$
\begin{aligned}
\left| (Y_{s,t} \otimes Y_{t,u})^2 - Y_{s,u}^2 \right| &\leqslant |R_3| + |R_4| \\
&\leqslant CM \left(\omega(s,u)^{3/p} + \omega(s,u)^{\gamma/p} \right),
\end{aligned}
$$

for some universal constant C. By the assumption $p < \gamma$, we may then conclude that Y is an almost rough path. ∎

Definition 5.2.1 (Under Condition 5.2.1) *Let $X \in \Omega_p(V)$. Then the integral of the one-form α against the rough path X, denoted by $\int \alpha(X)\,\mathrm{d}X$, is the unique rough path with roughness p in $T^{(2)}(W)$ associated with the almost rough path $Y \in C_0(\triangle, T^{(2)}(W))$, where*

$$
\begin{aligned}
Y_{s,t}^1 &= \alpha^1(X_s)(X_{s,t}^1) + \alpha^2(X_s)(X_{s,t}^2)\,, \\
Y_{s,t}^2 &= \alpha^1(X_s) \otimes \alpha^1(X_s)(X_{s,t}^2)\,,
\end{aligned}
$$

for all $(s,t) \in \triangle$. The integration operator $\int \alpha$ is defined to be the map from $\Omega_p(V)$ into $\Omega_p(W)$ which sends a rough path X into $\int \alpha(X)\,\mathrm{d}X$.

Remark 5.2.1 Following the tradition, we use $\int_s^t \alpha(X)\,\mathrm{d}X$ to denote $\left(\int \alpha(X)\,\mathrm{d}X \right)_{s,t}$, and we use $\int_s^t \alpha(X)\,\mathrm{d}X^i$ to denote the ith component of $\int_s^t \alpha(X)\,\mathrm{d}X$. We should emphasize that the integral $\int \alpha(X)\,\mathrm{d}X$ is a rough path which includes the 'iterated path integrals' up to degree $[p]$.

Theorem 5.2.2 (Under Condition 5.2.1) *The integration operator $\int \alpha$ is a continuous map from $\Omega_p(V)$ to $\Omega_p(W)$ in p-variation topology.*

Proof Let us use the notation from the proof of Theorem 5.2.1. Then it is clear that $X \to Y$ is a continuous operator from $\Omega_p(V)$ into $C_{0,p}(\triangle, T^{(2)}(W))$ and, by Theorem 3.2.2, $Y \to \int \alpha(X)\,\mathrm{d}X$ is continuous in p-variation topology. Therefore, $\int \alpha$ is continuous in p-variation topology. ∎

Theorem 5.2.3 (Under Condition 5.2.1) *Let $X \in \Omega_p(V)$ and let X be controlled by ω, namely*

$$
\left| X_{s,t}^i \right| \leqslant \frac{\omega(s,t)^{i/p}}{\beta\,(i/p)!}\,, \quad i = 1, 2, \quad \forall (s,t) \in \triangle\,,
$$

where β is a constant satisfying (3.34). Then there is a constant K_M depending only on $M, \max \omega, p, \gamma, \theta$, such that

$$\left| \int_s^t \alpha(X) \, \mathrm{d}X^i \right| \leqslant \frac{(K_M \omega(s,t))^{i/p}}{\beta \, (i/p)!} \, ,$$

for all $(s,t) \in \Delta$, $i = 1,2$.

5.3 Lipschitz continuity of integration

In the previous section, we have shown that the integration operator $\int \alpha$ is continuous in p-variation topology on $\Omega_p(V)$ if the Condition 5.2.1 is satisfied. In this section we shall provide an improvement of this continuity result when the one-form α satisfies a stronger requirement.

Theorem 5.3.1 *Let* $\alpha \in C^3(E; L(E,H))$, *and let* $E^{\otimes 2}$ *and* $H^{\otimes 2}$ *be their projective tensor products. Let* $X, \hat{X} \in \Omega_p(E)$, $\varepsilon \geqslant 0$, *and let* ω *be a control such that*

$$\left| X^i_{s,t} \right|, \, \left| \hat{X}^i_{s,t} \right| \leqslant \omega(s,t)^{i/p} \, , \quad i = 1,2, \quad \forall (s,t) \in \Delta \qquad (5.26)$$

and

$$\left| X^i_{s,t} - \hat{X}^i_{s,t} \right| \leqslant \varepsilon \omega(s,t)^{i/p} \, , \quad i = 1,2, \quad \forall (s,t) \in \Delta \, . \qquad (5.27)$$

Assume

$$\left| \mathrm{d}^i \alpha(\xi) \right|_{L(E \times \cdots \times E, H)} \leqslant M(1 + |\xi|) \, , \quad i = 0,1,2,3, \quad \forall \xi \in E \, .$$

Then

$$\left| \int_s^t \alpha(X) \, \mathrm{d}X^i - \int_s^t \alpha(\hat{X}) \, \mathrm{d}\hat{X}^i \right| \leqslant K \varepsilon \omega(s,t)^{i/p} \, ,$$

for all $(s,t) \in \Delta$ *and* $i = 1,2$, *where* K *is a constant depending only on* $M, \max \omega, p$.

However, when we use Picard's iteration for differential equations, in many cases, the projective tensor norm is too big and is not appropriate. Actually, we will use the following version of the above continuity theorem.

Theorem 5.3.2 *Let* $f \in C^3(W; L(V,W))$ *be a function satisfying*

$$\left| \mathrm{d}^i f(x) \right|_{L(V \times \cdots \times V, W)} \leqslant M(1 + |x|) \, , \quad i = 0,1,2,3, \quad \forall x \in V \, ,$$

and let $\alpha = \hat{f} : V \oplus W \to L(V \oplus W, V \oplus W)$ *($V \oplus W$ is the direct sum of V and W) be the induced vector field of f, namely*

$$\hat{f}(x,y)(v,w) = (v, f(y)v) \, , \quad \forall (x,y), (v,w) \in V \oplus W \, .$$

Let $(V \oplus W)^{\otimes 2}$ *be the Banach tensor product with the tensor norm* $|\cdot|_m$:

$$|v|_m = |\pi_{V^{\otimes 2}}(v)|_\vee + |\pi_{W \otimes V}(v)|_\vee + |\pi_{V \otimes W}(v)|_\vee + |\pi_{W^{\otimes 2}}(v)|_\vee \, ,$$

for $v \in (V \oplus W)^{\otimes 2}$. Then the same conclusion as in the previous theorem holds. Naturally, if $X, \hat{X} \in \Omega_p(V \oplus W)$, $\varepsilon \geqslant 0$, and if ω is a control such that

$$\left| X_{s,t}^i \right|, \left| \hat{X}_{s,t}^i \right| \leqslant \omega(s,t)^{i/p}, \quad i = 1, 2, \quad \forall (s,t) \in \Delta$$

and

$$\left| X_{s,t}^i - \hat{X}_{s,t}^i \right| \leqslant \varepsilon \omega(s,t)^{i/p}, \quad i = 1, 2, \quad \forall (s,t) \in \Delta,$$

then

$$\left| \int_s^t \alpha(X) \, \mathrm{d}X^i - \int_s^t \alpha(\hat{X}) \, \mathrm{d}\hat{X}^i \right| \leqslant K \varepsilon \omega(s,t)^{i/p},$$

for all $(s,t) \in \Delta$ and $i = 1, 2$, where K is a constant depending only on $M, \max \omega, p$.

The proof is quite long and we will divide it into several steps. First we need several estimates on the one-form α.

Several estimates about one-forms

In this part, we collect several estimates on the one-form which we need in the proof of the theorems. These estimates involve tensor product norms.

Recall that, in general, if $\beta \in C^{n+1}(U, H)$, then we have the following Taylor's formula (see the appendix of this chapter):

$$\beta(\xi + h) = \sum_{k=0}^n \frac{1}{k!} \mathrm{d}^k \beta(\xi) h^k + R_{n+1}(\xi, \xi + h), \tag{5.28}$$

where

$$R_{n+1}(\xi, \xi + h) = \int_0^1 \frac{(1-s)^n}{n!} \mathrm{d}^{n+1} \beta(\xi + h) h^{n+1} \, \mathrm{d}s.$$

Let $\alpha \in C^3(E, \boldsymbol{L}(E, H))$. Applying the above Taylor's formula to $\beta = \alpha^1$ and $n = 1$, we have ($\eta = \xi + h$)

$$\alpha^1(\eta) = \alpha^1(\xi) + \alpha^2(\xi) h + R_2(\xi, \eta).$$

More precisely,

$$\alpha^1(\eta)v = \alpha^1(\xi)v + \alpha^2(\xi)(h \otimes v) + R_2(\xi, \eta)v, \quad \forall v \in E, \tag{5.29}$$

and

$$R_2(\xi, \eta) = \int_0^1 (1-s)\alpha^3(\xi + sh) h^{\otimes 2} \, \mathrm{d}s, \tag{5.30}$$

$$R_2(\xi, \eta)v = \int_0^1 (1-s)\alpha^3(\xi + sh)(h^{\otimes 2} \otimes v) \, \mathrm{d}s, \quad \forall v \in E. \tag{5.31}$$

Similarly, we have

$$\alpha^2(\eta) = \alpha^2(\xi) + R_1(\xi, \eta),$$

that is, for any $v \in E^{\otimes_a 2}$,

$$\alpha^2(\eta)v = \alpha^2(\xi)v + R_1(\xi, \eta)v, \tag{5.32}$$

and

$$R_1(\xi, \eta) = \int_0^1 \alpha^3(\xi + sh)h \, ds. \tag{5.33}$$

Finally,

$$\alpha^3(\eta) = \alpha^3(\xi) + R_0(\xi, \eta) \tag{5.34}$$

with

$$R_0(\xi, \eta) = \int_0^1 \alpha^4(\xi + sh)h \, ds. \tag{5.35}$$

Assume that all of the derivatives α^i are at most linear in growth, namely

$$\left| \alpha^i(\xi) \right| \leqslant M(1 + |\xi|), \quad \forall \xi \in V, \quad 1 \leqslant i \leqslant 4. \tag{5.36}$$

In order to obtain the estimates needed later, we consider, for two pairs (ξ, η) and $(\hat{\xi}, \hat{\eta})$, the difference

$$
\begin{aligned}
R_2(\hat{\xi}, \hat{\eta}) - R_2(\xi, \eta) &= \int_0^1 (1 - s) \left[\alpha^3(\hat{\xi} + s\hat{h})\hat{h}^{\otimes 2} - \alpha^3(\xi + sh)h^{\otimes 2} \right] ds \\
&= \int_0^1 (1 - s) \left[\alpha^3(\hat{\xi} + s\hat{h}) - \alpha^3(\xi + sh) \right] \hat{h}^{\otimes 2} \, ds \\
&\quad + \int_0^1 (1 - s)\alpha^3(\xi + sh)\left(\hat{h}^{\otimes 2} - h^{\otimes 2} \right) ds \\
&= \int_0^1 (1 - s)R_0(\xi + sh, \hat{\xi} + s\hat{h})\hat{h}^{\otimes 2} \, ds \\
&\quad + \int_0^1 (1 - s)\alpha^3(\xi + sh)\left(\hat{h}^{\otimes 2} - h^{\otimes 2} \right) ds. \tag{5.37}
\end{aligned}
$$

Similarly,

$$
\begin{aligned}
R_1(\hat{\xi}, \hat{\eta}) - R_1(\xi, \eta) &= \int_0^1 R_0(\xi + sh, \hat{\xi} + s\hat{h})\hat{h} \, ds \\
&\quad + \int_0^1 \alpha^3(\xi + sh)(\hat{h} - h) \, ds. \tag{5.38}
\end{aligned}
$$

For R_0 we need the following estimate:

$$\left| R_0(\xi + sh, \hat{\xi} + s\hat{h}) \right| \leqslant M\left(|\hat{\xi} - \xi| + |\hat{h} - h| \right)\left(1 + |\xi| + |\hat{\xi}| + |h| + |\hat{h}| \right), \tag{5.39}$$

which follows from (5.35) and (5.36). From the above formulae, we may deduce that

$$\left|R_1(\hat{\xi},\hat{\eta})(v) - R_1(\xi,\eta)(v)\right|_W \leqslant M\left(1 + |\xi| + |h| + |\hat{\xi} - \xi| + |h - \hat{h}|\right)$$
$$\times \left(|\hat{\xi} - \xi| + |h - \hat{h}|\right)|\hat{h}||v|_V\,,$$

(5.40)

for any $v \in E^{\otimes_a 2}$. This estimate is obtained for the projective tensor norm on $E^{\otimes_a 2}$.

In the case where $E = H = V \oplus W$, and $\alpha = \hat{f}$ is the induced one-form of a function $f : W \to L(V, W)$, which is C^3 and

$$|d^k f(x)| \leqslant M(1 + |x|)\,, \quad \text{for any} \quad x \in W\,, \quad k = 0, 1, 2, 3\,,$$

(5.41)

we use the Banach tensor product $(V \oplus W)^{\otimes 2}$ with the norm $|\cdot|_m$ rather than the projective norm on $(V \oplus W)^{\otimes_a 2}$. Then $\hat{f}$ is Lip (γ) for any $\gamma < 3$. Write $\hat{f}^1 = \hat{f}$,

$$\hat{f}^1(\eta) = \hat{f}^1(\xi) + \hat{f}^2(\xi)(\eta - \xi) + R_2(\xi, \eta)\,,$$

and

$$\hat{f}^2(\eta) = \hat{f}^2(\xi) + R_1(\xi, \eta)\,,$$

where R_1 and R_2 are given by (5.31) and (5.33). We introduce the following notation:

$$\hat{\xi} = (\hat{\xi}_1, \hat{\xi}_2)\,,\ \hat{\eta} = (\hat{\eta}_1, \hat{\eta}_2)\,,\ \hat{h} = (\hat{h}_1, \hat{h}_2)\quad \text{with}\quad \hat{h}_1 = \hat{\eta}_1 - \hat{\xi}_1\,,\ \hat{h}_2 = \hat{\eta}_2 - \hat{\xi}_2\,,$$
$$\xi = (\xi_1, \xi_2)\,,\ \eta = (\eta_1, \eta_2)\,,\ h = (h_1, h_2)\quad \text{with}\quad h_1 = \eta_1 - \xi_1\,,\ h_2 = \eta_2 - \xi_2\,,$$

which are all in $V \oplus W$. Then, for any $v = \sum_i (v_i^2, w_i^2) \otimes (v_i^1, w_i^1)$ in $(V \oplus W)^{\otimes_a 2}$,

$$R_1(\hat{\xi}, \hat{\eta})(v) - R_1(\xi, \eta)(v)$$

$$= \int_0^1 R_0(\xi + sh, \hat{\xi} + s\hat{h})(\hat{h})(v)\,ds + \int_0^1 \alpha^3(\xi + sh)(\hat{h} - h)(v)\,ds$$

$$= \int_0^1 \int_0^1 \alpha^4\{\xi + sh + r[\hat{\xi} - \xi + s(\hat{h} - h)]\}$$

$$\times [\hat{\xi} - \xi + s(\hat{h} - h)](\hat{h})(v)\,ds\,dr$$

$$+ \int_0^1 \alpha^3(\xi + sh)(\hat{h} - h)(v)\,ds$$

$$= \left(0, \sum_i \int_0^1 \int_0^1 d^3 f\{\xi_2 + sh_2 + r[\hat{\xi}_2 - \xi_2 + s(\hat{h}_2 - h_2)]\}\right.$$

$$\left.\times [\hat{\xi}_2 - \xi_2 + s(\hat{h}_2 - h_2)](w_i^2)v_i^1\,ds\,dr\right)$$

$$+ \left(0, \sum_i \int_0^1 d^2 f(\xi_2 + sh_2)(\hat{h}_2 - h_2)(w_i^2)v_i^1\right),$$

so that

$$\left|R_1(\hat\xi,\hat\eta)(\boldsymbol v) - R_1(\xi,\eta)(\boldsymbol v)\right|$$
$$\leqslant M|\pi_{W\otimes V}(\boldsymbol v)|_\vee\left(1 + |\xi_2| + |h_2| + |\hat\xi_2 - \xi_2| + |\hat h_2 - h_2|\right)$$
$$\times\left[\left(|\hat\xi_2 - \xi_2| + |\hat h_2 - h_2|\right)|\hat h_2| + |\hat h_2 - h_2|\right]$$
$$\leqslant M|\boldsymbol v|_{\mathrm m}\left(1 + |\xi_2| + |h_2| + |\hat\xi_2 - \xi_2| + |\hat h_2 - h_2|\right)$$
$$\times\left[\left(|\hat\xi_2 - \xi_2| + |\hat h_2 - h_2|\right)|\hat h_2| + |\hat h_2 - h_2|\right]. \quad (5.42)$$

It is easy to estimate $\left|R_2(\hat\xi,\hat\eta)v - R_2(\xi,\eta)v\right|$ as it does not involve any tensor product norms.

Next we give estimates about the term $\alpha^1(\xi)\otimes\alpha^1(\xi)$. The case when $E^{\otimes 2}$ and $H^{\otimes 2}$ are projective tensor products is simple, and therefore we discuss the case $\alpha = \hat f$. By definition, for any $\boldsymbol v = \sum_i(v_i^2, w_i^2)\otimes(v_i^1, w_i^1)$ in $(V\oplus W)^{\otimes_{\mathrm a}2}$,

$$\hat f^1(\xi)\otimes\hat f^1(\xi)(\boldsymbol v)$$
$$= \sum_i\hat f^1(\xi)(v_i^2, w_i^2)\otimes\hat f^1(\xi)(v_i^1, w_i^1)$$
$$= \sum_i(v_i^2, f(\xi_2)v_i^2)\otimes(v_i^1, f(\xi_2)v_i^1)$$
$$= \sum_i\left(v_i^2\otimes v_i^1, f(\xi_2)v_i^2\otimes v_i^1, v_i^2\otimes f(\xi_2)v_i^1, f(\xi_2)v_i^2\otimes f(\xi_2)v_i^1\right),$$

and therefore

$$\left|\hat f^1(\xi)\otimes\hat f^1(\xi)(\boldsymbol v)\right|_{\mathrm m}$$
$$= \left|\sum_i v_i^2\otimes v_i^1\right|_\vee + \left|\sum_i f(\xi_2)v_i^2\otimes v_i^1\right|_\vee$$
$$+ \left|\sum_i v_i^2\otimes f(\xi_2)v_i^1\right|_\vee + \left|\sum_i f(\xi_2)v_i^2\otimes f(\xi_2)v_i^1\right|_\vee$$
$$\leqslant\left(1 + 2|f(\xi_2)| + |f(\xi_2)|^2\right)\left|\sum_i v_i^2\otimes v_i^1\right|_\vee$$
$$= \left(1 + 2|f(\xi_2)| + |f(\xi_2)|^2\right)|\pi_{V^{\otimes 2}}(\boldsymbol v)|_\vee$$
$$\leqslant\left(1 + 2|f(\xi_2)| + |f(\xi_2)|^2\right)|\boldsymbol v|_{\mathrm m}. \quad (5.43)$$

Next let us consider the difference

$$\hat f^1(\eta)\otimes\hat f^1(\eta)(\boldsymbol v) - \hat f^1(\xi)\otimes\hat f^1(\xi)(\boldsymbol v),$$

which equals

$$\hat f^1(\xi)\otimes R_4(\xi,\eta)(\boldsymbol v) + R_4(\xi,\eta)\otimes\hat f^1(\xi)(\boldsymbol v) + R_4(\xi,\eta)\otimes R_4(\xi,\eta)(\boldsymbol v),$$

where we write

$$R_4(\xi, \eta) = \hat{f}^1(\eta) - \hat{f}^1(\xi)$$
$$= \int_0^1 d\hat{f}(\xi + sh) h \, ds.$$

Therefore

$$\left| \left(\hat{f}^1(\xi) \otimes R_4(\xi, \eta) \right)(v) \right|_m$$
$$= \left| \sum_i \hat{f}^1(\xi)(v_i^2, w_i^2) \otimes R_4(\xi, \eta)(v_i^1, w_i^1) \right|_m$$
$$= \left| \sum_i (v_i^2, f(\xi_2) v_i^2) \otimes \left(0, \int_0^1 df(\xi_2 + sh_2)(h_2) v_i^1 \, ds \right) \right|_m$$
$$= \left| \sum_i \left(0, v_i^2 \otimes \int_0^1 df(\xi_2 + sh_2)(h_2) v_i^1 \, ds, \, 0, \right. \right.$$
$$\left. \left. f(\xi_2) v_i^2 \otimes \int_0^1 df(\xi_2 + sh_2)(h_2) v_i^1 \, ds \right) \right|_m$$
$$\leqslant M^2 (1 + |\xi_2|)(1 + |\xi_2| + |h_2|)|h_2| \left| \sum_i v_i^2 \otimes v_i^1 \right|_\vee$$
$$= M^2 (1 + |\xi_2|)(1 + |\xi_2| + |h_2|)|h_2| |\pi_{V^{\otimes 2}}(v)|_\vee$$
$$\leqslant M^2 (1 + |\xi_2|)(1 + |\xi_2| + |h_2|)|h_2| |v|_m,$$

and similarly we have

$$| (R_4(\xi, \eta) \otimes R_4(\xi, \eta))(v)|_m$$
$$= \left| \sum_i \left(0, \int_0^1 df(\xi_2 + sh_2)(h_2) v_i^2 \, ds \right) \right.$$
$$\left. \otimes \left(0, \int_0^1 df(\xi_2 + sh_2)(h_2) v_i^1 \, ds \right) \right|_m$$
$$= \left| \left(0, 0, 0, \int_0^1 df(\xi_2 + sh_2)(h_2) v_i^2 \, ds \right) \right.$$
$$\left. \otimes \int_0^1 df(\xi_2 + sh_2)(h_2) v_i^1 \, ds \right|_m$$
$$\leqslant M^2 (1 + |\xi_2| + |h_2|)^2 |h_2|^2 |\pi_{V^{\otimes 2}}(v)|_\vee$$
$$\leqslant M^2 (1 + |\xi_2| + |h_2|)^2 |h_2|^2 |v|_m.$$

Putting the two estimates together, we obtain

$$\left| \hat{f}^1(\eta) \otimes \hat{f}^1(\eta)(\boldsymbol{v}) - \hat{f}^1(\xi) \otimes \hat{f}^1(\xi)(\boldsymbol{v}) \right|_{\mathrm{m}}$$
$$\leqslant 2M^2(1+|\xi_2|)(1+|\xi_2|+|h_2|)|h_2||\pi_{V^{\otimes 2}}(\boldsymbol{v})|_{\vee}$$
$$+ M^2(1+|\xi_2|+|h_2|)^2|h_2|^2|\pi_{V^{\otimes 2}}(\boldsymbol{v})|_{\vee}$$
$$\leqslant 2M^2(1+|\xi_2|)(1+|\xi_2|+|h_2|)|h_2||\boldsymbol{v}|_{\mathrm{m}}$$
$$+ M^2(1+|\xi_2|+|h_2|)^2|h_2|^2|\boldsymbol{v}|_{\mathrm{m}}, \qquad (5.44)$$

for all $\boldsymbol{v} \in (V \oplus W)^{\otimes 2}$.

The proofs of the theorems

The proofs of the two theorems use the facts that α is Lip (γ) for any $\gamma < 3$, and estimates (5.40) (or (5.42)), (5.43), and (5.44), which are valid under the conditions of either Theorems 5.3.1 or 5.3.2. Hence we only present the proof for Theorem 5.3.2. That is, $\alpha = \hat{f}$ is the induced one-form of a function $f \in C^3(W; \boldsymbol{L}(V,W))$ which has at most linear growth in its Fréchet derivatives, and $E^{\otimes 2} = (V \oplus W)^{\otimes 2}$ is the Banach tensor product with tensor norm $|\cdot|_{\mathrm{m}}$. We will omit the subscript m from the norms.

Let us begin with some notation. Define two almost rough paths Y and $\hat{Y}$, both in the space $C_{0,p}(\Delta, T^{(2)}(V \oplus W))$, by

$$Y_{s,t}^1 = \alpha^1(X_s)(X_{s,t}^1) + \alpha^2(X_s)(X_{s,t}^2),$$
$$Y_{s,t}^2 = \alpha^1(X_s) \otimes \alpha^1(X_s)(X_{s,t}^2),$$

and

$$\hat{Y}_{s,t}^1 = \alpha^1(\hat{X}_s)(\hat{X}_{s,t}^1) + \alpha^2(\hat{X}_s)(\hat{X}_{s,t}^2),$$
$$\hat{Y}_{s,t}^2 = \alpha^1(\hat{X}_s) \otimes \alpha^1(\hat{X}_s)(\hat{X}_{s,t}^2).$$

Then $Z = \int \alpha(X)\,\mathrm{d}X$ is the unique rough path associated with Y, and the same conclusion is true for $\hat{Y}$. Set, for any finite partition D,

$$Y(D)_{s,t}^1 = \sum_l Y_{t_{l-1},t_l}^1$$

and

$$Y(D)_{s,t}^2 = \sum_l Y_{t_{l-1},t_l}^2 + \sum_l Z_{s,t_{l-1}}^1 \otimes Z_{t_{l-1},t_l}^1. \qquad (5.45)$$

Then $Z_{s,t}^1 = \lim_{m(D)\to 0} Y(D)_{s,t}^1$ and $Z_{s,t}^2 = \lim_{m(D)\to 0} Y(D)_{s,t}^2$, see Theorem 3.2.1. Moreover, there is a constant K_1 depending only on $M, \max \omega, p$ such that

$$\left| Z_{s,t}^i \right| \leqslant K_1 \omega(s,t)^{i/p}, \quad \forall i = 1, 2, \quad (s,t) \in \Delta \qquad (5.46)$$

and

$$\left| Z_{s,t}^i - Y_{s,t}^i \right| \leqslant K_1 \omega(s,t)^{3/p}, \quad \forall i = 1, 2, \quad (s,t) \in \Delta. \qquad (5.47)$$

The same estimates also hold for $\hat{Z}$ and $\hat{Y}$.

Next we use the same trick as before, and choose a point t_l in a finite division

$$D = \{s = t_0 < t_1 < \cdots < t_r = t\}$$

of $[s,t]$, such that

$$\omega(t_{l-1}, t_{l+1}) \leqslant \frac{2}{r-2}\omega(s,t), \quad \text{if} \quad r \geqslant 3.$$

Let $D' = D - \{t_l\}$. Then, by definition,

$$\begin{aligned}
Y(D)^1_{s,t} - Y(D')^1_{s,t} &= Y^1_{t_{l-1},t_l} + Y^1_{t_l,t_{l+1}} - Y^1_{t_{l-1},t_{l+1}} \\
&= \alpha^1(X_{t_{l-1}})(X^1_{t_{l-1},t_l}) + \alpha^2(X_{t_{l-1}})(X^2_{t_{l-1},t_l}) \\
&\quad + \alpha^1(X_{t_l})(X^1_{t_l,t_{l+1}}) + \alpha^2(X_{t_l})(X^2_{t_l,t_{l+1}}) \\
&\quad - \alpha^1(X_{t_{l-1}})(X^1_{t_{l-1},t_{l+1}}) - \alpha^2(X_{t_{l-1}})(X^2_{t_{l-1},t_{l+1}}).
\end{aligned}$$

Then move X_{t_l} to $X_{t_{l-1}}$ via Taylor's theorem,

$$\alpha^1(X_{t_l}) = \alpha^1(X_{t_{l-1}}) + \alpha^2(X_{t_{l-1}})(X^1_{t_{l-1},t_l}) + R_2(X_{t_{l-1}}, X_{t_l}), \quad (5.48)$$

$$\alpha^2(X_{t_l}) = \alpha^2(X_{t_{l-1}}) + R_1(X_{t_{l-1}}, X_{t_l}). \quad (5.49)$$

Therefore,

$$Y(D)^1_{s,t} - Y(D')^1_{s,t} = R_2(X_{t_{l-1}}, X_{t_l})(X^1_{t_l,t_{l+1}}) + R_1(X_{t_{l-1}}, X_{t_l})(X^2_{t_l,t_{l+1}}). \quad (5.50)$$

Similarly,

$$\hat{Y}(D)^1_{s,t} - \hat{Y}(D')^1_{s,t} = R_2(\hat{X}_{t_{l-1}}, \hat{X}_{t_l})(\hat{X}^1_{t_l,t_{l+1}}) + R_1(\hat{X}_{t_{l-1}}, \hat{X}_{t_l})(\hat{X}^2_{t_l,t_{l+1}}). \quad (5.51)$$

We want to estimate the difference

$$I(D, D') \equiv Y(D)^1_{s,t} - Y(D')^1_{s,t} - \left(\hat{Y}(D)^1_{s,t} - \hat{Y}(D')^1_{s,t}\right). \quad (5.52)$$

By (5.37) we have

$$\begin{aligned}
&R_2(\hat{X}_{t_{l-1}}, \hat{X}_{t_l}) - R_2(X_{t_{l-1}}, X_{t_l}) \\
&\quad = \int_0^1 (1-s)R_0\big(X^1_{t_{l-1}} + sX^1_{t_{l-1},t_l}, \hat{X}_{t_{l-1}} + s\hat{X}^1_{t_{l-1},t_l}\big)\big(\hat{X}^1_{t_{l-1},t_l}\big)^{\otimes 2}\, ds \\
&\qquad + \int_0^1 (1-s)\alpha^3\big(X_{t_{l-1}} + sX^1_{t_{l-1},t_l}\big)\big[\big(\hat{X}^1_{t_{l-1},t_l}\big)^{\otimes 2} - \big(X^1_{t_{l-1},t_l}\big)^{\otimes 2}\big]\, ds,
\end{aligned}$$

so that, by (5.26), (5.27), (5.36), and (5.39),

$$\big|R_2(\hat{X}_{t_{l-1}}, \hat{X}_{t_l}) - R_2(X_{t_{l-1}}, X_{t_l})\big| \leqslant K_2 \varepsilon \omega(t_{l-1}, t_l)^{2/p}, \quad (5.53)$$

for some constant K_2 depending only on $M, \max\omega, p$. For example,

$$K_2 = 2M(1 + 4\omega(0,T)^{1/p})^2$$

will do. By (5.42),

$$\left| R_1(\hat{X}_{t_{l-1}}, \hat{X}_{t_l}) - R_1(X_{t_{l-1}}, X_{t_l}) \right| \leqslant K_3 \varepsilon \omega(t_{l-1}, t_l)^{1/p}, \tag{5.54}$$

where

$$K_3 = 2M\omega(0,T)^{1/p}(1 + 4\omega(0,T)^{1/p}).$$

Furthermore, we can easily check the following estimate:

$$\left| R_j(X_{t_{l-1}}, X_{t_l}) \right| \leqslant K_4 \omega(t_{l-1}, t_l)^{j/p}, \tag{5.55}$$

where

$$K_4 = M(1 + 2\omega(0,T)^{1/p}).$$

With these estimates, (5.50)–(5.52), and the choice of t_l, we deduce that

$$\begin{aligned}
\left| I(D, D') \right| &\leqslant \left| R_2(\hat{X}_{t_{l-1}}, \hat{X}_{t_l}) - R_2(X_{t_{l-1}}, X_{t_l}) \right| \left| X^1_{t_l, t_{l+1}} \right| \\
&\quad + \left| R_2(X_{t_{l-1}}, X_{t_l}) \right| \left| X^1_{t_l, t_{l+1}} - \hat{X}^1_{t_l, t_{l+1}} \right| \\
&\quad + \left| R_1(\hat{X}_{t_{l-1}}, \hat{X}_{t_l}) - R_1(X_{t_{l-1}}, X_{t_l}) \right| \left| X^2_{t_l, t_{l+1}} \right| \\
&\quad + \left| R_1(X_{t_{l-1}}, X_{t_l}) \right| \left| X^2_{t_l, t_{l+1}} - \hat{X}^2_{t_l, t_{l+1}} \right| \\
&\leqslant K_5 \varepsilon \left(\frac{2}{r-2} \right)^{3/p} \omega(s,t)^{3/p},
\end{aligned}$$

where

$$K_5 = 8M(1 + 4\omega(0,T)^{1/p})^2.$$

Repeating the above argument until all of the inside points $t_l \in D$ have been deleted, we obtain

$$\left| Y(D)^1_{s,t} - Y^1_{s,t} - (\hat{Y}(D)^1_{s,t} - \hat{Y}^1_{s,t}) \right| \leqslant K_6 \varepsilon \omega(s,t)^{3/p},$$

where

$$K_6 = K_5 \left[1 + \sum_{r=3}^{\infty} \left(\frac{2}{r-2} \right)^{3/p} \right].$$

Letting $m(D) \to 0$ we may then establish that

$$\left| Z^1_{s,t} - Y^1_{s,t} - (\hat{Z}^1_{s,t} - \hat{Y}^1_{s,t}) \right| \leqslant K_6 \varepsilon \omega(s,t)^{3/p}, \quad \forall (s,t) \in \Delta. \tag{5.56}$$

By a similar argument, we can bound $|Y^1_{s,t} - \hat{Y}^1_{s,t}|$. In fact, we have

$$\left| Y^1_{s,t} - \hat{Y}^1_{s,t} \right| \leqslant K_7 \varepsilon \omega(s,t)^{1/p}, \quad \forall (s,t) \in \Delta, \tag{5.57}$$

where K_7 can be chosen to be

$$K_7 = 3M\left(1 + 4\omega(0,T)^{1/p}\right)^3.$$

Indeed,

$$
\begin{aligned}
\left|Y_{s,t}^1 - \hat{Y}_{s,t}^1\right| &\leqslant \left|\alpha^1(X_s) - \alpha^1(\hat{X}_s)\right|\left|X_{s,t}^1\right| \\
&\quad + \left|\alpha^1(\hat{X}_s)\right|\left|X_{s,t}^1 - \hat{X}_{s,t}^1\right| \\
&\quad + \left|\alpha^2(X_s) - \alpha^2(\hat{X}_s)\right|\left|X_{s,t}^2\right| \\
&\quad + \left|\alpha^2(\hat{X}_s)\right|\left|X_{s,t}^2 - \hat{X}_{s,t}^2\right|,
\end{aligned}
$$

and

$$\alpha^i(X_s) - \alpha^i(\hat{X}_s) = \int_0^1 \alpha^{1+i}(\hat{X}_s + r(X_s - \hat{X}_s))(X_s - \hat{X}_s)\,\mathrm{d}r,$$

so that

$$
\begin{aligned}
\left|\alpha^i(X_s) - \alpha^i(\hat{X}_s)\right| &\leqslant M\left(1 + 3\omega(0,s)^{1/p}\right)\varepsilon\omega(0,s)^{1/p} \\
&\leqslant M\varepsilon\omega(0,T)^{1/p}\left(1 + 3\omega(0,T)^{1/p}\right),
\end{aligned}
$$

for all $i = 1, 2, 3$, $s \in [0,T]$, and

$$\left|\alpha^i(X_s)\right| \leqslant M\left(1 + \omega(0,T)^{1/p}\right), \quad \forall\, 1 \leqslant i \leqslant 4, \quad s \in [0,T].$$

Hence,

$$
\begin{aligned}
\left|Y_{s,t}^1 - \hat{Y}_{s,t}^1\right| &\leqslant M\varepsilon\omega(0,T)^{1/p}\left(1 + 3\omega(0,T)^{1/p}\right)\left(\omega(s,t)^{1/p} + \omega(s,t)^{2/p}\right) \\
&\quad + M\varepsilon\left(1 + \omega(0,T)^{1/p}\right)\left(\omega(s,t)^{1/p} + \omega(s,t)^{2/p}\right) \\
&\leqslant K_7\varepsilon\omega(s,t)^{1/p}.
\end{aligned}
$$

Therefore,

$$\left|Z_{s,t}^1 - \hat{Z}_{s,t}^1\right| \leqslant K_8\varepsilon\omega(s,t)^{1/p}, \quad \forall(s,t) \in \Delta, \tag{5.58}$$

where

$$K_8 = K_7 + K_6\omega(0,T)^{2/p}.$$

Next we are going to consider the difference

$$J(D,D') = Y(D)_{s,t}^2 - Y(D')_{s,t}^2 - \left(\hat{Y}(D)_{s,t}^2 - \hat{Y}(D')_{s,t}^2\right).$$

The treatment is similar, but requires a more careful inspection. First, it is easily seen that

$$
\begin{aligned}
Y(D)_{s,t}^2 &- Y(D')_{s,t}^2 \\
&= Y_{t_{l-1},t_l}^2 + Y_{t_l,t_{l+1}}^2 - Y_{t_{l-1},t_{l+1}}^2 + Z_{t_{l-1},t_l}^1 \otimes Z_{t_l,t_{l+1}}^1 \\
&= \alpha^1(X_{t_{l-1}}) \otimes \alpha^1(X_{t_{l-1}})(X_{t_{l-1},t_l}^2 + X_{t_l,t_{l+1}}^2 - X_{t_{l-1},t_{l+1}}^2) \\
&\quad + \left(\alpha^1(X_{t_l}) \otimes \alpha^1(X_{t_l}) - \alpha^1(X_{t_{l-1}}) \otimes \alpha^1(X_{t_{l-1}})\right)(X_{t_l,t_{l+1}}^2) \\
&\quad + Z_{t_{l-1},t_l}^1 \otimes Z_{t_l,t_{l+1}}^1
\end{aligned}
$$

$$= -\alpha^1(X_{t_{l-1}}) \otimes \alpha^1(X_{t_{l-1}})(X^1_{t_{l-1},t_l} \otimes X^1_{t_l,t_{l+1}})$$
$$+ \left(\alpha^1(X_{t_l}) \otimes \alpha^1(X_{t_l}) - \alpha^1(X_{t_{l-1}}) \otimes \alpha^1(X_{t_{l-1}})\right)(X^2_{t_l,t_{l+1}})$$
$$+ Z^1_{t_{l-1},t_l} \otimes Z^1_{t_l,t_{l+1}}. \tag{5.59}$$

For simplicity, we set

$$T_1 = Z^1_{t_{l-1},t_l} \otimes Z^1_{t_l,t_{l+1}} - \alpha^1(X_{t_{l-1}}) \otimes \alpha^1(X_{t_{l-1}})(X^1_{t_{l-1},t_l} \otimes X^1_{t_l,t_{l+1}}), \tag{5.60}$$
$$T_2 = \left(\alpha^1(X_{t_l}) \otimes \alpha^1(X_{t_l}) - \alpha^1(X_{t_{l-1}}) \otimes \alpha^1(X_{t_{l-1}})\right)(X^2_{t_l,t_{l+1}}). \tag{5.61}$$

Similar formulae apply to the same quantities for $\hat{X}$, and we shall use the same notation with hat signs.

By using (5.43) and (5.44) we can show that

$$|T_2 - \hat{T}_2| \leqslant K_9 \varepsilon \left(\frac{2}{r-2}\right)^{3/p} \omega(s,t)^{3/p}, \tag{5.62}$$

for a constant K_9 depending only on $M, \max \omega, p$, namely

$$K_9 = 2MK_4\omega(0,T)^{1/p}\left(1 + 3\omega(0,T)^{1/p}\right)$$
$$+ 2MK_3\left(1 + \omega(0,T)^{1/p}\right) + 2K_4K_3\omega(0,T)^{1/p}.$$

Next we are going to show that $T_1 - \hat{T}_1$ satisfies a similar inequality. To this end, we notice that

$$T_1 = (Z^1_{t_{l-1},t_l} - Y^1_{t_{l-1},t_l}) \otimes Z^1_{t_l,t_{l+1}}$$
$$+ \alpha^1(X_{t_{l-1}})(X^1_{t_{l-1},t_l}) \otimes (Z^1_{t_l,t_{l+1}} - Y^1_{t_l,t_{l+1}})$$
$$+ \alpha^1(X_{t_{l-1}})(X^1_{t_{l-1},t_l}) \otimes \alpha^1(X_{t_l})(X^1_{t_l,t_{l+1}})$$
$$+ \alpha^1(X_{t_{l-1}})(X^1_{t_{l-1},t_l}) \otimes \alpha^2(X_{t_l})(X^2_{t_l,t_{l+1}})$$
$$+ \alpha^2(X_{t_{l-1}})(X^2_{t_{l-1},t_l}) \otimes Z^1_{t_l,t_{l+1}}$$
$$- \alpha^1(X_{t_{l-1}}) \otimes \alpha^1(X_{t_{l-1}})(X^1_{t_{l-1},t_l} \otimes X^1_{t_l,t_{l+1}}), \tag{5.63}$$

and then moving the point X_{t_l} to the point $X_{t_{l-1}}$ we get that

$$T_1 = (Z^1_{t_{l-1},t_l} - Y^1_{t_{l-1},t_l}) \otimes Z^1_{t_l,t_{l+1}}$$
$$+ \alpha^1(X_{t_{l-1}})(X^1_{t_{l-1},t_l}) \otimes (Z^1_{t_l,t_{l+1}} - Y^1_{t_l,t_{l+1}})$$
$$+ \alpha^1(X_{t_{l-1}})(X^1_{t_{l-1},t_l}) \otimes \alpha^2(X_{t_{l-1}})(X^1_{t_{l-1},t_l} \otimes X^1_{t_l,t_{l+1}})$$
$$+ \alpha^1(X_{t_{l-1}})(X^1_{t_{l-1},t_l}) \otimes R_2(X_{t_{l-1}}, X_{t_l})(X^1_{t_l,t_{l+1}})$$
$$+ \alpha^1(X_{t_{l-1}})(X^1_{t_{l-1},t_l}) \otimes \alpha^2(X_{t_l})(X^2_{t_l,t_{l+1}})$$
$$+ \alpha^2(X_{t_{l-1}})(X^2_{t_{l-1},t_l}) \otimes Z^1_{t_l,t_{l+1}}$$
$$\equiv S_1 + S_2 + S_3 + S_4 + S_5 + S_6. \tag{5.64}$$

We have the same equality for $\hat{T}_1$, replacing X, Z by $\hat{X}, \hat{Z}$. Then we may estimate the norm of the difference $T_1 - \hat{T}_1$ by controlling the differences term by term in (5.64). For example, by (5.46), (5.56), and (5.58),

$$
\begin{aligned}
|\hat{S}_1 - S_1| &= \left| \left(\hat{Z}^1_{t_{l-1},t_l} - \hat{Y}^1_{t_{l-1},t_l} \right) \otimes \hat{Z}^1_{t_l,t_{l+1}} - \left(Z^1_{t_{l-1},t_l} - Y^1_{t_{l-1},t_l} \right) \otimes Z^1_{t_l,t_{l+1}} \right| \\
&\leqslant \left| \left(\hat{Z}^1_{t_{l-1},t_l} - \hat{Y}^1_{t_{l-1},t_l} \right) - \left(Z^1_{t_{l-1},t_l} - Y^1_{t_{l-1},t_l} \right) \right| \left| \hat{Z}^1_{t_l,t_{l+1}} \right| \\
&\quad + \left| Z^1_{t_{l-1},t_l} - Y^1_{t_{l-1},t_l} \right| \left| \hat{Z}^1_{t_l,t_{l+1}} - Z^1_{t_l,t_{l+1}} \right| \\
&\leqslant K_{10}\varepsilon \left(\frac{2}{r-2} \right)^{3/p} \omega(s,t)^{3/p},
\end{aligned}
\tag{5.65}
$$

where $K_{10} = K_6(K_1 + K_8)\omega(0,T)^{1/p}$. Similarly,

$$
\begin{aligned}
|\hat{S}_2 - S_2| &= \left| \alpha^1(\hat{X}_{t_{l-1}})(\hat{X}^1_{t_{l-1},t_l}) \otimes \left(\hat{Z}^1_{t_l,t_{l+1}} - \hat{Y}^1_{t_l,t_{l+1}} \right) \right. \\
&\quad \left. - \alpha^1(X_{t_{l-1}})(X^1_{t_{l-1},t_l}) \otimes \left(Z^1_{t_l,t_{l+1}} - Y^1_{t_l,t_{l+1}} \right) \right| \\
&\leqslant K_{11}\varepsilon \left(\frac{2}{r-2} \right)^{3/p} \omega(s,t)^{3/p},
\end{aligned}
\tag{5.66}
$$

with $K_{11} = (1 + K_1 + K_6)\omega(0,T)^{1/p}(1 + 3\omega(0,T)^{1/p})$. It is easy to control the remaining terms in (5.64). In fact, as the arguments are similar, we only write down the following estimates:

$$
|\hat{S}_i - S_i| \leqslant B_i\varepsilon \left(\frac{2}{r-2} \right)^{3/p} \omega(s,t)^{3/p}, \quad \forall i = 3, 4, 5, 6,
$$

where B_i are constants depending only on $M, \max\omega, p$. The precise values can be (of course, not the optimal ones)

$$
\begin{aligned}
B_3 &= 2 + M\left(1 + \omega(0,T)^{1/p}\right)\left(2 + 3\omega(0,T)^{1/p}\right), \\
B_4 &= M\left(K_4\omega(0,T)^{1/p} + 2K_4 + K_2\right)\left(1 + 3\omega(0,T)^{1/p}\right), \\
B_5 &= M\left[1 + M\left(1 + \omega(0,T)^{1/p}\right)\omega(0,T)^{1/p}\right]\left(1 + 3\omega(0,T)^{1/p}\right) \\
&\quad + 2M^2\left(1 + \omega(0,T)^{1/p}\right)^2, \\
B_6 &= MK_3\left(1 + 3\omega(0,T)^{1/p}\right)^2 + K_8M\left(1 + \omega(0,T)^{1/p}\right).
\end{aligned}
$$

Thus,

$$
|T_1 - \hat{T}_1| \leqslant K_{12}\varepsilon \left(\frac{2}{r-2} \right)^{3/p} \omega(s,t)^{3/p},
$$

so that

$$
|J(D,D')| \leqslant K_{13}\varepsilon \left(\frac{2}{r-2} \right)^{3/p} \omega(s,t)^{3/p},
$$

for some constants K_{12}, K_{13} depending only on $M, \max\omega, p$. Therefore, by the same trick as we have used many times, we get that

$$\left|Z_{s,t}^2 - Y_{s,t}^2 - \left(\hat{Z}_{s,t}^2 - \hat{Y}_{s,t}^2\right)\right| \leqslant K_{14}\varepsilon\omega(s,t)^{3/p}, \quad \forall(s,t) \in \Delta. \tag{5.67}$$

However, it is obvious that

$$\left|Y_{s,t}^2 - \hat{Y}_{s,t}^2\right| \leqslant K_{15}\varepsilon\omega(s,t)^{2/p}, \quad \forall(s,t) \in \Delta.$$

All of the constants K_i depend only on $M, \max\omega, p$. We have thus completed the proof.

From the proof and eqn (5.67), we also have the following corollary.

Corollary 5.3.1 *Under the same assumptions as in the previous theorems, and with $Y, \hat{Y}$ defined as in their proof (both are almost rough paths), we have*

$$\left|\int_s^t \alpha(X)\,\mathrm{d}X^i - Y_{s,t}^i - \left(\int_s^t \alpha(\hat{X})\,\mathrm{d}\hat{X}^i - \hat{Y}_{s,t}^i\right)\right| \leqslant K\varepsilon\omega(s,t)^{3/p},$$

for all $i = 1, 2$ and $(s,t) \in \Delta$, where K is a constant depending only on $M, \max\omega$, and p.

Remark 5.3.1 There is no significance in the precise values of the constants K_i appearing in the above proof, except for the fact that these constants depend only on $\max\omega$, the Lipschitz constant M of the one-form α, and the roughness p of the paths. The reason that we present some precise values of these constants (far from the optimal ones) is to exhibit their interesting structure, namely they are polynomials of $\max\omega^{1/p}$ and M.

5.4 Itô's formula and stochastic integration

In this section, we first deduce an Itô formula for rough paths, and then we show that the path integration, which we have defined in the previous section, reduces to Stratonovich's integrals if the rough paths are the canonical geometric rough paths associated with a standard Brownian motion.

5.4.1 *Itô's formula*

Let us first give the following version of the fundamental theorem of the calculus.

Theorem 5.4.1 *Let $X \in G\Omega_p(\mathbb{R}^m)$, $2 \leqslant p < 3$, and let f be a $\mathrm{Lip}\,(\gamma)$ function, where $\gamma > p + 1$ (i.e. $\mathrm{d}f$ is a $\mathrm{Lip}\,(\gamma - 1)$ one-form). Then,*

$$f(X_t) - f(X_s) = \int_s^t \mathrm{d}f(X)\,\mathrm{d}X^1, \quad \forall(s,t) \in \Delta. \tag{5.68}$$

Remark 5.4.1 We will see later that (5.68) cannot be true for non-geometric rough paths.

Proof The proof is very simple. First, (5.68) certainly holds for any Lipschitz paths. Then, by continuity of the two sides, (5.68) holds for the path X in p-variation topology, and therefore remains true for any geometric rough path. ∎

We now turn to an Itô formula.

Lemma 5.4.1 *Let* $X \in \Omega_p(V)$, *and let* $Y_{s,t} = (1, X^1_{s,t}, X^2_{s,t} + \phi_{s,t})$. *Then* $Y \in \Omega_p(V)$ *if and only if* ϕ *is additive (so that* $\phi_{s,t} = \phi_t - \phi_s$ *), and possesses finite* $p/2$-*variation. In this case we use* $X(\phi)$ *to denote* Y.

Theorem 5.4.2 (Under Condition 5.2.1) *Let* $X \in \Omega_p(V)$, *and let* ϕ *be a continuous path in* $V^{\otimes 2}$ *which possesses finite* $p/2$-*variation. Then,*

$$\int_s^t \alpha(X(\phi)) \, \mathrm{d}X(\phi)^1 = \int_s^t \alpha(X) \, \mathrm{d}X^1 + \int_s^t \alpha^2(X_r)(\mathrm{d}\phi_r) \qquad (5.69)$$

and

$$\int_s^t \alpha(X(\phi)) \, \mathrm{d}X(\phi)^2 = \int_s^t \alpha(X) \, \mathrm{d}X^2 + \int_s^t \alpha^1(X_r) \otimes \alpha^1(X_r)(\mathrm{d}\phi_r)$$

$$+ \int_s^t \left[\int_s^r \alpha^2(X_u)(\mathrm{d}\phi_r) \right] \otimes \mathrm{d}Z_r$$

$$+ \int_s^t Z_{s,r} \otimes \alpha^2(X_r)(\mathrm{d}\phi_r) , \qquad (5.70)$$

where the integrals involving ϕ *on the right-hand sides are Young's integrals, and* $Z_{s,t} = \int_s^t \alpha(X) \, \mathrm{d}X^1$.

Proof By definition,

$$\int_s^t \alpha(X(\phi)) \, \mathrm{d}X(\phi)^1 = \lim_{m(D) \to 0} \sum_l \alpha^1(X_{t_{l-1}})(X^1_{t_{l-1}, t_l})$$

$$+ \alpha^2(X_{t_{l-1}})(X^2_{t_{l-1}, t_l} + \phi_{t_{l-1}, t_l})$$

$$= \int_s^t \alpha(X) \, \mathrm{d}X^1 + \int_s^t \alpha^2(X_r)(\mathrm{d}\phi_r) ,$$

and the proof of (5.70) is similar. ∎

5.4.2 *Stochastic integration*

Let $W_t = (W^1_t, \ldots, W^d_t)$ be a standard, d-dimensional Brownian motion, and let $X_{s,t} = (1, X^1_{s,t}, X^2_{s,t})$ be the associated geometric rough path(s). Therefore,

$$X^1_{s,t} = (W^1_t - W^1_s, \ldots, W^d_t - W^d_s)$$

and

$$X^2_{s,t} = \left(\int_{s < t_1 < t_2 < t} \circ \, \mathrm{d}W^i_{t_1} \otimes \circ \, \mathrm{d}W^j_{t_2} \right)_{i,j \leqslant d} .$$

Let $\alpha = (\alpha_i) : \mathbb{R}^d \to L(\mathbb{R}^d, \mathbb{R}^N)$ be a one-form on $\mathbb{R}^d$. Without loss of generality, we may and will assume that $N = 1$. Then, according to the definition of the integral along a rough path,

$$\int_0^t \alpha(X)\,\mathrm{d}X^1 = \lim_{m(D)\to 0} \sum_l \sum_{i=1}^d \alpha_i(W_{t_{l-1}})(W_{t_l}^i - W_{t_{l-1}}^i)$$
$$+ \sum_l \alpha^2(W_{t_{l-1}})(X^2_{t_{l-1},t_l})$$

$$= \int_0^t \alpha(W_u)\,\mathrm{d}W_u \quad (\text{Itô's integral})$$
$$+ \lim_{m(D)\to 0} \sum_l \alpha^2(X_{t_{l-1}})(X^2_{t_{l-1},t_l}).$$

However,

$$X^2_{t_{l-1},t_l} = \int_{t_{l-1}}^{t_l} (W_r^i - W_{t_{l-1}}^i)\circ\mathrm{d}W_r^j,$$

so that

$$\lim_{m(D)\to 0} \sum_l \alpha^2(W_{t_{l-1}})(X^2_{t_{l-1},t_l})$$

$$= \lim_{m(D)\to 0} \sum_{i,j=1}^d \sum_l \frac{\partial \alpha_i}{\partial x^j}(W_{t_{l-1}}) \int_{t_{l-1}}^{t_l} (W_r^i - W_{t_{l-1}}^i)\circ\mathrm{d}W_r^j$$

$$= \sum_{i,j=1}^d \lim_{m(D)\to 0} \sum_l \frac{\partial \alpha_i}{\partial x^j}(W_{t_{l-1}}) \frac{1}{2}(W_{t_l}^i - W_{t_{l-1}}^i)(W_{t_l}^j - W_{t_{l-1}}^j)$$

$$= \sum_{i=1}^d \frac{1}{2}\langle W^i, \alpha_i(W)\rangle_t.$$

Therefore, we have

$$\int_0^t \alpha(X)\,\mathrm{d}X^1 = \int_0^t \alpha(W_s)\circ\mathrm{d}W_s \quad (\text{Stratonovich integral}) \text{ a.s.}$$

Thus we have proved the following theorem.

Theorem 5.4.3 *Let W_t be a d-dimensional Brownian motion, and let $X_{s,t} = (1, X^1_{s,t}, X^2_{s,t})$ be the canonical geometric rough path(s) associated with W. Then,*

$$Z_{s,t} \equiv \int_s^t \alpha(X)\,\mathrm{d}X^1 = \int_s^t \alpha(W_r)\circ\mathrm{d}W_r \quad (\text{Stratonovich integral}) \text{ a.s.}$$

and

$$\int_s^t \alpha(X)\,\mathrm{d}X^2 = \int_{s<t_1<t_2<t} \circ\,\mathrm{d}Z_{t_1}\otimes\circ\,\mathrm{d}Z_{t_2} \quad (\text{Stratonovich integral}) \text{ a.s.}$$

5.5 Integration against geometric rough paths

In this section we shall establish an integration theory for any geometric rough path with any roughness p. Actually, we shall prove that the integration operator $\int \alpha$ from $\Omega_1(V)$ into $\Omega_1(W)$ is continuous in p-variation topology (by embedding $\Omega_1(V)$ into $\Omega_p(V)$, for any $p \geqslant 1$, using Theorem 3.1.2), and therefore $\int \alpha$ can be extended to all geometric rough paths. However, rather than directly proving this continuity result, the approach we adopt here is the following. First, we define the integral $\int \alpha(X)\,\mathrm{d}X$ for any geometric rough path X. Then we show that $X \to \int \alpha(X)\,\mathrm{d}X$ is indeed a continuous map from $\Omega_p(V)$ to $\Omega_p(W)$ for any p, and that it coincides with the integral defined previously if X is a Lipschitz path (together with its higher-degree iterated integrals).

In this section, we enforce the following condition.

Condition 5.5.1 Let $\alpha : V \to L(V, W)$ be a one-form. Given $p < \gamma \leqslant [p] + 1$ and given Banach tensor products $V^{\otimes j}$ and $W^{\otimes j}$ $(j = 1, \ldots, [p])$, we assume that $(\alpha, V^{\otimes j}, W^{\otimes j} : 1 \leqslant j \leqslant [p])$ is admissible.

Before we give the formal definition for the integration of one-forms along a geometric rough path, let us describe some facts about geometric rough paths which are needed in the following.

The first thing we should mention is that the Taylor formula (5.4) remains true for any geometric rough path X (but may not be true for non-geometric rough paths).

Suppose $X \in \Omega_p(V)$ is a smooth rough path. This means that the projection X_t is a continuous path with finite variation, and $X_{s,t}^i$ is the usual ith iterated path integral. That is,

$$X_{s,t}^1 = X_t - X_s\,,$$

and

$$
\begin{aligned}
X_{s,t}^k &= \lim_{m(D)\to 0} \sum_{i=1}^{k-1} \sum_{l} X_{s,t_{l-1}}^i \otimes X_{t_{l-1},t_l}^{k-i} \\
&= \lim_{m(D)\to 0} \sum_{l} X_{s,t_{l-1}}^{k-1} \otimes X_{t_{l-1},t_l}^1\,.
\end{aligned}
\tag{5.71}
$$

This immediately yields that

$$\mathrm{d}X_{s,t}^k = X_{s,t}^{k-1} \otimes \mathrm{d}X_{s,t}^1\,, \tag{5.72}$$

where $\mathrm{d}X_{s,t}^k$ means the differential of the function $t \to X_{s,t}^k$ as s is fixed.

If $X \in \Omega_p(V)$ is only a rough path, then its differential should include the higher-order terms of its increments up to degree $[p]$. Therefore, the differential of $t \to \int_s^t \alpha(X)\,\mathrm{d}X$ (which we are going to try to define to be a rough path) should be

$$\sum_{l=1}^{[p]} \alpha^l(X_s)(X_{s,t}^l)\,.$$

This means that $\int_s^t \alpha(X)\,\mathrm{d}X^1$ should be the integral of the 'almost multiplicative one'

$$\int_s^t \sum_{l=1}^{[p]} \alpha^l(X_s)(\mathrm{d}X_{s,u}^l)$$

and its iterated path integral $\int_s^t \alpha(X)\,\mathrm{d}X^i$ should be formally the integral of the almost multiplicative one

$$\sum_{\substack{l=(l_1,\ldots,l_i) \\ 1 \leqslant l_j \leqslant [p]}} \int_{s<u_1<\cdots<u_i<t} \sum_{l_1=1}^{[p]} \alpha^{l_1}(X_s)(\mathrm{d}X_{s,u_1}^{l_1}) \otimes \cdots \otimes \sum_{l_i=1}^{[p]} \alpha^{l_i}(X_s)(\mathrm{d}X_{s,u_i}^{l_i}). \quad (5.73)$$

We are going to show that (5.73) makes sense for any geometric rough path.

For any $n \in \mathbb{N}$, we use Π_n to denote the set of all permutations of $\{1, 2, \ldots, n\}$. Given $\boldsymbol{l} = (l_1, \ldots, l_i)$, let $|\boldsymbol{l}| = l_1 + \cdots + l_i$, and use Π_l to denote all those permutations $\pi \in \Pi_{|l|}$ such that

$$\pi(1) < \cdots < \pi(l_1),$$
$$\pi(l_1 + 1) < \cdots < \pi(l_1 + l_2),$$
$$\vdots$$
$$\pi(l_1 + \cdots + l_{i-1} + 1) < \cdots < \pi(|\boldsymbol{l}|).$$

Lemma 5.5.1 *If $X \in \Omega_p(V)$ is a smooth rough path, then, for any $\boldsymbol{l} = (l_1, \ldots, l_i)$, $l_j \geqslant 1$, we have*

$$\int_{s<u_1<\cdots<u_i<t} \mathrm{d}X_{s,u_1}^{l_1} \otimes \cdots \otimes \mathrm{d}X_{s,u_i}^{l_i} = \sum_{\pi \in \Pi_l} \pi^{-1} X_{s,t}^{|l|}, \quad (5.74)$$

for all $(s,t) \in \Delta$.

Proof Since the path X_t is piecewise smooth, we have

$$\int_{s<u_1<\cdots<u_i<t} \mathrm{d}X_{s,u_1}^{l_1} \otimes \cdots \otimes \mathrm{d}X_{s,u_i}^{l_i} = \int_\sigma \dot{X}_{u_{1,1}} \otimes \cdots \otimes \dot{X}_{u_{i,l_i}}\,\mathrm{d}u_{1,1} \cdots \mathrm{d}u_{i,l_i},$$

where the domain σ is the set of all those points $(u_{1,1}, \ldots, u_{i,l_i})$ in $\mathbb{R}^{|l|}$ such that

$$s < u_1 < \cdots < u_i < t,$$
$$s < u_{1,1} < \cdots < u_{1,l_1} = u_1 < t,$$
$$s < u_{2,1} < \cdots < u_{2,l_2} = u_2 < t,$$
$$\vdots$$
$$s < u_{i,1} < \cdots < u_{i,l_i} = u_i < t.$$

However,

$$\sigma = \cup_{\pi \in \Pi_l} \left\{ (u_1, \ldots, u_{|l|}) \in \mathbb{R}^{|l|} : s < u_{\pi(1)} < \cdots < u_{\pi(|l|)} < t \right\}$$

and all these sets in the union are disjoint. Therefore,

$$\int_{s < u_1 < \cdots < u_i < t} \mathrm{d}X_{s,u_1}^{l_1} \otimes \cdots \otimes \mathrm{d}X_{s,u_i}^{l_i}$$

$$= \sum_{\pi \in \Pi_l} \pi^{-1} \int_{s < u_{\pi(1)} < \cdots < u_{\pi(|l|)} < t} \dot{X}_{u_{\pi(1)}} \otimes \cdots \otimes \dot{X}_{u_{\pi(|l|)}} \, \mathrm{d}u_{\pi(1)} \cdots \mathrm{d}u_{\pi(|l|)}$$

$$= \sum_{\pi \in \Pi_l} \pi^{-1} X_{s,t}^{|l|}.$$

Thus we have completed the proof. ∎

Definition 5.5.1 *If $X \in \Omega_p(V)$, then its quasi α-differential Y is defined to be a function in $C_{0,p}(\Delta, T^{([p])}(W))$ with ith component*

$$Y_{s,t}^i = \sum_{\substack{l=(l_1,\ldots,l_i) \\ 1 \leqslant l_j \leqslant [p]}} \alpha^{l_1}(X_s) \otimes \cdots \otimes \alpha^{l_i}(X_s) \left(\sum_{\pi \in \Pi_l} \pi^{-1} X_{s,t}^{|l|} \right),$$

for all $(s,t) \in \Delta$ and $i = 1, \ldots, [p]$.

Therefore, if $X \in \Omega_p(V)$ is a smooth rough path, then its quasi α-differential is

$$Y_{s,t}^i = \sum_{\substack{l=(l_1,\ldots,l_i) \\ 1 \leqslant l_j \leqslant [p]}} \alpha^{l_1}(X_s) \otimes \cdots \alpha^{l_i}(X_s) \left(\int_{s < u_1 < \cdots < u_i < t} \mathrm{d}X_{s,u_1}^{l_1} \otimes \cdots \otimes \mathrm{d}X_{s,u_i}^{l_i} \right)$$

$$= \int_{s < u_1 < \cdots < u_i < t} \sum_{l_1=1}^{[p]} \alpha^{l_1}(X_s)(\mathrm{d}X_{s,u_1}^{l_1}) \otimes \cdots \otimes \sum_{l_i=1}^{[p]} \alpha^{l_i}(X_s)(\mathrm{d}X_{s,u_i}^{l_i})$$

$$= \sum_{\substack{l=(l_1,\ldots,l_i) \\ 1 \leqslant l_j \leqslant [p]}} \int_{s < u_1 < \cdots < u_i < t} \alpha^{l_1}(X_s)(\mathrm{d}X_{s,u_1}^{l_1}) \otimes \cdots \otimes \alpha^{l_i}(X_s)(\mathrm{d}X_{s,u_i}^{l_i}).$$

$$(5.75)$$

Our next goal is to prove that if $X \in \Omega_p(V)$ is geometric, then its quasi α-differential Y, defined as above, is an almost rough path.

Lemma 5.5.2 *If $X \in \Omega_p(V)$ is a smooth rough path, then, for any $(s,t),(t,u) \in \Delta$,*

$$\sum_{l=1}^{[p]} \alpha^l(X_s)(\mathrm{d}X_{s,u}^l) = \sum_{l=1}^{[p]} (\alpha^l(X_t) - R_l(X_s, X_t))(\mathrm{d}X_{t,u}^l). \qquad (5.76)$$

Proof By (5.72) and the fact that $\mathrm{d}X_{s,u}^1 = \mathrm{d}X_{t,u}^1$ (as $X_{s,u}^1 = X_{s,t}^1 + X_{t,u}^1$), we have

$$
\begin{aligned}
\mathrm{d}X_{s,u}^k &= X_{s,u}^{k-1} \otimes \mathrm{d}X_{s,u}^1 \\
&= \sum_{i=0}^{k-1} X_{s,t}^i \otimes X_{t,u}^{k-i-1} \otimes \mathrm{d}X_{t,u}^1 \\
&= \sum_{i=0}^{k-1} X_{s,t}^i \otimes \mathrm{d}X_{t,u}^{k-i} \,.
\end{aligned}
$$

Therefore,

$$
\begin{aligned}
\sum_{l=1}^{[p]} \alpha^l(X_s)(\mathrm{d}X_{s,u}^l) &= \sum_{l=1}^{[p]} \sum_{i=0}^{l-1} \alpha^l(X_s)\big(X_{s,t}^i \otimes \mathrm{d}X_{t,u}^{l-i}\big) \\
&= \sum_{j=1}^{[p]} \sum_{i=0}^{[p]-j} \alpha^{i+j}(X_s)(X_{s,t}^i)(\mathrm{d}X_{t,u}^j) \\
&= \sum_{j=1}^{[p]} \big(\alpha^j(X_t) - R_j(X_s, X_t)\big)(\mathrm{d}X_{t,u}^j) \,.
\end{aligned}
$$

Thus we have proved the lemma. ∎

If $X \in \Omega_p(V)$ is a smooth rough path, then $Y_{s,u}^i$ equals

$$
\sum_{\substack{l=(l_1,\ldots,l_i) \\ 1\leqslant l_j \leqslant [p]}} \int_{s<u_1<\cdots<u_i<u} \alpha^{l_1}(X_s)(\mathrm{d}X_{s,u_1}^{l_1}) \otimes \cdots \otimes \alpha^{l_i}(X_s)(\mathrm{d}X_{s,u_i}^{l_i})
$$

$$
= \sum_{r=1}^{i} \sum_{\substack{l=(l_{r+1},\ldots,l_i) \\ 1\leqslant l_j \leqslant [p]}} \int_{t<u_{r+1}<\cdots<u_i<u} Y_{s,t}^r
$$

$$
\otimes\, \alpha^{l_{r+1}}(X_s)(\mathrm{d}X_{s,u_{r+1}}^{l_{r+1}}) \otimes \cdots \otimes \alpha^{l_i}(X_s)(\mathrm{d}X_{s,u_i}^{l_i})
$$

$$
= \sum_{r:r+j=i} Y_{s,t}^r
$$

$$
\otimes \sum_{\substack{l=(l_1,\ldots,l_j) \\ 1\leqslant l_k \leqslant [p]}} \int_{t<u_1<\cdots<u_j<u} \alpha^{l_1}(X_s)(\mathrm{d}X_{s,u_1}^{l_1}) \otimes \cdots \otimes \alpha^{l_j}(X_s)(\mathrm{d}X_{s,u_j}^{l_j})
$$

$$
= \sum_{r:r+j=i} Y_{s,t}^r \otimes \sum_{\substack{l=(l_1,\ldots,l_j) \\ 1\leqslant l_k \leqslant [p]}} \int_{t<u_1<\cdots<u_j<u} \big(\alpha^{l_1}(X_t) - R_{l_1}(X_s, X_t)\big)(\mathrm{d}X_{t,u_1}^{l_1}) \otimes
$$

$$
\cdots \otimes \big(\alpha^{l_j}(X_t) - R_{l_j}(X_s, X_t)\big)(\mathrm{d}X_{t,u_j}^{l_j})
$$

$$= \sum_{r:r+j=i} Y_{s,t}^r \otimes \sum_{\substack{l=(l_1,\dots,l_j) \\ 1 \leqslant l_k \leqslant [p]}} \left(\alpha^{l_1}(X_t) - R_{l_1}(X_s, X_t) \right) \otimes$$

$$\cdots \otimes \left(\alpha^{l_j}(X_t) - R_{l_j}(X_s, X_t) \right) \left(\sum_{\pi \in \Pi_l} \pi^{-1} X_{t,u}^{|l|} \right).$$

Thus we employ the following notation. If $X \in \Omega_p(V)$, then we define $M \in C_{0,p}(\Delta, T^{([p])}(W))$ having the jth component

$$M_{t,u}^j = \sum_{\substack{l=(l_1,\dots,l_j) \\ 1 \leqslant l_k \leqslant [p]}} \left(\alpha^{l_1}(X_t) - R_{l_1}(X_s, X_t) \right) \otimes$$

$$\cdots \otimes \left(\alpha^{l_j}(X_t) - R_{l_j}(X_s, X_t) \right) \left(\sum_{\pi \in \Pi_l} \pi^{-1} X_{t,u}^{|l|} \right),$$

for $j = 1, \dots, [p]$.

Lemma 5.5.3 *If $X \in \Omega_p(V)$ is a geometric rough path, then*

$$Y_{s,u} = Y_{s,t} \otimes M_{t,u}, \quad \forall (s,t), (t,u) \in \Delta$$

in $T^{([p])}(W)$.

Proof We have already seen that this is true for smooth rough paths. Since M and Y are obviously continuous in p-variation topology, the equality remains true for any geometric rough path.   ∎

Furthermore, we notice that

$$M_{t,u}^j = Y_{t,u}^j + N_{t,u}^j,$$

where $N_{t,u}^0 = 0$,

$$N_{t,u}^j = \sum_{\substack{l=(l_1,\dots,l_j) \\ 1 \leqslant l_k \leqslant [p]}} \sum \beta^{l_1}(X_s, X_t) \otimes \cdots \otimes \beta^{l_j}(X_s, X_t) \left(\sum_{\pi \in \Pi_l} \pi^{-1} X_{t,u}^{|l|} \right),$$

$j = 1, \dots, [p]$, and

$$\beta^{l_k}(X_s, X_t) = \alpha^{l_k}(X_t) \quad \text{or} \quad - R_{l_k}(X_s, X_t).$$

The second summation is taken over all of these possible choices such that at least one of them is $-R_{l_k}(X_s, X_t)$. Clearly, $N \in C_{0,p}(\Delta, T^{([p])}(W))$. We thus have the following lemma.

Lemma 5.5.4 *Let $X \in \Omega_p(V)$ be a geometric rough path. Then, using the above notation,*

$$Y_{s,u} = Y_{s,t} \otimes Y_{t,u} + Y_{s,t} \otimes N_{t,u},$$

for all $(s,t), (t,u) \in \Delta$, where $\otimes$ is the product associated with the algebra $T^{([p])}(W)$.

Theorem 5.5.1 (Under Condition 5.5.1) *Let $X \in \Omega_p(V)$ be a geometric rough path. Then its quasi α-differential $Y \in C_{0,p}(\Delta, T^{([p])}(W))$ is an almost rough path, where*

$$Y_{s,t}^i = \sum_{\substack{l=(l_1,\ldots,l_i) \\ 1 \leqslant l_j \leqslant [p]}} \alpha^{l_1}(X_s) \otimes \cdots \otimes \alpha^{l_i}(X_s) \left(\sum_{\pi \in \Pi_l} \pi^{-1} X_{s,t}^{|l|} \right),$$

for all $(s,t) \in \Delta$ and $i = 1, \ldots, [p]$.

Proof We continue to use the notation established above. We may choose a control ω such that

$$\left| X_{s,t}^i \right| \leqslant \omega(s,t)^{i/p}, \quad \forall 1 \leqslant i \leqslant [p] \quad \text{and} \quad (s,t) \in \Delta.$$

Then

$$\left| Y_{s,t}^i \right| \leqslant C_1 \omega(s,t)^{i/p},$$

for all $(s,t) \in \Delta$ and $1 \leqslant i \leqslant [p]$, where C_1 is a constant depending only on the Lipschitz constant of α, $\max \omega, p, \gamma$. Notice that

$$\left| \beta^{l_1}(X_s, X_t) \otimes \cdots \otimes \beta^{l_j}(X_s, X_t) \left(\sum_{\pi \in \Pi_l} \pi^{-1} X_{t,u}^{|l|} \right) \right| \leqslant C \omega(s,t)^{(\gamma - l_k)/p} \omega(t,u)^{|l|/p}$$

$$\leqslant C \omega(s,u)^{\gamma/p},$$

for some l_k, where C are constants depending only on the Lipschitz constant of α, $\max \omega, p, \gamma$, which may be different from line to line. Now, since

$$Y_{s,u} = Y_{s,t} \otimes (Y_{t,u} + N_{t,u})$$
$$= Y_{s,t} \otimes Y_{t,u} + Y_{s,t} \otimes N_{t,u},$$

we have

$$Y_{s,u}^k - (Y_{s,t} \otimes Y_{t,u})^k = (Y_{s,t} \otimes N_{t,u})^k$$
$$= \sum_{\substack{i+j=k \\ i \geqslant 0, j \geqslant 1}} Y_{s,t}^i \otimes N_{t,u}^j.$$

Therefore,

$$\left| Y_{s,u}^k - (Y_{s,t} \otimes Y_{t,u})^k \right| \leqslant C \omega(s,u)^{\gamma/p}, \tag{5.77}$$

for any $(s,t), (t,u) \in \Delta$ and $k = 1, \ldots, [p]$, where C is a constant depending only on $M, \max \omega, p, \gamma$. Since $\gamma/p > 1$, Y is an almost rough path in $T^{([p])}(W)$. Thus we have completed the proof. ∎

Definition 5.5.2 (Under Condition 5.5.1) *Let* $X \in \Omega_p(V)$ *be a geometric rough path. Then the integral* $\int \alpha(X) \, dX$ *is the unique rough path in* $\Omega_p(W)$ *associated with the quasi* α-*differential* $Y \in C_{0,p}(\Delta, T^{([p])}(W))$ *which is an almost rough path, where*

$$Y^i_{s,t} = \sum_{\substack{l=(l_1,\ldots,l_i) \\ 1 \leqslant l_j \leqslant [p]}} \alpha^{l_1}(X_s) \otimes \cdots \otimes \alpha^{l_i}(X_s) \left(\sum_{\pi \in \Pi_l} \pi^{-1} X^{|l|}_{s,t} \right),$$

for all $(s,t) \in \Delta$ *and* $i = 1, \ldots, [p]$.

Proposition 5.5.1 (Under the same assumptions as in Theorem 5.5.1) *Set*

$$Z^i_{s,t} = \sum_{\substack{l=(l_1,\ldots,l_i) \\ 1 \leqslant l_j, \, |l| \leqslant [p]}} \alpha^{l_1}(X_s) \otimes \cdots \otimes \alpha^{l_i}(X_s) \left(\sum_{\pi \in \Pi_l} \pi^{-1} X^{|l|}_{s,t} \right),$$

for all $(s,t) \in \Delta$ *and* $i = 1, \ldots, [p]$. *Then* $Z \in C_{0,p}(\Delta, T^{([p])}(W))$ *is an almost rough path such that, for some control* ω,

$$\left| Y^i_{s,t} - Z^i_{s,t} \right| \leqslant \omega(s,t)^{([p]+1)/p},$$

for all $(s,t) \in \Delta$ *and* $i = 1, \ldots, [p]$. *Therefore, both* Z *and* Y *define the same rough path* $\int \alpha(X) \, dX$.

This is obvious by definition and Theorem 3.1.2. This proposition shows that our definition for integration of a one-form, given here for a geometric rough path, coincides with the previous definition in the case when $p < 3$. However, the definition given here only makes sense for geometric rough paths or, more precisely, we are not able to show that the quasi α-differential Y of a non-geometric rough path is an almost rough path when $p > 3$.

Theorem 5.5.2 (Under the same assumptions on α as in Theorem 5.5.1) *The integration operator* $X \to \int \alpha(X) \, dX$ *is a continuous map in p-variation-topology from* $G\Omega_p(V)$ *into* $G\Omega_p(W)$. *Moreover, if* $X \in G\Omega_p(V)$ *and if* ω *is a control such that*

$$\left| X^i_{s,t} \right| \leqslant \omega(s,t)^{i/p}, \quad i = 1, \ldots, [p], \quad (s,t) \in \Delta,$$

then

$$\left| \int_s^t \alpha(X) \, dX^i \right| \leqslant (K\omega(s,t))^{i/p},$$

for all $(s,t) \in \Delta$ *and* $i = 1, \ldots, [p]$, *where the constant* K *depends only on* $M, p, \gamma, \max \omega$.

Proof By definition, we can easily see that the map which sends X to its quasi α-differential Y is continuous in p-variation topology, and Y satisfies the uniform, almost multiplicative condition (5.77). Therefore, by Theorem 3.2.1,

$X \to \int \alpha(X)\,\mathrm{d}X$ is continuous in p-variation topology. Moreover, if X is a smooth rough path, then again, by definition, $\int_s^t \alpha(X)\,\mathrm{d}X^1$ is the Riemann integral $\int_s^t \alpha(X_u)\,\mathrm{d}X_u$ and $\int_s^t \alpha(X)\,\mathrm{d}X^i$ is the ith iterated integral of $\int_s^t \alpha(X_u)\,\mathrm{d}X_u$. ∎

5.6 Appendix of Chapter 5

5.6.1 *Banach tensor products*

Given two Banach spaces, there are different Banach tensor products, while there is a unique Hilbert tensor product of two Hilbert spaces. In some sense, this is why life with Hilbert spaces is simpler than with Banach spaces. The theory of tensor products is trivial in the finite-dimensional case, as far as applications to the approach in this book are concerned. However, we must point out that understanding the dependence on dimension is one of the key issues in the local theory of Banach spaces and usually involves very hard analysis. In this respect let us cite the reference Pisier (1986).

Let V and W be two Banach spaces. Then we will use $V \otimes_{\mathrm{a}} W$ to denote the algebraic tensor product, which is the smallest linear space spanned by all elements $v \otimes w, v \in V, w \in W$. Therefore, a general element in $V \otimes_{\mathrm{a}} W$ is a finite linear combination $\sum_i v_i \otimes w_i$.

- The algebraic tensor product $V \otimes_{\mathrm{a}} W$ is a subspace of $\boldsymbol{L}(V^*, W)$, where

$$v \otimes w \in V \otimes_{\mathrm{a}} W \iff v \otimes w(v^*) \equiv \langle v, v^* \rangle w, \quad \forall v^* \in V^*.$$

  In particular, if $W = V$, then we will use $V^{\otimes_{\mathrm{a}} 2}$ to denote $V \otimes_{\mathrm{a}} V$, and it is a subspace of $\boldsymbol{L}(V^*, V)$.

- A norm $|\cdot|$ on the algebraic tensor product $V \otimes_{\mathrm{a}} W$ is a tensor norm if

$$|v \otimes w| \leqslant |v||w|, \quad \forall v \otimes w \in V \otimes_{\mathrm{a}} W, \qquad (5.78)$$

  and it is called a cross-norm if the inequality in (5.78) becomes an equality. If we are given a tensor norm on the algebraic tensor product $V \otimes_{\mathrm{a}} W$, then we shall use $V \otimes W$ to denote the completion of the algebraic tensor product under the tensor norm. $V \otimes W$ is a Banach space, but it usually depends on the choice of tensor norm. In the case when confusion may arise, we will point out which tensor norm we are using, although we use the same notation for a tensor Banach space for simplicity.

- If $\boldsymbol{v} = \sum_{i=1}^n v_i \otimes w_i \in V \otimes_{\mathrm{a}} W$, then we define $\Phi(\boldsymbol{v}) \in \boldsymbol{L}(V^*, W)$ by

$$\Phi(\boldsymbol{v})v^* = \sum_{i=1}^n \langle v_i, v^* \rangle w_i, \quad \forall v^* \in V^*.$$

It is clear $\Phi : V \otimes_{\mathrm{a}} W \to \boldsymbol{L}(V^*, W)$ is injective, and $\Phi(\boldsymbol{v})^* \in \boldsymbol{L}(W^*, V)$ is defined by

$$\Phi(\boldsymbol{v})^* w^* = \sum_{i=1}^n \langle w_i, w^* \rangle v_i, \quad \forall w^* \in W^*.$$

There are two typical cross-tensor norms on the algebraic tensor product of two Banach spaces.

- **Injective tensor norm** If $v = \sum_{i=1}^{n} v_i \otimes w_i$ is an element in $V \otimes_a W$, then

$$|v|_\wedge = \sup\left\{ \left| \sum_{i=1}^{n} \langle v_i, v^* \rangle \langle w_i, w^* \rangle \right| : \right.$$

$$\left. \forall v^* \in V^*,\, w^* \in W^*,\, |v^*| \leqslant 1,\, |w^*| \leqslant 1 \right\},$$

  which is the least cross-norm.
- **Projective tensor norm** For $v \in V \otimes_a W$, then the projective norm is

$$|v|_\vee = \inf \left\{ \sum_{i=1}^{n} |\xi_i^1| |\xi_i^2| : \text{ if } v = \sum_{i=1}^{n} v_i \otimes w_i \right\}.$$

This is the largest cross-norm on $V \otimes_a W$.

Theorem 5.6.1 *The map* $\Phi : (V \otimes_a W, |\cdot|_\wedge) \to L(V^*, W)$ *defined above is an isometry, and therefore* Φ *is extended to an isometry of* $(V \otimes W, |\cdot|_\wedge)$ *into* $L(V^*, W)$.

Suppose $|\cdot|_\beta$ is a cross-norm on $V \otimes_a W$, then we define a norm (called an adjoint-norm) $|\cdot|_{\beta^*}$ on $V^* \otimes_a W^*$ by

$$|v^*|_{\beta^*} = \sup \left\{ \langle v, v^* \rangle : v \in V \otimes_a W,\, |v|_\beta \leqslant 1 \right\}, \quad \forall v^* \in V^* \otimes_a W^*,$$

where if $v^* = \sum_{j=1}^{m} v_j^* \otimes w_j^*$ and $v = \sum_{i=1}^{n} v_i \otimes w_i$, then

$$\langle v, v^* \rangle = \sum_{i,j=1}^{m,n} \langle v_i, v_j^* \rangle \langle w_i, w_j^* \rangle.$$

Theorem 5.6.2 *Let* $|\cdot|_\beta$ *be a cross-norm on* $V \otimes_a W$. *Then the adjoint-norm* $|\cdot|_{\beta^*}$ *is a cross-norm on* $V^* \otimes_a W^*$ *if and only if*

$$|\cdot|_\wedge \leqslant |\cdot|_\beta \leqslant |\cdot|_\vee.$$

The following is called the lifting theorem, which is the most important property about projective tensor norms.

Theorem 5.6.3 *Let V and W be two Banach spaces, and let $(V \otimes W, |\cdot|_\vee)$ be the projective tensor product. Let F be another Banach space, and let $\phi : V \times W \to F$ be a bounded bilinear operator. Then there is a unique $\phi_0 \in L(V \otimes W, F)$ such that*

$$\phi_0(v \otimes w) = \phi(v, w), \quad \text{and} \quad |\phi_0| = |\phi|.$$

5.6.2 *Differentiation, Taylor's theorem*

We recall Taylor's formula for functions on Banach spaces.

Definition 5.6.1 *Let V, H be two Banach spaces, let $U \subseteq V$ be an open subset, and let $f : U \to H$ be a function. f is Fréchet differentiable at ξ_0 if there is a bounded linear operator $A \in L(V, H)$ such that in a neighborhood of ξ_0 we have*

$$|f(\xi) - f(\xi_0) - A(\xi - \xi_0)| = o(|\xi - \xi_0|) \,. \tag{5.79}$$

In this case we write $A = \mathrm{d}f(\xi_0)$, and $\mathrm{d}f(\xi_0)$ is called the Fréchet derivative.

If f is Fréchet differentiable at any $\xi \in U$, and $\mathrm{d}f : U \to L(V, H)$ is Fréchet differentiable on U, then we use $\mathrm{d}^2 f$ to denote $\mathrm{d}(\mathrm{d}f)$ which is a map from U into $L(V, L(V, H))$. We use induction to define $\mathrm{d}^{k+1} f = \mathrm{d}(\mathrm{d}^k f)$, which is a map from U into $L(V, L(V, L(V, \ldots, H)))$.

If $V_1, \ldots, V_k, H$ are Banach spaces, then we use $L(V_1, \ldots, V_k; H)$ to denote the Banach space of all continuous multilinear maps $g : V_1 \times \cdots \times V_k \to H$, and $|g|$ is the smallest constant K such that

$$g(\xi_1, \ldots, \xi_k) \leqslant K|\xi_1| \cdots |\xi_k| \,, \quad \xi_i \in V_i \,, \quad i = 1, \ldots, k \,.$$

Lemma 5.6.1 *The Banach spaces $L(V_1, V_2; H)$ and $L(V_1, L(V_2, H))$ are identical up to a linear isometry, namely*

$$\alpha(\xi_1, \xi_2) = \alpha(\xi_1)\xi_2 \,, \quad \forall \xi_1 \in V_1 \,, \quad \xi_2 \in V_2 \,. \tag{5.80}$$

A map $\alpha \in L(V, \ldots, V; H)$ is called symmetric if, for any permutation π of $(1, \ldots, k)$, we have $\alpha = \alpha \circ \pi$, and we will use $L_s(V, \ldots, V; H)$ to denote the Banach subspace of all symmetric $\alpha \in L(V, \ldots, V; H)$.

By the above lemma, we may regard $\mathrm{d}^k f$ as a function from U into $L_s(V, \ldots, V; H)$. For simplicity, we write $\mathrm{d}^k f(\xi)\eta^k$ to denote $\mathrm{d}^k f(\xi)(\eta, \ldots, \eta)$.

Theorem 5.6.4 (Taylor's theorem) *Suppose $f \in C^{n+1}(U, H)$, $U \subseteq V$ is an open subset, and there is a line segment $[\xi, \xi + h] \subset U$. Then*

$$f(\xi + h) = \sum_{k=0}^{n} \frac{1}{k!} \mathrm{d}^k f(\xi) h^k + R_{n+1}(\xi, \xi + h) \,,$$

where

$$R_{n+1}(\xi, \xi + h) = \int_0^1 \frac{(1-s)^n}{n!} \mathrm{d}^{n+1} f(\xi + sh) h^{n+1} \, \mathrm{d}s \,.$$

5.7 Comments and notes on Chapter 5

There are many good accounts available about the development of the theory of stochastic integration for Brownian motion and, more generally, for semimartingales. The reader may find a complete picture in Dellacherie and Meyer

(1978), Ikeda and Watanabe (1981), Karatzas and Shreve (1988), Revuz and Yor (1991), and Malliavin (1997).

In the 1940s K. Itô defined his theory of stochastic integration for Brownian motion by using the martingale characteristic of the Brownian motion. The main property which Itô has used for the Brownian motion $(b_t)_{t\geqslant 0}$ is that the process $(b_t^2 - t)_{t\geqslant 0}$ is again a martingale, which contributes to the fact that $\sum_l |b_{t_l} - b_{t_{l-1}}|^2$ converges to t in probability. Therefore, the usual 'Riemannian sum' to define integrals works in the probability sense (i.e. convergence happens at the distribution level). Later in his book, Doob (1953) pointed out that Itô's stochastic integration theory can be established for any continuous and square-integrable martingale $(M_t)_{t\geqslant 0}$, provided that there is a continuous and adapted increasing process $(\langle M \rangle_t)_{t\geqslant 0}$ such that $M_t^2 - \langle M \rangle_t$ is a martingale. Doob conjectured that such an increasing process $(\langle M \rangle_t)_{t\geqslant 0}$ always exists. The latter conjecture was proved to be true by Meyer (1962, 1963), and is now known as the Doob–Meyer decomposition (for non-negative super-martingales). Based on the resolution of Doob's conjecture, Kunita and Watanabe (1967) established a stochastic integration theory for square-integrable martingales, and completed the work that Doob had begun. Stochastic analysis for semi-martingales underwent a revolutionary development from the 1960s to the 1980s, especially by the French school, and has now become the classical part of stochastic analysis, see Dellacherie and Meyer (1978), Jacod (1979), and He *et al.* (1992).

In an independent development, Skorohod (1961) established his integration theory, now named after him as Skorohod's integration.

Another direction is that of stochastic calculus for Markov processes, notably developed by Stroock and Varadhan (1979) under the name of the martingale method, and the theory of Dirichlet forms by Silverstein (1974, 1976), Fukushima (1980), etc., see also Ma and Röckner (1992).

Section 5.1. The definition of Lipschitz functions is borrowed from Stein (1970).

Section 5.2. All results are taken from Lyons (1998) and Lyons and Qian (1998).

Section 5.3. The Lipschitz continuity of integration operation was proved in Lyons and Qian (1998). However, the idea follows from Lyons (1998).

Section 5.4. The reader may find a discussion in Lyons and Qian (1996) regarding the Itô formula for rough paths.

Section 5.5. The material is taken from Lyons (1998) with some modifications.

6

UNIVERSAL LIMIT THEOREM

In this chapter we present the central result of this book. It is the universal limit theorem, the name offered by P. Malliavin (1997) to the main result in T. Lyons (1998), which states that the Itô map defined via a differential equation is continuous in p-variation topology.

In the first part of the chapter we shall, in order to exhibit the main idea, consider the case in which the driving path is a rough path of roughness p, where $2 \leqslant p < 3$, and we shall prove a slightly stronger result under a strong regular requirement on the vector fields which govern differential equations. Moreover, in this case we are able to solve differential equations driven by any, not necessarily geometric, rough path. By this, we cover both Itô- and Stratonovich-type stochastic differential equations. Actually, what we shall prove is that Itô's map is Lipschitz continuous in p-variation topology on the space $\Omega_p(V)$ of rough paths for $p < 3$, if the vector fields are C^3. In the remainder of this chapter, we prove the universal limit theorem for any rough path other than a geometric rough path.

6.1 Introduction

We describe the framework in which we are going to discuss differential equations driven by rough paths. Let V and W be two Banach spaces together with their Banach tensor products $V^{\otimes k}$, $W^{\otimes k}$ up to some degree. Let $f : W \to L(V, W)$ be a function, which can be viewed as a map sending vectors of V linearly to vector fields on W. Thus, we will call such functions (a family of) vector fields on W. Consider the following differential equation (initial value problem):

$$\mathrm{d}Y_t = f(Y_t)\,\mathrm{d}X_t, \quad Y_0 = y_0. \tag{6.1}$$

If f is sufficiently regular (for example, f is Lipschitz continuous), then we may solve (6.1) for any Lipschitz path X (see Chapter 2) and obtain the unique solution Y. We may write this unique solution as $Y = F(y_0, X)$. The map $X \to F(y_0, X)$ is called the Itô map (defined through the differential eqn (6.1)). As we have pointed out before, in general, the Itô map $X \to F(y_0, X)$ is not continuous with respect to the topology of uniform convergence. The main result which we shall show in this chapter then states that the Itô map $X \to F(y_0, X)$ is continuous under the p-variation topology.

To establish this universal limit theorem, we first solve (6.1) for any geometric rough path X, and then we prove that the solution is continuous in X and y_0 as well. In fact, we shall prove that the solution map $X \to F(y_0, X)$ is smooth in a sense that will be described later.

148

Since the integral $\int f(Y)\, dX$, for rough paths X, Y, makes no sense generally, we are not able to iterate the differential eqn (6.1) to obtain the unique solution directly. To overcome this difficulty, the idea is to combine X and Y together as a new path. Let us describe this idea in a little detail. Rather than writing (6.1) as a single equation, instead we view this equation as

$$
\begin{aligned}
dX_t &= dX_t, \\
dY_t &= f(Y_t)\, dX_t, \quad Y_0 = y_0.
\end{aligned}
\tag{6.2}
$$

The initial condition of X is irrelevant, and therefore we simply set $X_0 = 0$. Define $\hat{f} : V \oplus W \to L(V \oplus W, V \oplus W)$ by

$$
\hat{f}(x,y)(v,w) = (v, f(y+y_0)v), \quad \forall (x,y), (v,w) \in V \oplus W.
\tag{6.3}
$$

Then eqn (6.2) can be written in the following more appealing form:

$$
dZ_t = \hat{f}(Z_t)\, dZ_t, \quad Z_0 = 0.
\tag{6.4}
$$

It is eqn (6.4) which we are going to solve, since it makes very good sense for geometric rough paths. Naturally, given a geometric rough path X in V, we say that a geometric rough path Z in $V \oplus W$ is a solution to (6.4) if $\pi_V(Z) = X$ and

$$
Z = \int \hat{f}(Z)\, dZ.
\tag{6.5}
$$

6.2   Itô maps: rough paths with $2 \leqslant p < 3$

In this section, we consider the differential eqn (6.1) driven by a rough path with $2 \leqslant p < 3$. Let us first describe the setting in which we are working in this section. Let $T^{(2)}(V)$ be constructed with the projective tensor product $V^{\otimes 2}$, and let X be a rough path in $T^{(2)}(V)$ with roughness $2 \leqslant p < 3$. Let $f \in C^3(W; L(V,W))$ be a vector field. Assume that the derivatives of f are at most linear in growth, namely

$$
|d^i f(x)| \leqslant M(1 + |x|), \quad \forall x \in W, \quad i = 0, 1, 2, 3.
$$

Therefore, $\hat{f}$, as defined by (6.3), belongs to $C^3(V \oplus W)$, where $V \oplus W$ is the direct sum of two Banach spaces with the direct sum norm. The tensor products $(V \oplus W)^{\otimes k}$ will be the completion of the algebraic tensor product $(V \oplus W)^{\otimes_a k}$ under the tensor norm $|v|_m$. We are mainly interested in $(V \oplus W)^{\otimes 2}$, which has the following direct decomposition:

$$
(V \oplus W)^{\otimes 2} = V^{\otimes 2} \oplus (V \otimes W) \oplus (W \otimes V) \oplus W^{\otimes 2},
$$

where $V^{\otimes 2}$, $(V \otimes W)$, $(W \otimes V)$, and $W^{\otimes 2}$ are all projective tensor products.

Any function $K \in C_0(\Delta, T^{(2)}(V \oplus W))$ possesses the corresponding decomposition

$$
\begin{aligned}
K^1_{s,t} &= (X^1_{s,t}, Y^1_{s,t}), \\
K^2_{s,t} &= (X^2_{s,t}, K^{1,0}_{s,t}, K^{0,1}_{s,t}, Y^2_{s,t}),
\end{aligned}
\tag{6.6}
$$

such that $X \in C_0(\Delta, T^{(2)}(V))$ and $Y \in C_0(\Delta, T^{(2)}(W))$. The following property is obvious and follows by a simple calculation.

Lemma 6.2.1 *If $K \in C_0(\triangle, T^{(2)}(V \oplus W))$ is multiplicative, that is K satisfies Chen's identity, and if (6.6) is its decomposition, then $X \in C_0(\triangle, T^{(2)}(V))$ and $Y \in C_0(\triangle, T^{(2)}(W))$ are multiplicative also. In addition,*

$$K \in C_{0,p}(\triangle, T^{(2)}(V \oplus W))$$

if and only if

$$X \in C_{0,p}(\triangle, T^{(2)}(V)),$$
$$Y \in C_{0,p}(\triangle, T^{(2)}(W)),$$
$$K_{s,t}^{1,0} \in C_{0,p/2}(\triangle, T^{(1)}(V \otimes W)),$$

and

$$K_{s,t}^{0,1} \in C_{0,p/2}(\triangle, T^{(1)}(W \otimes V)).$$

Let us discuss several special features of the one-form $\hat{f}$ which are not shared by a general one-form. Without lose of the generality, we may assume that $y_0 = 0$.

Lemma 6.2.2 *(i) We have*

$$\hat{f}^1(\boldsymbol{x})(v, w) = \hat{f}(\boldsymbol{x})(v, 0), \quad \forall \boldsymbol{x}, (v, w) \in V \oplus W,$$
$$\hat{f}^1(\boldsymbol{x})(0, w) = 0, \qquad \forall \boldsymbol{x} \in V \oplus W, \quad w \in W.$$

(ii) Define $\hat{f}^2 \equiv \mathrm{d}\hat{f} : V \oplus W \to \boldsymbol{L}((V \oplus W)^{\otimes_a 2}, V \oplus W)$. Then

$$\hat{f}^2(\boldsymbol{x})(\boldsymbol{v}) = \left(0, \sum_j \mathrm{d}f(y + y_0)(w_2^j)v_1^j\right),$$

for all $\boldsymbol{x} \in V \oplus W$ and $\boldsymbol{v} = \sum_j (v_2^j, w_2^j) \otimes (v_1^j, w_1^j)$ in $(V \oplus W)^{\otimes_a 2}$, and

$$\left|\hat{f}^2(x, y)(\boldsymbol{v})\right| \leqslant |\mathrm{d}f(y + y_0)| \, |\pi_{W \otimes V}(\boldsymbol{v})|_\vee$$
$$\leqslant |\mathrm{d}f(y + y_0)| \, |\boldsymbol{v}|_m.$$

In particular, for any $\boldsymbol{x} \in V \oplus W$, we have

$$\hat{f}^2(\boldsymbol{x}) : V^{\otimes 2} \oplus (V \otimes W) \oplus W^{\otimes 2} \to \{0\}, \tag{6.7}$$

and

$$\pi_{V^{\otimes 2}}(\hat{f}^2(\boldsymbol{x})(\boldsymbol{v})) = 0, \quad \forall \boldsymbol{x} \in V \oplus W, \quad \forall \boldsymbol{v} \in (V \oplus W)^{\otimes 2}. \tag{6.8}$$

Here, and in what follows, we use $\pi_{V^{\otimes i} \otimes W^{\otimes j}}$ to denote the natural projection in the direct decomposition

$$(V \otimes W)^{\otimes k} = \sum_{i+j=k} \oplus (V^{\otimes i} \otimes W^{\otimes j}).$$

Another fact that we need is contained in the following lemma.

Lemma 6.2.3 *For any $x = (x, y) \in V \oplus W$ and $v = \sum_j (v_1^j, w_1^j) \otimes (v_2^j, w_2^j)$ in $(V \oplus W)^{\otimes_a 2}$, then*

$$\hat{f}^1(x) \otimes \hat{f}^1(x)(v) = \sum_j \hat{f}(x)(v_1^j, w_1^j) \otimes \hat{f}(x)(v_2^j, w_2^j)$$

$$= \hat{f}^1(x) \otimes \hat{f}^1(x)(\pi_{V^{\otimes 2}}(v))$$

and

$$\hat{f}^1(x) \otimes \hat{f}^1(x)(v) = \Big(\pi_{V^{\otimes 2}}(v), (f(y) \otimes \mathbf{1})(\pi_{V^{\otimes 2}}(v)),$$

$$(\mathbf{1} \otimes f(y))(\pi_{V^{\otimes 2}}(v)), (f(y) \otimes f(y))(\pi_{V^{\otimes 2}}(v)) \Big).$$

In particular, we have

$$\pi_{V^{\otimes 2}}(\hat{f}^1(x) \otimes \hat{f}^1(x)(v)) = \pi_{V^{\otimes 2}}(v), \quad \forall v \in (V \oplus W)^{\otimes 2}. \qquad (6.9)$$

We also need a scaling property related to path integrals for the one-form $\hat{f}$. For any two real numbers ϵ, δ, the map

$$A_{\epsilon,\delta}(\xi, \eta) = (\epsilon \xi, \delta \eta)$$

defines a bounded linear operator on $V \oplus W$. Its second quantization operator $\Gamma(A_{\epsilon,\delta})$ (see Remark 3.3.3) is an operator from $T^{(2)}(V \oplus W)$ into itself. For simplicity, we shall denote $\Gamma(A_{\epsilon,\delta})$ by $\Gamma_{\epsilon,\delta}$.

Lemma 6.2.4 *For any $\epsilon, \delta \in \mathbb{R}$, if $Z \in C_0(\Delta, T^{(2)}(V \oplus W))$ is multiplicative, then so is $\Gamma_{\epsilon,\delta} Z$. Moreover, the map $Z \to \Gamma_{\epsilon,\delta} Z$ is continuous on $\Omega_p(V \oplus W)$.*

Lemma 6.2.5 *For any $\epsilon, \delta \in \mathbb{R}$, where $\delta \neq 0$, we have*

$$\Gamma_{1,\delta^{-1}} \circ \Gamma_{\epsilon,\delta} = \Gamma_{\epsilon,1}.$$

The proofs of these two lemmas are immediate, so we omit them.

Lemma 6.2.6 *If $Z \in \Omega_p(V \oplus W)$, where $2 \leqslant p < 3$, then*

$$\Gamma_{\epsilon,\epsilon} \int_s^t \hat{f}(Z) \, dZ = \int_s^t \hat{f}(\Gamma_{\epsilon,1} Z) \, d\Gamma_{\epsilon,1} Z, \quad \forall (s, t) \in \Delta.$$

Proof Let $X = \pi_V(Z)$ and $Y = \pi_W(Z)$. Then $\int \hat{f}(\Gamma_{\epsilon,1} Z) \, d\Gamma_{\epsilon,1} Z$ is, by definition, the unique rough path associated with the following almost rough path $Z(\epsilon)$:

$$Z(\epsilon)_{s,t}^1 = \hat{f}^1(\Gamma_{\epsilon,1} Z_s)(\Gamma_{\epsilon,1} Z_{s,t}^1) + \hat{f}^2(\Gamma_{\epsilon,1} Z_s)(\Gamma_{\epsilon,1} Z_{s,t}^2)$$

$$= \big(\pi_V(\Gamma_{\epsilon,1} Z_{s,t}^1), f^1(\pi_W(\Gamma_{\epsilon,1} Z_s)) \pi_V(\Gamma_{\epsilon,1} Z_{s,t}^1) \big)$$

$$\quad + \big(0, f^2(\pi_W(\Gamma_{\epsilon,1} Z_s)) \pi_{W \otimes V}(\Gamma_{\epsilon,1} Z_{s,t}^2) \big)$$

$$= \big(\epsilon X_{s,t}^1, \epsilon f^1(Y_s) X_{s,t}^1 + \epsilon f^2(Y_s) \pi_{W \otimes V}(Z_{s,t}^2) \big)$$

$$= \epsilon \big(X_{s,t}^1, f^1(Y_s) X_{s,t}^1 + f^2(Y_s) \pi_{W \otimes V}(Z_{s,t}^2) \big)$$

and

$$\begin{aligned}
Z(\epsilon)_{s,t}^2 &= \hat{f}^1(\Gamma_{\epsilon,1} Z_s) \otimes \hat{f}^1(\Gamma_{\epsilon,1} Z_s)(\Gamma_{\epsilon,1} Z_{s,t}^2) \\
&= \hat{f}^1(\Gamma_{\epsilon,1} Z_s) \otimes \hat{f}^1(\Gamma_{\epsilon,1} Z_s)(\pi_{V^{\otimes 2}}(\Gamma_{\epsilon,1} Z_{s,t}^2)) \\
&= \hat{f}^1(Y_s) \otimes \hat{f}^1(Y_s)(\epsilon^2 X_{s,t}^2, 0, 0, 0) \\
&= \epsilon^2 \hat{f}^1(Y_s) \otimes \hat{f}^1(Y_s)(X_{s,t}^2, 0, 0, 0) \, .
\end{aligned}$$

Therefore, $Z(\epsilon) = \Gamma_{\epsilon,\epsilon} K$,

$$K_{s,t}^1 = \big(X_{s,t}^1, f^1(Y_s) X_{s,t}^1 + f^2(Y_s) \pi_{W \otimes V}(Z_{s,t}^2) \big) \, ,$$
$$K_{s,t}^2 = \hat{f}^1(Y_s) \otimes \hat{f}^1(Y_s)(X_{s,t}^2, 0, 0, 0) \, .$$

However, by the same calculation we know that $\int \hat{f}(Z) \, \mathrm{d}Z$ is the rough path associated to the almost rough path K, and this yields the conclusion. ∎

The following is an easy but useful estimate.

Lemma 6.2.7 *Let* $Z \in \Omega_p(V \oplus W)$ *and let* $X = \pi_V(Z)$, *which belongs to* $\Omega_p(V)$. *If*

$$\left| X_{s,t}^i \right| \leqslant 2^{-1} \omega(s,t)^{i/p} \, , \quad \left| Z_{s,t}^i \right| \leqslant (K \omega(s,t))^{i/p} \, ,$$

for all $(s,t) \in \Delta$, $i = 1, 2$, *then*

$$\left| \Gamma_{1,\delta} Z_{s,t}^i \right| \leqslant (K_\delta \omega(s,t))^{i/p} \, , \quad i = 1, 2 \quad and \quad \forall (s,t) \in \Delta \, ,$$

where

$$K_\delta = \max \big\{ (2^{-1} + \delta K^{1/p})^p \, , (2^{-1} + 3\delta^2 K^{2/p})^{p/2} \big\} \, .$$

In particular, if we choose δ *such that* $K_\delta \leqslant 1$, *then*

$$\left| \Gamma_{1,\delta} Z_{s,t}^i \right| \leqslant \omega(s,t)^{i/p} \, ,$$

for all $(s,t) \in \Delta$, $i = 1, 2$.

Next we are going to solve eqn (6.1) or, more precisely, the following integral equation:

$$Z = \int \hat{f}(Z) \, \mathrm{d}Z \, , \tag{6.10}$$

with the one-form $\hat{f}$ defined as before, such that $\pi_V(Z) = X$.

Let $X \in \Omega_p(V)$ (remember that $2 \leqslant p < 3$). The Lipschitz constant for both f and $\hat{f}$ is $M \geqslant 1$. Let ω be a control such that

$$\left| X_{s,t}^i \right| \leqslant \frac{1}{2} \omega(s,t)^{i/p} \, , \quad i = 1, 2 \quad and \quad \forall (s,t) \in \Delta \, . \tag{6.11}$$

6.2.1   *The Picard iteration*

We shall use the Picard iteration procedure for eqn (6.4). That is, $Z(0)_{s,t} = (1, Z(0)_{s,t}^1, Z(0)_{s,t}^2)$, where

$$Z(0)_{s,t}^1 = (X_{s,t}^1, 0), \quad Z(0)_{s,t}^2 = (X_{s,t}^2, 0, 0, 0), \tag{6.12}$$

and

$$Z(n) = \int \hat{f}(Z(n-1)) \, dZ(n-1), \quad n = 1, 2, \dots.$$

Then $Z(n) \in \Omega_p(V \oplus W)$, for all n. We will eventually show that the sequence $Z(n)$ converges to a rough path Z, which will be the solution of (6.4).

To this end, we also define a sequence of almost rough paths $K(n)$, where $K(0) = Z(0)$ and

$$K(n+1)_{s,t}^1 = \hat{f}^1(Z(n)_s)(Z(n)_{s,t}^1) + \hat{f}^2(Z(n)_s)(Z(n)_{s,t}^2),$$
$$K(n+1)_{s,t}^2 = \hat{f}^1(Z(n)_s) \otimes \hat{f}^1(Z(n)_s)(Z(n)_{s,t}^2).$$

By definition, we know that $\int Z(n) \, dZ(n)$ is the unique rough path associated with the almost rough path $K(n+1)$. Therefore, $Z(n+1)$ is the unique rough path corresponding to the almost rough path $K(n+1)$.

Lemma 6.2.8   *(i) For each n, we have*

$$\pi_V(Z(n)) = \pi_V(K(n)) = X.$$

(ii) For each n, we have

$$K(n+1)_{s,t}^2 = \hat{f}^1(Z(n)_s) \otimes \hat{f}^1(Z(n)_s) (X_{s,t}^2, 0, 0, 0).$$

(iii) We have

$$\hat{f}^1(x)(Z(n+1)_{s,t}^1 - Z(n)_{s,t}^1) = 0,$$

for all $x \in V \oplus W$ and for all n.

Proof   (i) It is clear that $\pi_V(Z(0)) = \pi_V(K(0)) = X$. We prove that $\pi_V(K(1)) = X$. In fact, by definition,

$$\pi_V(K(1)_{s,t}^1) = \pi_V\left[\hat{f}^1(Z(0)_s)(Z(0)_{s,t}^1) + \hat{f}^2(Z(0)_s)(Z(0)_{s,t}^2)\right]$$
$$= \pi_V\left[X_{s,t}^1, f^1(y_0 + \pi_W Z(0)_s)X_{s,t}^1 + f^2(y_0 + \pi_W Z(0)_s)X_{s,t}^2\right]$$
$$= X_{s,t}^1$$

and

$$\pi_{V^{\otimes 2}}(K(1)_{s,t}^2) = \pi_{V^{\otimes 2}}\left[\hat{f}^1(Z(0)_s) \otimes \hat{f}^1(Z(0)_s)(Z(0)_{s,t}^2)\right]$$
$$= X_{s,t}^2.$$

We now use induction. Suppose that the result is true for n, that is $\pi_V(Z(n)) = \pi_V(K(n)) = X$. Then

$$\pi_V(K(n+1)^1_{s,t})$$
$$= \pi_V\big[\hat{f}^1(Z(n)_s)(Z(n)^1_{s,t}) + \hat{f}^2(Z(n)_s)(Z(n)^2_{s,t})\big]$$
$$= \pi_V\big[X^1_{s,t}, f^1(y_0 + \pi_W Z(n)_s)X^1_{s,t} + f^2(y_0 + \pi_W Z(n)_s)X^2_{s,t}\big]$$
$$= X^1_{s,t}$$

and

$$\pi_{V^{\otimes 2}}(K(n+1)^2_{s,t}) = \pi_{V^{\otimes 2}}\big[\hat{f}^1(Z(n)_s) \otimes \hat{f}^1(Z(n)_s)(Z(n)^2_{s,t})\big]$$
$$= \pi_{V^{\otimes 2}}\big[\hat{f}^1(Z(n)_s) \otimes \hat{f}^1(Z(n)_s)(X^2_{s,t})\big]$$
$$= X^2_{s,t}.$$

Moreover, by definition,

$$Z(n+1)^1_{s,t} = \lim_{m(D)\to 0} \sum_l K(n+1)^1_{t_{l-1},t_l},$$

so that

$$\pi_V(Z(n+1)^1_{s,t}) = \lim_{m(D)\to 0} \sum_l \pi_V(K(n+1)^1_{t_{l-1},t_l})$$
$$= \lim_{m(D)\to 0} \sum_l X^1_{t_{l-1},t_l}$$
$$= X^1_{s,t}.$$

Similarly,

$$\pi_{V^{\otimes 2}}(Z(n+1)^2_{s,t}) = \lim_{m(D)\to 0} \sum_l \pi_{V^{\otimes 2}}(K(n+1)^2_{t_{l-1},t_l})$$
$$\qquad + \pi_{V^{\otimes 2}}\big[K(n+1)^1_{s,t_{l-1}} \otimes K(n+1)^1_{t_{l-1},t_l}\big]$$
$$= \lim_{m(D)\to 0} \sum_l X^2_{t_{l-1},t_l} + X^1_{s,t_{l-1}} \otimes X^1_{t_{l-1},t_l}$$
$$= X^2_{s,t},$$

where we have used the fact that

$$\pi_{V^{\otimes 2}}\big[K(n+1)^1_{s,t_{l-1}} \otimes K(n+1)^1_{t_{l-1},t_l}\big] = X^1_{s,t_{l-1}} \otimes X^1_{t_{l-1},t_l}.$$

(ii) The conclusion follows from the fact that

$$K(n+1)^2_{s,t} = \hat{f}^1(Z(n)_s) \otimes \hat{f}^1(Z(n)_s)(Z(n)^2_{s,t})$$
$$= \hat{f}^1(Z(n)_s) \otimes \hat{f}^1(Z(n)_s)(\pi_{V^{\otimes 2}}(Z(n)^2_{s,t}))$$
$$= \hat{f}^1(Z(n)_s) \otimes \hat{f}^1(Z(n)_s)(X^2_{s,t}, 0, 0, 0).$$

(iii) By definition,

$$\hat{f}(x)(Z(n+1)^1_{s,t}) = \big(\pi_V(Z(n+1)^1_{s,t}), f^1(y+y_0)\pi_V(Z(n+1)^1_{s,t})\big)$$
$$= \big(X^1_{s,t}, f^1(y+y_0)X^1_{s,t}\big),$$

which is independent of n. ∎

6.2.2 *Basic estimates*

By Theorem 5.3.2, we have the following conclusion: there exists a constant $K_1 \geqslant 1$, depending only on $M(1+|y_0|)$ and p, such that, if $\tilde{Z} \in \Omega_p(V \oplus W)$ and if $\tilde{\omega}$ is a control such that $\max \tilde{\omega} \leqslant 1$ and

$$\big|\tilde{Z}^i_{s,t}\big| \leqslant \tilde{\omega}(s,t)^{i/p}, \quad \forall i = 1,2 \quad \text{and} \quad \forall (s,t) \in \Delta,$$

then

$$\left|\int_s^t \hat{f}(\tilde{Z})\, \mathrm{d}\tilde{Z}^i\right| \leqslant (K_1\tilde{\omega}(s,t))^{i/p}, \tag{6.13}$$

for all $(s,t) \in \Delta$ and $i = 1,2$.

With this constant K_1, we choose a constant $\delta \leqslant 1$ such that

$$K_\delta = \max\left\{\big(2^{-1} + \delta K_1^{1/p}\big)^p, \big(2^{-1} + 3\delta^2 K_1^{2/p}\big)^{p/2}\right\} = 1.$$

Let $\epsilon = \delta^{-1} \geqslant 1$. Then

$$\Gamma_{\epsilon,\epsilon}Z(0) = \Gamma_{\epsilon,1}Z(0)$$

and

$$\big|\Gamma_{\epsilon,1}Z(0)^i_{s,t}\big| = \epsilon^i \big|X^i_{s,t}\big| \leqslant \frac{1}{2}\,(\epsilon\omega(s,t))^{i/p}$$
$$\leqslant (\epsilon\omega(s,t))^{i/p},$$

for all $(s,t) \in \Delta$ and $i = 1,2$. Choose $T_1 > 0$ such that $\epsilon \max \omega \leqslant 1$. Then again T_1 depends only on $M, \max \omega, p$.

In the following we replace Δ by Δ_{T_1}.

Lemma 6.2.9 *Under the above assumptions, we have, for all n,*

$$\big|\Gamma_{\epsilon,1}Z(n)^i_{s,t}\big| \leqslant (\epsilon\omega(s,t))^{i/p},$$

for all $0 \leqslant s \leqslant t \leqslant T_1$ and $i = 1,2$.

Proof It is true for $n = 0$. We now use induction. Suppose that it is true for n. Then, since $\epsilon \max \omega \leqslant 1$, we have

$$\left|\int_s^t \hat{f}(\Gamma_{\epsilon,1}Z(n))\, \mathrm{d}\Gamma_{\epsilon,1}Z(n)^i\right| \leqslant (K_1\epsilon\omega(s,t))^{i/p},$$

for all $0 \leqslant s \leqslant t \leqslant T_1$ and $i = 1, 2$, and therefore, by Lemma 6.2.7, we have

$$\left| \Gamma_{1,\delta} \int_s^t \hat{f}(\Gamma_{\epsilon,1} Z(n)) \, \mathrm{d}\Gamma_{\epsilon,1} Z(n)^i \right| \leqslant (\epsilon \omega(s,t))^{i/p} \,,$$

for all $0 \leqslant s \leqslant t \leqslant T_1$ and $i = 1, 2$. However, by Lemma 6.2.6,

$$\int_s^t \hat{f}(\Gamma_{\epsilon,1} Z(n)) \, \mathrm{d}\Gamma_{\epsilon,1} Z(n)^i = \Gamma_{\epsilon,\epsilon} \int_s^t \hat{f}(Z(n)) \, \mathrm{d}Z(n)^i$$

and therefore

$$\Gamma_{\epsilon,1} Z(n+1) = \Gamma_{\epsilon,1} \int_s^t \hat{f}(Z(n)) \, \mathrm{d}Z(n)$$

$$= \Gamma_{1,\delta} \circ \Gamma_{\epsilon,\epsilon} \int_s^t \hat{f}(Z(n)) \, \mathrm{d}Z(n)$$

$$= \Gamma_{1,\delta} \int_s^t \hat{f}(\Gamma_{\epsilon,1} Z(n)) \, \mathrm{d}\Gamma_{\epsilon,1} Z(n) \,.$$

Hence

$$\left| \Gamma_{\epsilon,1} Z(n+1)^i_{s,t} \right| \leqslant (\epsilon \omega(s,t))^{i/p} \,,$$

for all $0 \leqslant s \leqslant t \leqslant T_1$ and $i = 1, 2$. This completes the proof. ∎

Corollary 6.2.1 *There is a constant K_0 depending only on δ (and therefore only on K_1) such that*

$$\left| Z(n)^i_{s,t} \right| \leqslant (K_0 \omega(s,t))^{i/p} \,,$$

for all $0 \leqslant s \leqslant t \leqslant T_1$ and $i = 1, 2$.

By Corollary 5.3.1, there exists a constant K_2 depending only on M, p such that, if $\tilde{Z}, \hat{Z} \in \Omega_p(V \oplus W)$ and if $\tilde{\omega}$ is a control such that $\max \tilde{\omega} \leqslant 1$,

$$\left| \tilde{Z}^i_{s,t} \right|, \left| \hat{Z}^i_{s,t} \right| \leqslant \tilde{\omega}(s,t)^{i/p} \,, \qquad \forall i = 1, 2, \quad \forall (s,t) \in \Delta \,,$$

and

$$\left| \tilde{Z}^i_{s,t} - \hat{Z}^i_{s,t} \right| \leqslant \rho \tilde{\omega}(s,t)^{i/p} \,, \qquad \forall i = 1, 2, \quad \forall (s,t) \in \Delta \,,$$

then

$$\left| \int_s^t \hat{f}(\tilde{Z}) \, \mathrm{d}\tilde{Z}^i - \int_s^t \hat{f}(\hat{Z}) \, \mathrm{d}\hat{Z}^i - (\tilde{K}^i_{s,t} - \hat{K}^i_{s,t}) \right| \leqslant K_2 \rho \tilde{\omega}(s,t)^{3/p} \,, \qquad (6.14)$$

for all $(s,t) \in \Delta$ and $i = 1, 2$, where

$$\tilde{K}^1_{s,t} = \hat{f}^1(\tilde{Z}_s)(\tilde{Z}^1_{s,t}) + \hat{f}^2(\tilde{Z}_s)(\tilde{Z}^2_{s,t}) \,,$$

$$\tilde{K}^2_{s,t} = \hat{f}^1(\tilde{Z}_s) \otimes \hat{f}^1(\tilde{Z}_s)(\tilde{Z}^2_{s,t}) \,,$$

and similar for $\hat{K}$.

Lemma 6.2.10 *With the constant* K_2 *and* $T_1 \geqslant T_2 > 0$ *such that*

$$\left[(M^2 + M)(3 + |y_0|) + K_2 \right] K_0 \omega(0, T_2) < 1 \,,$$

we have

$$\left| Z(n+1)_{s,t}^i - Z(n)_{s,t}^i \right| \leqslant h_n(T_2) \left(K_0 \omega(s,t) \right)^{i/p} \,,$$

for all $0 \leqslant s \leqslant t \leqslant T_2$ *and* $i = 1, 2$, *where*

$$h_n(T_2) = 2 \left[(M^2 + M)(3 + |y_0|) + K_2 \right]^{n-1} \left(K_0 \omega(0, T_2)^{1/p} \right)^{n-1} \,.$$

Proof It is true for $n = 0$ as

$$\left| Z(1)_{s,t}^i - Z(0)_{s,t}^i \right| \leqslant 2 \left(K_0 \omega(s,t) \right)^{i/p} \,,$$

for all $(s,t) \in \Delta$ and $i = 1, 2$. Suppose that the inequality is true for n. Since

$$
\begin{aligned}
K(n+1)_{s,t}^1 - K(n)_{s,t}^1 &= \left[\hat{f}^1(Z(n)_s) - \hat{f}^1(Z(n-1)_s) \right] \left(Z(n)_{s,t}^1 \right) \\
&\quad + \left[\hat{f}^2(Z(n)_s) - \hat{f}^2(Z(n-1)_s) \right] \left(Z(n)_{s,t}^2 \right) \\
&\quad + \hat{f}^1(Z(n-1)_s) \left(Z(n)_{s,t}^1 - Z(n-1)_{s,t}^1 \right) \\
&\quad + \hat{f}^2(Z(n-1)_s) \left(Z(n)_{s,t}^2 - Z(n-1)_{s,t}^2 \right) \,,
\end{aligned}
$$

therefore

$$
\begin{aligned}
\left| K(n+1)_{s,t}^1 - K(n)_{s,t}^1 \right| & \\
&\leqslant M \left| Z(n)_s - Z(n-1)_s \right| \left| Z(n)_{s,t}^1 \right| \\
&\quad + M \left| Z(n)_s - Z(n-1)_s \right| \left| Z(n)_{s,t}^2 \right| \\
&\quad + M \left(1 + |y_0| + \omega(0, T_2)^{1/p} \right) \left| Z(n)_{s,t}^2 - Z(n-1)_{s,t}^2 \right| \\
&\quad + M \left(1 + |y_0| + \omega(0, T_2)^{1/p} \right) \left| Z(n)_{s,t}^1 - Z(n-1)_{s,t}^1 \right| \\
&\leqslant 2M \left(3 + |y_0| \right) K_0 \omega(0, T_2)^{1/p} h_{n-1}(T_2) \left(K_0 \omega(s,t) \right)^{1/p} \,.
\end{aligned}
$$

By Lemma 6.2.8,

$$
\begin{aligned}
K(n+1)_{s,t}^2 &- K(n)_{s,t}^2 \\
&= \left[\hat{f}^1(Z(n)_s) \otimes \hat{f}^1(Z(n)_s) \right. \\
&\qquad \left. - \hat{f}^1(Z(n-1)_s) \otimes \hat{f}^1(Z(n-1)_s) \right] (X_{s,t}^2, 0, 0, 0) \,,
\end{aligned}
$$

and therefore

$$
\begin{aligned}
\left| K(n+1)_{s,t}^2 - K(n)_{s,t}^2 \right| & \\
&\leqslant 2M \left(1 + |y_0| + \omega(0, T_2)^{1/p} \right) \left| Z(n)_s - Z(n-1)_s \right| \left| X_{s,t}^2 \right| \\
&\leqslant 2M^2 \left(1 + |y_0| + \omega(0, T_2)^{1/p} \right) K_0 \omega(0, T_2)^{1/p} \omega(s,t)^{2/p} \\
&\leqslant 2M^2 \left(3 + |y_0| \right) K_0 \omega(0, T_2)^{1/p} \left(K_0 \omega(s,t) \right)^{2/p} \,.
\end{aligned}
$$

Hence, by (6.13), we have

$$
\begin{aligned}
\left| Z(n+1)^i_{s,t} - Z(n)^i_{s,t} \right| &\leqslant \left| K(n+1)^i_{s,t} - K(n)^i_{s,t} \right| + K_2 \big(K_0 \omega(s,t) \big)^{3/p} \\
&\leqslant h_n(T_2) \big(K_0 \omega(s,t) \big)^{i/p} .
\end{aligned}
$$

Thus we have completed the proof. ∎

Therefore, the limit $Z \equiv \lim_{n \to \infty} Z(n)$ exists on $\triangle_{T_2}$, which is again in $\Omega_p(V \oplus W)$. Hence, by the continuity of the integration operator, we may conclude that Z is a solution of eqn (6.4) on the shrinking interval $[0, T_2]$.

6.2.3 *Lipschitz continuity*

Furthermore, we can prove that the solution produced in the previous section is Lipschitz continuous in the driving rough path X.

Let $X, \hat{X} \in \Omega_p(V)$. We shall continue to use the notation of the previous section, and also apply it to X and $\hat{X}$. Thus $\hat{Z}(n)$ is the iteration sequence for $\hat{X}$, that is

$$
\hat{Z}(0)^1_{s,t} = (\hat{X}^1_{s,t}, 0), \quad \hat{Z}(0)^2_{s,t} = (\hat{X}^2_{s,t}, 0, 0, 0),
$$

and

$$
\hat{Z}(n) = \int \hat{f}(\hat{Z}(n-1)) \, d\hat{Z}(n-1), \quad n = 1, 2, \dots .
$$

Suppose that ω is a control such that

$$
\left| X^i_{s,t} \right|, \left| \hat{X}^i_{s,t} \right| \leqslant \frac{1}{2} \omega(s,t)^{i/p}, \quad \forall i = 1, 2, \quad \forall (s,t) \in \triangle,
$$

and, for some $\varepsilon \in [0, 1)$, we have

$$
\left| X^i_{s,t} - \hat{X}^i_{s,t} \right| \leqslant \frac{\varepsilon}{2} \omega(s,t)^{i/p}, \quad \forall i = 1, 2, \quad \forall (s,t) \in \triangle.
$$

Then, as we have shown,

$$
\left| Z(n)^i_{s,t} \right|, \left| \hat{Z}(n)^i_{s,t} \right| \leqslant \big(K_0 \omega(s,t) \big)^{i/p},
$$

for all $0 \leqslant s \leqslant t \leqslant T_2$ and $i = 1, 2$.

Proposition 6.2.1 *Under the above notation and assumptions, we choose* $0 < T_3 \leqslant T_2 \leqslant T_1$ *such that*

$$
\big[(2M^2 + M)(4 + |y_0|) + K_2 \big] \big(K_0 \omega(0, T_3) \big)^{1/p} \leqslant 1 .
$$

Then we have

$$
\left| Z(n)^i_{s,t} - \hat{Z}(n)^i_{s,t} \right| \leqslant \varepsilon \big(K_0 \omega(s,t) \big)^{i/p},
$$

for all $0 \leqslant s \leqslant t \leqslant T_3$ *and* $i = 1, 2$.

Proof By our choice of T_3, we have $K_0\omega(0, T_3) \leqslant 1$. It is clear that the inequality is true for $n = 0$ as $K_0 \geqslant 1$. Suppose that it is true for n. Then notice that

$$K(n+1)^1_{s,t} = \hat{f}^1(Z(n)_s)((X^1_{s,t}, 0)) + \hat{f}^2(Z(n)_s)(Z(n)^2_{s,t}),$$
$$\hat{K}(n+1)^1_{s,t} = \hat{f}^1(\hat{Z}(n)_s)((X^1_{s,t}, 0)) + \hat{f}^2(\hat{Z}(n)_s)(\hat{Z}(n)^2_{s,t}),$$

so that

$$
\begin{aligned}
K(n+1)^1_{s,t} - \hat{K}(n+1)^1_{s,t} &= \left(\hat{f}^1(Z(n)_s) - \hat{f}^1(\hat{Z}(n)_s)\right)((X^1_{s,t}, 0)) \\
&\quad + \left(\hat{f}^2(Z(n)_s) - \hat{f}^2(\hat{Z}(n)_s)\right)(Z(n)^2_{s,t}) \\
&\quad + \hat{f}^2(\hat{Z}(n)_s)\left(Z(n)^2_{s,t} - \hat{Z}(n)^2_{s,t}\right).
\end{aligned}
$$

Therefore

$$\left|K(n+1)^1_{s,t} - \hat{K}(n+1)^1_{s,t}\right| \leqslant \varepsilon M \left(K_0\omega(0, T_3)\right)^{1/p} (4 + |y_0|) \left(K_0\omega(s,t)\right)^{1/p}.$$

Similarly, since

$$
\begin{aligned}
K(n+1)^2_{s,t} &- \hat{K}(n+1)^2_{s,t} \\
&= \left(\hat{f}^1(Z(n)_s) \otimes \hat{f}^1(Z(n)_s) - \hat{f}^1(\hat{Z}(n)_s) \otimes \hat{f}^1(\hat{Z}(n)_s)\right)(X^2_{s,t}, 0, 0, 0),
\end{aligned}
$$

we have the following estimate:

$$\left|K(n+1)^2_{s,t} - \hat{K}(n+1)^2_{s,t}\right| \leqslant \varepsilon M^2 (2 + |y_0|) \left(K_0\omega(0, T_3)\right)^{1/p} \omega(s,t)^{2/p}.$$

Then, together with (6.14), we can deduce that

$$
\begin{aligned}
\left|Z(n+1)^i_{s,t} - \hat{Z}(n+1)^i_{s,t}\right| &\leqslant K_2\varepsilon\left(K_0\omega(s,t)\right)^{3/p} + \left|K(n+1)^i_{s,t} - \hat{K}(n+1)^i_{s,t}\right| \\
&\leqslant \varepsilon\left(K_0\omega(s,t)\right)^{i/p},
\end{aligned}
$$

for all $0 \leqslant s \leqslant t \leqslant T_3$ and $i = 1, 2$. ∎

Therefore, we have

$$\left|Z^i_{s,t} - \hat{Z}^i_{s,t}\right| \leqslant \varepsilon\left(K_0\omega(s,t)\right)^{i/p},$$

for all $0 \leqslant s \leqslant t \leqslant T_3$ and $i = 1, 2$, where $Z = \lim_{n\to\infty} Z(n)$ and $\hat{Z} = \lim_{n\to\infty} \hat{Z}(n)$ are solutions of (6.4) with driving rough paths X and $\hat{X}$, respectively.

6.2.4 *Uniqueness*

Suppose that $\hat{Z}$ is a solution of the integration eqn (6.4) on an interval $[0, T_4]$, for some $T_4 > 0$, such that $\pi_V(\hat{Z}) = X$. We want to prove that, on some small interval $[0, T_5]$, $\hat{Z}$ coincides with Z, which is the limit of our Picard iteration sequence $Z(n)$. We may assume that $T_4 \leqslant T_3$. Define an almost rough path $\hat{K}$ by

$$\hat{K}^1_{s,t} = \hat{f}^1(\hat{Z}_s)(\hat{Z}^1_{s,t}) + \hat{f}^2(\hat{Z}_s)(\hat{Z}^2_{s,t}),$$
$$\hat{K}^2_{s,t} = \hat{f}^1(\hat{Z}_s) \otimes \hat{f}^1(\hat{Z}_s)(\hat{Z}^2_{s,t}).$$

We can choose a control ω such that

$$\left|\hat{Z}^i_{s,t}\right|, \left|\hat{K}^i_{s,t}\right|, \left|Z(n)^i_{s,t}\right| \leqslant \omega(s,t)^{i/p}$$

and

$$\left|X^i_{s,t}\right| \leqslant \frac{1}{2}\omega(s,t)^{i/p},$$

for all $(s,t) \in \Delta_{T_4}$, $n \in \mathbb{N}$, and $i = 1, 2$. If necessary, we may shrink the interval $[0, T_4]$ so that $\max_{[0,T_4]} \omega \leqslant 1$.

Lemma 6.2.11 *We have*

$$\left|Z(n+1)^i_{s,t} - \hat{Z}^i_{s,t}\right| \leqslant \hat{h}_n(T_4)\omega(s,t)^{i/p},$$

for all $0 \leqslant s \leqslant t \leqslant T_4$ and $i = 1, 2$, where

$$\hat{h}_n(T_4) = 2\big[\left(M^2 + M\right)(4 + |y_0|) + K_2\big]^n \omega(0, T_4)^{n/p}.$$

Proof It is true for $n = 0$ as

$$\left|Z(1)^i_{s,t} - \hat{Z}^i_{s,t}\right| \leqslant 2\omega(s,t)^{i/p}, \quad \forall i = 1, 2, \quad \forall (s,t) \in \Delta.$$

Suppose that the inequality is true for n. Since

$$K(n+1)^1_{s,t} - \hat{K}^1_{s,t} = \left(\hat{f}^1(Z(n)_s) - \hat{f}^1(\hat{Z}_s)\right)(X^1_{s,t}, 0)$$
$$+ \left(\hat{f}^2(Z(n)_s) - \hat{f}^2(\hat{Z}_s)\right)(Z(n)^2_{s,t})$$
$$+ \hat{f}^2(\hat{Z}_s)\left(Z(n)^2_{s,t} - \hat{Z}^2_{s,t}\right),$$

then

$$\left|K(n+1)^1_{s,t} - \hat{K}^1_{s,t}\right| \leqslant M(4 + |y_0|)\,\omega(0, T_4)^{1/p}\hat{h}_{n-1}(T_4)\omega(s,t)^{1/p}.$$

Similarly, by Lemma 6.2.8,

$$K(n+1)^2_{s,t} - \hat{K}^2_{s,t}$$
$$= \left(\hat{f}^1(Z(n)_s) \otimes \hat{f}^1(Z(n)_s) - \hat{f}^1(\hat{Z}_s) \otimes \hat{f}^1(\hat{Z}_s)\right)(X^2_{s,t}, 0, 0, 0),$$

and therefore

$$\left|K(n+1)_{s,t}^2 - \hat{K}_{s,t}^2\right| \leqslant M^2 \left(4 + |y_0|\right) \omega(0,T_4)^{1/p} \hat{h}_{n-1}(T_4) \omega(s,t)^{2/p}.$$

Hence, by (6.13), we have

$$\left|Z(n+1)_{s,t}^i - \hat{Z}_{s,t}^i\right| \leqslant \left|K(n+1)_{s,t}^i - \hat{K}_{s,t}^i\right| + K_2 \omega(s,t)^{3/p}$$
$$\leqslant \hat{h}_n(T_4) \omega(s,t)^{i/p}.$$

Thus the conclusion may be established. ∎

Therefore, if we choose $T_4 \geqslant T_5 > 0$ such that

$$\left[(M^2 + M)\left(4 + |y_0|\right) + K_2\right] \omega(0,T_5)^{1/p} < 1,$$

then $\lim_{n \to \infty} Z(n) = \hat{Z}$ on $\triangle_{T_5}$, and therefore on $\triangle_{T_5}$ we have $Z = \hat{Z}$.

Let us summarize what we have proved. Suppose that $V^{\otimes 2}$, $W^{\otimes 2}$, $V \otimes W$, and $W \otimes V$ are projective tensor products, and let $(V \oplus W)^{\otimes 2}$ be the direct sum

$$(V \oplus W)^{\otimes 2} = V^{\otimes 2} \oplus (V \otimes W) \oplus (W \otimes V) \oplus W^{\otimes 2}$$

with the norm $|\cdot|_{\mathrm{m}}$. Let $X \in \Omega_p(V)$, $2 \leqslant p < 3$, and let ω be a control such that

$$\left|X_{s,t}^i\right| \leqslant \frac{1}{2}\omega(s,t)^{i/p}, \quad \forall i = 1,2, \quad \forall (s,t) \in \triangle_T.$$

Suppose that $f : W \to \boldsymbol{L}(V,W)$ is C^3, and suppose that M is a positive constant such that

$$\left|f^i(\xi)\right| \leqslant M(1 + |\xi|), \quad i = 0,1,2,3, \quad \forall \xi \in V,$$

and

$$\left|f^i(\xi) - f^i(\eta)\right| \leqslant M|\xi - \eta|, \quad i = 0,1,2,3, \quad \forall \xi, \eta \in V.$$

Then there are S_1, $T \geqslant S_1 > 0$, and a constant M_1, which depends only on M, $|y_0|$, p, and $\max_{[0,T_4]} \omega$, such that the following conditions hold.

- There is a unique solution Z on $\triangle_{S_1}$, that is $Z \in \Omega_p(V \oplus W)$ such that $\pi_V(Z) = X$ and $Z = \int \hat{f}(Z) \, \mathrm{d}Z$.
- We have the following estimate:

$$\left|Z_{s,t}^i\right| \leqslant \left(M_1 \omega(s,t)\right)^{i/p}, \quad i = 1,2, \quad \forall (s,t) \in \triangle_{S_1}.$$

- If $\hat{X} \in \Omega_p(V)$, and if

$$\left|\hat{X}_{s,t}^i\right| \leqslant \frac{1}{2}\omega(s,t)^{i/p}, \quad i = 1,2, \quad \forall (s,t) \in \triangle_T,$$

and

$$\left|X_{s,t}^i - \hat{X}_{s,t}^i\right| \leqslant \frac{\varepsilon}{2}\omega(s,t)^{i/p}, \quad i = 1,2, \quad \forall (s,t) \in \triangle_T,$$

then

$$\left|Z_{s,t}^i - \hat{Z}_{s,t}^i\right| \leqslant \varepsilon\left(M_1 \omega(s,t)\right)^{i/p}, \quad i = 1,2, \quad \forall (s,t) \in \triangle_{S_1},$$

where $\hat{Z}$ is the solution corresponding to $\hat{X}$.

6.2.5 *Continuity theorem*

Finally, we may extend the solution to the whole interval $[0, T]$, as the estimates we have established are universal in the sense that all constants depend only on $M, |y_0|, p$, and an upper bound of $\max_{[0,T]} \omega$. For example, we may solve the integral equation beyond S_1 by replacing the initial condition y_0 by $Y_{S_1} = \pi_W(Z)_{S_1}$, and the only thing we need to notice is that we can control $|Y_{S_1}|$ by $|y_0|$ and $M, p, \max_{[0,T]} \omega$ due to the above estimate for Z. Thus we have proved the following theorem.

Theorem 6.2.1 *We retain the same convention on the tensor products about* $V^{\otimes 2}$, $W^{\otimes 2}$, *and* $(V \oplus W)^{\otimes 2}$. *Let* $f : W \to \boldsymbol{L}(V, W)$ *be a* C^3 *vector field, and let* M *be a constant such that*

$$\left| f^i(\xi) \right| \leqslant M(1 + |\xi|), \quad i = 0, 1, 2, 3, \quad \forall \xi \in V,$$

and

$$\left| f^i(\xi) - f^i(\eta) \right| \leqslant M|\xi - \eta|, \quad i = 0, 1, 2, 3, \quad \forall \xi, \eta \in V.$$

Define $\hat{f} : V \oplus W \to \boldsymbol{L}(V \oplus W, V \oplus W)$ *by*

$$\hat{f}(x, y)(v, w) = (v, f(y + y_0)v), \quad \forall (x, y), (v, w) \in V \oplus W,$$

where y_0 *is the initial point of our differential equation. Then, for any* $X \in \Omega_p(V)$, *there is a unique* $Z \in \Omega_p(V \oplus W)$ *such that* $\pi_V(Z) = X$ *and* Z *satisfies the following integral equation:*

$$Z^i_{s,t} = \int_s^t \hat{f}(Z) \, dZ^i, \quad i = 1, 2, \quad \forall (s, t) \in \Delta_T.$$

Moreover, there is a constant C *depending only on* $M, p, |y_0|$, *and* $\max_{[0,T]} \omega$ *such that the following conditions hold.*

(i) If

$$\left| X^i_{s,t} \right| \leqslant \omega(s, t)^{i/p}, \quad i = 1, 2, \quad \forall (s, t) \in \Delta_T,$$

then

$$\left| Z^i_{s,t} \right| \leqslant \left(C\omega(s, t) \right)^{i/p}, \quad i = 1, 2, \quad \forall (s, t) \in \Delta_T.$$

(ii) If $X, \hat{X} \in \Omega_p(V)$,

$$\left| X^i_{s,t} \right|, \left| \hat{X}^i_{s,t} \right| \leqslant \omega(s, t)^{i/p}, \quad i = 1, 2, \quad \forall (s, t) \in \Delta_T,$$

and

$$\left| X^i_{s,t} - \hat{X}^i_{s,t} \right| \leqslant \varepsilon \omega(s, t)^{i/p}, \quad i = 1, 2, \quad \forall (s, t) \in \Delta_T,$$

then

$$\left| Z^i_{s,t} - \hat{Z}^i_{s,t} \right| \leqslant \varepsilon \left(C\omega(s, t) \right)^{i/p}, \quad i = 1, 2, \quad \forall (s, t) \in \Delta_T.$$

Therefore we give the following definition.

Definition 6.2.1 *Let* $f : W \to L(V, W)$ *be a vector field on* W. *Let* $y_0 \in W$, *and let* $X \in \Omega_p(V)$, *for* $2 \leqslant p < 3$. *Then we say that a rough path* $Y \in \Omega_p(W)$ *is a solution to the following initial value problem:*

$$\mathrm{d}Y_t = f(Y_t)\,\mathrm{d}X_t\,, \quad Y_0 = y_0 \tag{6.15}$$

if there is a rough path $Z \in \Omega_p(V \oplus W)$ *such that* $\pi_V(Z) = X$, $\pi_W(Z) = Y$, *and*

$$Z = \int \hat{f}(Z)\,\mathrm{d}Z\,, \tag{6.16}$$

where $\hat{f} : V \oplus W \to L(V \oplus W, V \oplus W)$ *is defined by*

$$\hat{f}(x, y)(v, w) = (v, f(y_0 + y)v)\,, \quad \forall (x, y), (v, w) \in V \oplus W\,.$$

According to this definition and Theorem 6.2.1 we have the following corollary.

Corollary 6.2.2 *If* $f \in C^3(W, L(V, W))$ *satisfies the conditions in Theorem 6.2.1, then, for any* $X \in \Omega_p(V)$, *where* $2 \leqslant p < 3$, *there exists a unique solution* Y *to eqn (6.15).*

We will use $F(y_0, X)$ to denote this unique solution Y, and call the map $X \to F(y_0, X)$ Itô's map defined via the differential eqn (6.15). Thus, by Theorem 6.2.1, we have the following theorem.

Theorem 6.2.2 *If* $f \in C^3(W, L(V, W))$ *satisfies the conditions in Theorem 6.2.1, then the Itô map* $X \to F(y_0, X)$ *from* $\Omega_p(V)$ *to* $\Omega_p(W)$, *where* $2 \leqslant p < 3$, *defined by eqn (6.15) is Lipschitz continuous in the following sense. If* $X, \hat{X} \in \Omega_p(V)$, *and if* ω *is a control such that*

$$\left|X_{s,t}^i\right|, \left|\hat{X}_{s,t}^i\right| \leqslant \omega(s, t)^{i/p}\,, \quad i = 1, 2\,, \quad \forall (s, t) \in \Delta$$

and

$$\left|X_{s,t}^i - \hat{X}_{s,t}^i\right| \leqslant \varepsilon\omega(s, t)^{i/p}\,, \quad i = 1, 2\,, \quad \forall (s, t) \in \Delta\,,$$

then there exists a constant C *depending only on* $M, \max \omega, p$, *and* $|y_0|$ *such that*

$$\left|F(y_0, X)_{s,t}^i\right| \leqslant \left(C\omega(s, t)\right)^{i/p}\,, \quad i = 1, 2\,, \quad \forall (s, t) \in \Delta$$

and

$$\left|F(y_0, X)_{s,t}^i - F(y_0, \hat{X})_{s,t}^i\right| \leqslant \varepsilon\left(C\omega(s, t)\right)^{i/p}\,,$$

for all $(s, t) \in \Delta$ *and* $i = 1, 2$.

Proposition 6.2.2 *We retain the same assumptions as in Theorem 6.2.1. Suppose that $X \in \Omega_p(V)$, where $2 \leqslant p < 3$, is a smooth rough path, and therefore we have a unique solution $t \to \tilde{Y}_t$ to the ordinary differential equation*

$$\mathrm{d}Y_t = f(Y_t)\,\mathrm{d}X_t\,, \quad Y_0 = y_0 \tag{6.17}$$

in the sense that

$$\tilde{Y}_t = y_0 + \int_0^t f(\tilde{Y}_s)\,\mathrm{d}X_s\,, \tag{6.18}$$

the integral is the Riemannian integral, and $X_s = X_{0,s}^1$. Then

$$\tilde{Y}_t = F(y_0, X)_{0,t}^1 + y_0 \tag{6.19}$$

and $F(y_0, X)_{s,t}^i$ is the ith iterated integral of $\tilde{Y}_t$, that is

$$F(y_0, X)_{s,t}^i = \int_{s<t_1<\cdots<t_i<t} \mathrm{d}\tilde{Y}_{t_1} \otimes \cdots \otimes \mathrm{d}\tilde{Y}_{t_i}\,.$$

Proof The conclusion of this proposition is quite natural and obvious, but we still need a proof. We argue as follows. Let $\tilde{Y}_t$ be the unique solution to the differential eqn (6.17). Then $\tilde{Y}_t$ is a continuous and piecewise-smooth path in W. Set $Z_{s,t} = (1, Z_{s,t}^1, Z_{s,t}^2)$, where

$$Z_{s,t}^1 = (X_{s,t}^1, \tilde{Y}_{s,t}^1)\,,$$

$$Z_{s,t}^2 = \left(X_{s,t}^2, \int_s^t X_{s,u}^1\,\mathrm{d}\tilde{Y}_{s,u}^1, \int_s^t \tilde{Y}_{s,u}^1\,\mathrm{d}X_{s,u}^1, \tilde{Y}_{s,t}^2\right).$$

Then $Z \in \Omega_p(V \oplus W)$ is a smooth rough path. By the definition of the integration of one-forms, we have

$$Z_{s,t}^1 = \int_s^t \hat{f}(Z)\,\mathrm{d}Z^1\,,$$

which follows from (6.18) and $\pi_V(Z) = X$, and the integral is the usual Young integral. Similarly,

$$Z_{s,t}^2 = \int_s^t \hat{f}(Z)\,\mathrm{d}Z^2\,.$$

Hence, Z is the unique solution of (6.16) with $\pi_V(Z) = X$ and $\pi_W(Z) = \tilde{Y}$. Therefore, by the uniqueness, we have $\tilde{Y}_{s,t}^i = F(y_0, X)_{s,t}^i$. By identifying the initial value, we have (6.19). ∎

6.2.6 *Flows of diffeomorphisms*

To avoid complexity, we assume in this subsection that all Banach spaces are finite-dimensional. Moreover, to simplify our arguments, in this section we assume that the vector field $f : W \to \mathbf{L}(V, W)$ is C^∞ with all bounded derivatives, namely

$$\left| f^i(x) \right|_{\mathbf{L}(V \otimes W^i, W)} \leqslant M, \quad \forall x \in W, \quad i \geqslant 1.$$

We consider the following differential equation:

$$dY_t = f(Y_t)\, dX_t, \quad Y_0 = y, \tag{6.20}$$

which defines an Itô map $X \to F(y, X)$ from $\Omega_p(V)$ to $\Omega_p(W)$, for any $2 \leqslant p < 3$.

For an obvious reason, in this section we move away from our convention that the projection of a rough path X is the path $t \to X_{0,t}^1$. Instead, we say that the path $t \to F(y, X)_{0,t}^1 + y$ is the projection path of $F(y, X)$, and denote it by $F(y, X)_t$, that is

$$F(y, X)_t = y + F(y, X)_{0,t}^1, \quad \forall\, 0 \leqslant t \leqslant T,$$

for $X \in \Omega_p(V)$, where $2 \leqslant p < 3$. The advantage of using this convention is that, if X is a smooth rough path, then $F(y, X)_t$ solves the following ordinary differential equation:

$$dY_t = f(Y_t)\, dX_{0,t}^1, \quad Y_0 = y. \tag{6.21}$$

This section studies the function $y \to F(y, X)_t$ as a map from W to W, for a fixed rough path X and time t.

Lemma 6.2.12 *Let* $X \in C_0(\triangle_T, T^{([p])}(V))$, *and let* $0 \leqslant \tau < T$. *Define* $\theta_\tau X \in C_0(\triangle_{T-\tau}, T^{([p])}(V))$ *by*

$$(\theta_\tau X)_{s,t}^i = X_{s+\tau, t+\tau}^i,$$

for all $0 \leqslant s + \tau \leqslant t + \tau \leqslant T$ *and* $i = 1, \ldots, [p]$. *If* $X \in \Omega_p(V)$, *then so is* $\theta_\tau X$ *(with running time* $T - \tau$*), and if* X *is geometric, then so is* $\theta_\tau X$.

By uniqueness, we have the following lemma.

Lemma 6.2.13 *Let* $X \in \Omega_p(V)$, *where* $2 \leqslant p < 3$. *Then*

$$F(y, X)_{s+t} = F(F(y, X)_s, \theta_s X)_t, \quad \forall\, 0 \leqslant s + t \leqslant T.$$

Suppose that $X \in \Omega_p(V)$ is a smooth rough path. Then the solution $Y_t = F(y, X)_t$ of eqn (6.21) is smooth as a function of y when X, t are fixed. Moreover, the derivative $K_t = d_y F(y, X)_t$ (the Fréchet derivative in y, so that K_t is $\mathbf{L}(W, W)$-valued) satisfies the following differential equation:

$$dK_t = \partial f(Y_t)(dX_t) \circ K_t, \quad K_0 = \mathrm{id}, \tag{6.22}$$

where $\partial f : W \to \mathbf{L}(V, \mathbf{L}(W, W))$ is defined as follows. For any $x \in W$ and $\xi \in V$, we define

$$\partial f(x)(\xi)\eta = \lim_{h\to 0}\frac{f(x+h\eta)\xi - f(x)\xi}{h}$$
$$= \mathrm{d}f(x)(\eta)(\xi)$$
$$= f^1(x)(\eta\otimes\xi)\,,$$

and $\partial f(x)(\xi)\circ K$ is the composition of two operators in $\boldsymbol{L}(W,W)$. Thus

$$(\partial f(x)(\xi)\circ K)\,\eta = f^1(x)(K\eta\otimes\xi)\,.$$

Define $g: W\oplus\boldsymbol{L}(W,W)\to\boldsymbol{L}(V,\boldsymbol{L}(W,W))$ by

$$g(x,K)\xi = \partial f(x)(\xi)\circ K\,,\quad \forall(x,K)\in W\oplus\boldsymbol{L}(W,W)\,,\quad \forall\xi\in V\,.$$

Then (6.22) can be rewritten as

$$\mathrm{d}K_t = g(Y_t,K_t)\,\mathrm{d}X_t\,,\quad K_0 = \mathrm{id}\,,$$

which, together with (6.20), gives us the following system:

$$\begin{aligned} \mathrm{d}Y_t &= f(Y_t)\,\mathrm{d}X_t\,, & Y_0 &= y\,,\\ \mathrm{d}K_t &= g(Y_t,K_t)\,\mathrm{d}X_t\,, & K_0 &= \mathrm{id}\,. \end{aligned}\tag{6.23}$$

Let $E_t = (Y_t,K_t)$ and define $h: W\oplus\boldsymbol{L}(W,W)\to\boldsymbol{L}(V,W\oplus\boldsymbol{L}(W,W))$ by

$$h((x,K))\xi = (f(x)\xi, g(x,K)\xi)\,,\quad \forall(x,K)\in W\oplus\boldsymbol{L}(W,W)\,,\quad \xi\in V\,.$$

Then the system (6.23) becomes

$$\mathrm{d}E_t = h(E_t)\,\mathrm{d}X_t\,,\quad E_0 = (y,\mathrm{id})\,,\tag{6.24}$$

which makes sense for any rough path X. In other words, the solution Y_t, together with its space derivative K_t, is an Itô functional.

Furthermore, K_t is invertible and K_t^{-1} satisfies the following differential equation:

$$\mathrm{d}K_t^{-1} = -K_t^{-1}\circ\partial f(Y_t)\,\mathrm{d}X_t\,,\quad K_0^{-1} = \mathrm{id}\,,$$

i.e.

$$\mathrm{d}K_t^{-1} = \tilde{g}(Y_t,K_t^{-1})\,\mathrm{d}X_t\,,\quad K_0^{-1} = \mathrm{id}\,,$$

where $\tilde{g}: W\oplus\boldsymbol{L}(W,W)\to\boldsymbol{L}(V,\boldsymbol{L}(W,W))$ is given by

$$\tilde{g}(x,K)\xi = -K\circ\partial f(x)(\xi)\,,\quad \forall(x,K)\in W\oplus\boldsymbol{L}(W,W)\,,\quad \xi\in V\,.$$

By the same argument, $\tilde{E}_t = (Y_t,K_t^{-1})$ is the unique solution to the following differential equation:

$$\mathrm{d}\tilde{E}_t = \tilde{h}(\tilde{E}_t)\,\mathrm{d}X_t\,,\quad \tilde{E}_0 = (y,\mathrm{id})\,,\tag{6.25}$$

where $\tilde{h}: W\oplus\boldsymbol{L}(W,W)\to\boldsymbol{L}(V,W\oplus\boldsymbol{L}(W,W))$ is defined by

$$\tilde{h}((x,K))\xi = (f(x)\xi, \tilde{g}(x,K)\xi)\,,\quad \forall(x,K)\in W\oplus\boldsymbol{L}(W,W)\,,\quad \xi\in V\,.$$

We summary the above in the following lemma.

Lemma 6.2.14 *Let* $X \in \Omega_p(V)$ *be a smooth rough path (and so we are considering an ordinary differential equation). Under the above assumptions and notation,* $y \to F(y, X)_t$ *is smooth, and* $E_t = (F(y, X)_t, \mathrm{d}_y F(y, X)_t)$ *and* $\tilde{E}_t = (F(y, X)_t, \mathrm{d}_y F(y, X)_t^{-1})$ *are the unique solutions to the differential eqns (6.24) and (6.25), respectively. Moreover,* $(F(y, X)_t)$ *is a flow of diffeomorphisms.*

By the universal limit theorem, namely Theorem 6.2.2, we may extend this conclusion to any geometric rough path.

Theorem 6.2.3 *Let* $X \in \Omega_p(V)$, *where* $2 \leqslant p < 3$, *be a geometric rough path, let* $f \in C_b^\infty(W, \boldsymbol{L}(V, W))$, *and let* $F(y, X)$ *be the Itô map defined by the differential equation*

$$\mathrm{d}Y_t = f(Y_t)\,\mathrm{d}X_t, \quad Y_0 = y, \tag{6.26}$$

and $F(y, X)_t = F(y, X)_{0,t}^1 + y$. *Then, for any* $t \in [0, T]$, $y \to F(y, X)_t$ *is differentiable (and therefore smooth). Let* $E, \tilde{E}$ *be the solutions of the differential equations*

$$\mathrm{d}E_t = h(E_t)\,\mathrm{d}X_t, \quad E_0 = (y, \mathrm{id}), \tag{6.27}$$

$$\mathrm{d}\tilde{E}_t = \tilde{h}(\tilde{E}_t)\,\mathrm{d}X_t, \quad \tilde{E}_0 = (y, \mathrm{id}), \tag{6.28}$$

as rough paths in $T^{(2)}(W \oplus \boldsymbol{L}(W, W))$. *Then*

$$\mathrm{d}_y F(y, X)_t = \mathrm{id} + \pi_{\boldsymbol{L}(W,W)}(E_{0,t}^1),$$

and $\mathrm{d}_y F(y, X)_t^{-1}$ *exists and is given by*

$$\mathrm{d}_y F(y, X)_t^{-1} = \mathrm{id} + \pi_{\boldsymbol{L}(W,W)}(\tilde{E}_{0,t}^1).$$

Proof Since $X \in G\Omega_p(V)$, we can therefore choose a sequence of smooth rough paths $X(n) \in \Omega_p(V)$ and a control ω such that

$$\left|X(n)_{s,t}^i\right|, \left|X_{s,t}^i\right| \leqslant \omega(s,t)^{i/p}, \qquad i = 1, 2, \quad \forall (s, t) \in \triangle,$$

and

$$\left|X(n)_{s,t}^i - X_{s,t}^i\right| \leqslant n^{-1}\omega(s,t)^{i/p}, \quad i = 1, 2, \quad \forall (s, t) \in \triangle.$$

Let $E(n), \tilde{E}(n)$ denote the corresponding solutions of (6.27) and (6.28) with driving force $X(n)$. Then, by Theorem 6.2.2, there exists a constant C, depending only on M, $\max \omega$, and $|y|$, such that

$$\left|E(n)_{s,t}^i - E_{s,t}^i\right| \leqslant n^{-1}C\omega(s,t)^{i/p}, \quad i = 1, 2, \quad \forall (s, t) \in \triangle,$$

$$\left|\tilde{E}(n)_{s,t}^i - \tilde{E}_{s,t}^i\right| \leqslant n^{-1}C\omega(s,t)^{i/p}, \quad i = 1, 2, \quad \forall (s, t) \in \triangle,$$

and

$$|F(y, X(n))_t - F(y, X)_t| \leqslant n^{-1}C\omega(0,t)^{i/p}, \quad i = 1, 2.$$

However,

$$d_y F(y, X(n))_t = \mathrm{id} + E(n)_{0,t}^1, \quad \forall n,$$

and

$$d_y F(y, X(n))_t^{-1} = \mathrm{id} + \tilde{E}(n)_{0,t}^1, \quad \forall n.$$

In order to pass to the limit, the only thing we require is the fact that the differential operator d_y is a closed operator. Hence, $d_y F(y, X)_t$ exists and

$$d_y F(y, X)_t = \mathrm{id} + E_{0,t}^1,$$

by continuity. Thus we have completed the proof. ∎

If $X \in \Omega_p(V)$, where $2 \leqslant p < 3$, is a smooth rough path, and, for $T_1 \leqslant T$, $\hat{X} \in \Omega_p(V)$ is the time-reversal of X at time T_1 (as a geometric rough path), then, since $\hat{X}_t = X_{T_1-t}$, by the uniqueness of the solution to an ordinary differential equation, we have

$$F(y, X)_{T_1-t} = F(F(y, X)_{T_1}, \hat{X})_t, \quad \forall t \leqslant T_1. \tag{6.29}$$

By the continuity of Itô's map $X \to F(y, X)$ and the map $X \to \hat{X}$ (in p-variation topology), we know that (6.29) remains true for any geometric rough path X. Therefore, the map $y \to F(y, X)$ is a bijection. Together with Lemma 6.2.12, we have the following theorem.

Theorem 6.2.4 *If $f \in C_b^\infty(W, L(V, W))$, and if $X \in G\Omega_p(V)$, where $2 \leqslant p < 3$, is a geometric rough path, then $(F(\cdot, X)_t)$ is a flow of diffeomorphisms.*

6.3 The Itô map: geometric rough paths

We continue to study Itô's map defined by the differential equation

$$dY_t = f(Y_t)\, dX_t, \quad Y_0 = y_0. \tag{6.30}$$

However, in this section, the driving path X is a geometric rough path in V with roughness p, but $p \geqslant 1$ can be any number.

We still use the Picard iteration method to solve the corresponding integration equation, namely

$$Z_{s,t}^i = \int_s^t \hat{f}(Z)\, dZ^i, \quad i = 1, \dots, [p], \quad (s,t) \in \triangle, \tag{6.31}$$

such that $\pi_V(Z) = X$. Notice that the initial condition $Y_0 = y_0$ has been absorbed into the one-form $\hat{f}$, see (6.3). However, in this section, by a solution we mean a geometric rough path $Y \in \Omega_p(W)$ such that there is a geometric rough path $Z \in \Omega_p(V \oplus W)$ satisfying (6.31), $\pi_V(Z) = X$, and $\pi_W(Z) = Y$.

As a general principle, in this section we assume that all one-forms, together with the tensor products which we used, are admissible.

Several facts

We need the following lemma about Lip (γ) functions:

Lemma 6.3.1 *If* $f : W \to L(V,W)$ *is a* Lip (γ) *function, then there is a* Lip $(\gamma - 1)$ *function* $g : W \oplus W \to L(W \otimes V, W)$ *such that*

$$f(x)\xi - f(y)\xi = g(x,y)((x-y) \otimes \xi), \quad \forall x, y \in W, \quad \xi \in V.$$

The iteration procedure which we will use is the same as for the case when $p < 3$. Therefore,

$$Z(n) = \int \hat{f}(Z(n-1)) \, \mathrm{d}Z(n-1),$$
$$Z(0) = (X, 0). \tag{6.32}$$

However, if X is a smooth rough path, then this iteration procedure is equivalent to the following:

$$\mathrm{d}X = \mathrm{d}X,$$
$$\mathrm{d}Y(n+1) = f(Y(n)) \, \mathrm{d}X, \quad Y(n)_0 = y_0, \tag{6.33}$$

where $Y(0) = 0$. Our aim is to show that $Y(n)$ converges to a solution of eqn (6.30) uniformly in p-variation topology. Therefore, we consider the difference $D(n) = Y(n) - Y(n-1)$, with $D(1) = Y(1)$. By (6.33), we have

$$\mathrm{d}X = \mathrm{d}X,$$
$$\mathrm{d}Y(n+1) = f(Y(n)) \, \mathrm{d}X, \tag{6.34}$$
$$\mathrm{d}D(n+1) = g(Y(n), Y(n) - D(n))(D(n)) \, \mathrm{d}X.$$

The essential fact which we will use is the homogenous property, namely, for any $\beta \neq 0$, $\beta D(n+1)$ satisfies the following equation:

$$\mathrm{d}X = \mathrm{d}X,$$
$$\mathrm{d}Y(n+1) = f(Y(n)) \, \mathrm{d}X,$$
$$\mathrm{d}\beta D(n+1) = g(Y(n), Y(n) - \beta^{-1}\beta D(n))(\beta D(n)) \, \mathrm{d}X.$$

Therefore, let us define a one-form Φ_β, for any $\beta \neq 0$, such that $\Phi_\beta : V \oplus W \oplus W \to L(V \oplus W \oplus W, V \oplus W \oplus W)$, by

$$\Phi_\beta((x,y,d))(\xi, \eta, \zeta) = (\xi, f(y_0 + y)\xi, g(y, y - \beta^{-1}d)(d \otimes \xi)), \tag{6.35}$$

for $(x, y, d), (\xi, \eta, \zeta) \in V \oplus W \oplus W$, and set $\Phi = \Phi_1$.

We shall assume that f is a Lip (γ) function for some $\gamma > p$, and therefore g is a Lip $(\gamma - 1)$ function. Hence, for any $\beta \neq 0$, Φ_β is a Lip $(\gamma - 1)$ one-form on $V \oplus W \oplus W$ (with reasonable choices of their tensor norms). Our first key observation is given in the following lemma.

Lemma 6.3.2 *If M is a Lipschitz constant for Φ, then it also a Lipschitz constant for Φ_β, for all $\beta \geqslant 1$.*

Therefore, by Theorem 5.5.2, we have the following proposition.

Proposition 6.3.1 *There exists a constant C_1, depending only on the Lipschitz constant M, an upper bound of $\max \omega$, p and γ, such that, if $K \in G\Omega_p(V \oplus W \oplus W)$ and if*

$$\left| K^i_{s,t} \right| \leqslant \omega(s,t)^{i/p}, \quad i = 1, \ldots, [p], \quad \forall (s,t) \in \Delta,$$

then, for all $\beta \geqslant 1$, we have

$$\left| \int_s^t \Phi_\beta(K) \, \mathrm{d}K^i \right| \leqslant (C_1 \omega(s,t))^{i/p}, \tag{6.36}$$

for all $(s,t) \in \Delta$ and $i = 1, \ldots, [p]$.

We also need several scaling properties.

Lemma 6.3.3 *If $K \in G\Omega_p(V \oplus W \oplus W)$ is a geometric rough path, then so is $\Gamma_{\epsilon,\delta,\beta}K$, for any $\epsilon, \delta, \beta \in \mathbb{R}$. Moreover, the map $K \to \Gamma_{\epsilon,\delta,\beta}K$ is continuous in p-variation topology.*

Proof If K is a smooth rough path, then

$$(\Gamma_{\epsilon,\delta,\beta}K)^i_{s,t} = \int_{s<t_1<\cdots<t_i<t} \mathrm{d}A_{\epsilon,\delta,\beta}K_{t_1} \otimes \cdots \otimes \mathrm{d}A_{\epsilon,\delta,\beta}K_{t_i}$$

and therefore $\Gamma_{\epsilon,\delta,\beta}K$ is also a smooth rough path. Clearly, the map $K \to \Gamma_{\epsilon,\delta,\beta}K$ (as a map from $C_{0,p}(\Delta, T^{([p])}(V \oplus W \oplus W))$ to itself) is continuous in p-variation topology, and therefore we obtain all of the conclusions. ∎

Lemma 6.3.4 *For any $\epsilon, \rho, \beta \neq 0$, and a geometric rough path $K \in G\Omega_p(V \oplus W \oplus W)$, we have*

$$\Gamma_{\epsilon,\epsilon,\beta\epsilon} \int \Phi(K) \, \mathrm{d}K = \int \Phi_\beta(\Gamma_{\epsilon,1,\beta}K) \, \mathrm{d}\Gamma_{\epsilon,1,\beta}K \tag{6.37}$$

and

$$\Gamma_{1,1,\beta} \int \Phi_\rho(K) \, \mathrm{d}K = \int \Phi_{\rho\beta}(\Gamma_{1,1,\beta}K) \, \mathrm{d}\Gamma_{1,1,\beta}K. \tag{6.38}$$

Proof By the continuity of the integration operators (see Theorem 5.5.2), we only need to prove these identities for a smooth rough path K. In this case all integrals reduce to ordinary integrals and their iterated path integrals. Therefore, we only need to check (6.37) and (6.38) for the first degree components. Write $K = (X, Y, D)$ and $\int \Phi(K)\, dK = (\tilde{X}, \tilde{Y}, \tilde{D})$. Then

$$d\tilde{X} = dX\,,$$
$$d\tilde{Y} = f(Y)\, dX\,,$$
$$d\tilde{D} = g(Y, Y - D)(D)\, dX\,,$$

and therefore

$$d\epsilon\tilde{X} = d\epsilon X\,,$$
$$d\epsilon\tilde{Y} = f(Y)\, d\epsilon X\,,$$
$$d\beta\epsilon\tilde{D} = g(Y, Y - \beta^{-1}\beta D)(\beta D)\, d\epsilon X\,.$$

This gives (6.37). Similarly, if $K = (X, Y, D)$ and $\int \Phi_\rho(K)\, dK = (\tilde{X}, \tilde{Y}, \tilde{D})$, then

$$d\tilde{X} = dX\,,$$
$$d\tilde{Y} = f(Y)\, dX\,,$$
$$d\tilde{D} = g(Y, Y - \rho^{-1}D)(D)\, dX\,.$$

Therefore,

$$d\tilde{X} = dX\,,$$
$$d\tilde{Y} = f(Y)\, dX\,,$$
$$d\beta\tilde{D} = g(Y, Y - \rho^{-1}\beta^{-1}\beta D)(\beta D)\, dX\,,$$

which gives us (6.38). ∎

The following estimate is almost trivial.

Lemma 6.3.5 *If $K \in G\Omega_p(V \oplus W \oplus W)$ is a geometric rough path such that $\pi_V(K) = X$, and if ω is a control such that*

$$\left| X^i_{s,t} \right| \leqslant 2^{-1}\omega(s,t)^{i/p} \tag{6.39}$$

and

$$\left| K^i_{s,t} \right| \leqslant \left(C_2\omega(s,t) \right)^{i/p}, \tag{6.40}$$

for all $(s,t) \in \Delta$ and $i = 1, \ldots, [p]$, then there exists a constant $\delta \in (0,1]$ depending only on C_2 and $[p]$ such that, for any $\delta_1, \delta_2 \leqslant \delta$, we have

$$\left| (\Gamma_{1,\delta_1,\delta_2} K)^i_{s,t} \right| \leqslant \omega(s,t)^{i/p}, \tag{6.41}$$

for all $(s,t) \in \Delta$ and $i = 1, \ldots, [p]$.

Remark 6.3.1 However, δ also depends on the nature of the norms which define the involved tensor products and their direct sums.

Iteration procedure

We have defined the following iteration procedure:

$$Z(n+1) = \int \hat{f}(Z(n)) \, \mathrm{d}Z(n) \,,$$

$$Z(0) = (X, 0) \,.$$

Also, we have pointed out that we need to consider the following iteration program:

$$K(n+1) = \int \Phi(K(n)) \, \mathrm{d}K(n) \,. \tag{6.42}$$

Let us specify the first term $K(1)$. This should be defined by

$$\mathrm{d}X = \mathrm{d}X \,,$$
$$\mathrm{d}Y(1) = f(y_0) \, \mathrm{d}X \,, \tag{6.43}$$
$$\mathrm{d}D(1) = f(y_0) \, \mathrm{d}X \,,$$

which makes sense for any geometric rough path X. In fact, set

$$h : V \oplus W \oplus W \to \boldsymbol{L}(V \oplus W \oplus W, V \oplus W \oplus W)$$

by

$$h((x, y, d))(\xi, \eta, \zeta) = (\xi, f(y_0)\xi, f(y_0)\xi) \,,$$

for all $(x, y, d), (\xi, \eta, \zeta) \in V \oplus W \oplus W$, and let

$$K(1) = \int h((X, 0, 0)) \, \mathrm{d}(X, 0, 0) \,, \tag{6.44}$$

where $(X, 0, 0) \in G\Omega_p(V \oplus W \oplus W)$ is the rough path such that $\pi_V(X, 0, 0) = X$ and $\pi_{V\perp}(X, 0, 0) = 0$, that is

$$(X, 0, 0)^i_{s,t} = (X^i_{s,t}, 0, \ldots, 0) \,, \quad i = 1, \ldots, [p] \,, \quad (s, t) \in \Delta \,.$$

By definition, if $X \in G\Omega_p(V)$ is a smooth rough path, then all of the $K(n)$ are smooth rough paths and, if $K(n) = (X, Y(n), D(n))$, then

$$\mathrm{d}X = \mathrm{d}X \,,$$
$$\mathrm{d}Y(n+1) = f(Y(n)) \, \mathrm{d}X \,, \tag{6.45}$$
$$\mathrm{d}D(n+1) = g(Y(n), Y(n) - D(n))(D(n)) \, \mathrm{d}X \,.$$

Therefore, $\pi_V(K(n)) = X$ and $\pi_{V\oplus W}(K(n)) = Z(n)$, for any n. This conclusion is true for any geometric rough path by continuity. We present this as the following lemma.

Lemma 6.3.6 *If $X \in G\Omega_p(V)$, then we have*

$$\pi_V(K(n)) = X \quad and \quad \pi_{V \oplus W}(K(n)) = Z(n),$$

for all $n \geqslant 1$.

Next we choose any $\beta \neq 0$ (which will be determined later), and define a new sequence of geometric rough paths $H(n)$ by

$$H(n) = \Gamma_{1,1,\beta^{n-1}} K(n).$$

Notice that, if X is a smooth rough path, then

$$H(n) = (X, Y(n), \beta^{n-1} D(n)), \quad \forall n \geqslant 1.$$

The key part of our argument is to show that the sequence $H(n)$ is uniformly bounded (in p-variation topology, see below for the precise meaning), and therefore we can conclude that $D(n)$ tends to zero geometrically fast.

Lemma 6.3.7 *For any $\beta \neq 0$, we have*

$$H(n+1) = \Gamma_{1,1,\beta} \int \Phi_{\beta^{n-1}}(H(n)) \, \mathrm{d}H(n). \tag{6.46}$$

Proof By definition and eqns (6.37) and (6.38),

$$\begin{aligned}
H(n+1) &= \Gamma_{1,1,\beta^n} \int \Phi(K(n)) \, \mathrm{d}K(n) \\
&= \int \Phi_{\beta^n}(\Gamma_{1,1,\beta^n} K(n)) \, \mathrm{d}\Gamma_{1,1,\beta^n} K(n) \\
&= \int \Phi_{\beta^n}(\Gamma_{1,1,\beta} H(n)) \, \mathrm{d}\Gamma_{1,1,\beta} H(n) \\
&= \Gamma_{1,1,\beta} \int \Phi_{\beta^{n-1}}(H(n)) \, \mathrm{d}H(n).
\end{aligned}$$

Thus we have completed the proof. ∎

Proposition 6.3.2 *Choose $\beta > 1$ and let $\epsilon = \delta^{-1}\beta$. Choose a control (which is always possible by our assumption) ω such that*

$$\left| X_{s,t}^i \right| \leqslant 2^{-1} \omega(s,t)^{i/p}$$

and

$$\left| \Gamma_{\epsilon,1,1} K(1)_{s,t}^i \right| \leqslant \left(\epsilon^p \omega(s,t) \right)^{i/p},$$

for all $(s,t) \in \triangle$ and $i = 1, \ldots, [p]$. Finally, choose $T_1 > 0$ such that we have $\epsilon^p \max_{[0,T_1]} \omega \leqslant 1$. Then, for all $n \geqslant 1$, we have

$$\left| \Gamma_{\epsilon,1,1} H(n)_{s,t}^i \right| \leqslant \left(\epsilon^p \omega(s,t) \right)^{i/p}, \tag{6.47}$$

for all $(s,t) \in \triangle$ and $i = 1, \ldots, [p]$.

Proof We use induction. By the choice of ω, (6.47) holds for $n = 1$. Suppose that it is true for n. Then, by (6.36), we have

$$\left| \int_s^t \Phi_{\beta^{n-1}}(\Gamma_{\epsilon,1,1}H(n)) \, d\Gamma_{\epsilon,1,1}H(n)^i \right| \leqslant \left(C_1 \epsilon^p \omega(s,t) \right)^{i/p},$$

for all $(s,t) \in \Delta_{T_1}$ and $i = 1, \ldots, [p]$, where C_1 depends only on M, p, γ, and an upper bound of $\epsilon\omega$ (which is one). Therefore, by (6.37),

$$\left| \Gamma_{\epsilon,\epsilon,\epsilon} \int_s^t \Phi_{\beta^{n-1}}(H(n)) \, dH(n)^i \right| \leqslant \left(C_1 \epsilon^p \omega(s,t) \right)^{i/p},$$

for all $(s,t) \in \Delta_{T_1}$ and $i = 1, \ldots, [p]$. However,

$$\pi_V \Gamma_{\epsilon,\epsilon,\epsilon} \int_s^t \Phi_{\beta^{n-1}}(H(n)) \, dH(n)^i = \epsilon^i X_{s,t}^i,$$

and therefore

$$\left| \pi_V \Gamma_{\epsilon,\epsilon,\epsilon} \int_s^t \Phi_{\beta^{n-1}}(H(n)) \, dH(n)^i \right| \leqslant 2^{-1} \left(\epsilon^p \omega(s,t) \right)^{i/p},$$

for all $(s,t) \in \Delta_{T_1}$ and $i = 1, \ldots, [p]$. Thus, by (6.41), we have

$$\left| \Gamma_{1,\delta_1,\delta_2} \circ \Gamma_{\epsilon,\epsilon,\epsilon} \int_s^t \Phi_{\beta^{n-1}}(H(n)) \, dH(n)^i \right| \leqslant \left(\epsilon^p \omega(s,t) \right)^{i/p},$$

for all $(s,t) \in \Delta_{T_1}$ and $i = 1, \ldots, [p]$, where $\delta_1 = \beta^{-1}\delta$ and $\delta_2 = \delta$. Hence

$$\left| \Gamma_{\epsilon,1,\delta_2\epsilon} \int_s^t \Phi_{\beta^{n-1}}(H(n)) \, dH(n)^i \right| \leqslant \left(\epsilon^p \omega(s,t) \right)^{i/p},$$

and therefore

$$\left| \Gamma_{\epsilon,1,\delta_2\epsilon} \circ \Gamma_{1,1,\beta^{-1}} \circ \Gamma_{1,1,\beta} \int_s^t \Phi_{\beta^{n-1}}(H(n)) \, dH(n)^i \right| \leqslant \left(\epsilon^p \omega(s,t) \right)^{i/p},$$

which gives us that

$$\left| \Gamma_{\epsilon,1,1} \circ \Gamma_{1,1,\beta} \int_s^t \Phi_{\beta^{n-1}}(H(n)) \, dH(n)^i \right| \leqslant \left(\epsilon^p \omega(s,t) \right)^{i/p}.$$

By (6.46), we then have

$$\left| \Gamma_{\epsilon,1,1}H(n+1)_{s,t}^i \right| \leqslant \left(\epsilon^p \omega(s,t) \right)^{i/p},$$

for all $(s,t) \in \Delta_{T_1}$ and $i = 1, \ldots, [p]$. The induction is completed. ∎

Corollary 6.3.1 *Under the above notation, there is a constant C_3, depending only on p and ϵ, such that*

$$\left|H(n)^i_{s,t}\right| \leqslant \left(C_3 \epsilon^p \omega(s,t)\right)^{i/p},$$

for all $(s,t) \in \triangle_{T_1}$ and $i = 1, \ldots, [p]$.

On the other hand, we know by definition that

$$K(n) = \Gamma_{1,1,\beta^{-(n-1)}} H(n),$$

and therefore

$$\left|\hat{Z}(n)^i_{s,t} - K(n)^i_{s,t}\right| \leqslant C_4 \beta^{-(n-1)} \left(C_3 \epsilon^p \omega(s,t)\right)^{i/p}, \tag{6.48}$$

for all $(s,t) \in \triangle_{T_1}$ and $i = 1, \ldots, [p]$, where the constant C_4 depends only on $[p]$, and $\hat{Z}(n) = (Z(n), 0)$. Equation (6.48) follows from Corollary 6.3.1 and the fact that $\pi_{V \oplus W} \hat{Z}(n) = \pi_{V \oplus W} K(n) = Z(n)$.

Finally, we consider the map $\phi : V \oplus W \oplus W \to V \oplus W$ defined by

$$\phi(x, y, d) = (x, y - d).$$

Then ϕ is smooth, and its first derivative

$$d\phi : V \oplus W \oplus W \to \boldsymbol{L}(V \oplus W \oplus W, V \oplus W)$$

is a smooth one-form on $V \oplus W \oplus W$. Define

$$\Psi(K) = \int d\phi(K) \, dK, \quad \forall K \in G\Omega_p(V \oplus W \oplus W),$$

which is the integration operator on $G\Omega_p(V \oplus W \oplus W)$. It is clear that

$$d\phi((x, y, d))(\xi, \eta, \zeta) = (\xi, \eta - \zeta), \quad \forall (x, y, d), (\xi, \eta, \zeta) \in V \oplus W \oplus W.$$

Therefore, the derivative $\alpha = d\phi$ is a constant-valued one-form. By definition, $\Psi(K)$ is the unique rough path associated with the almost rough path $\tilde{K}$, namely

$$\tilde{K}^i_{s,t} = \underbrace{\alpha(K_s) \otimes \cdots \otimes \alpha(K_s)}_{i}(K^i_{s,t})$$

$$= \underbrace{\alpha \otimes \cdots \otimes \alpha}_{i}(K^i_{s,t}),$$

so that

$$\Psi(K)^i_{s,t} = \underbrace{\alpha \otimes \cdots \otimes \alpha}_{i}(K^i_{s,t}),$$

for all $(s,t) \in \triangle$ and $i = 1, \ldots, [p]$. Hence we have the following continuity lemma.

Lemma 6.3.8 *Let $K, \hat{K} \in G\Omega_p(V \oplus W \oplus W)$, and let ω be a control such that*

$$\left| K^i_{s,t} - \hat{K}^i_{s,t} \right| \leqslant \varepsilon \omega(s,t)^{i/p} ,$$

for all $(s,t) \in \Delta$ and $i = 1, \ldots, [p]$. Then

$$\left| \Psi(K)^i_{s,t} - \Psi(\hat{K})^i_{s,t} \right| \leqslant C_5 \varepsilon \omega(s,t)^{i/p} ,$$

for all $(s,t) \in \Delta$ and $i = 1, \ldots, [p]$, where C_5 is a constant such that $|\alpha| \leqslant C_5$.

If X is a smooth rough path (and so also is $K(n)$ for any n), then, by (6.34),

$$\Psi(K(n)) = Z(n-1) ,$$
$$\Psi(\hat{Z}(n)) = Z(n) .$$

This remains true for any geometric rough path X by the continuity of Ψ. Therefore, by the above lemma, we have

$$\left| Z(n)^i_{s,t} - Z(n-1)^i_{s,t} \right| \leqslant C_5 C_4 \beta^{-(n-1)} \left(C_3 \varepsilon \omega(s,t) \right)^{i/p} ,$$

for all $(s,t) \in \Delta_1$ and $i = 1, \ldots, [p]$. However,

$$\sum_{n=1}^{\infty} \beta^{-(n-1)} < \infty ,$$

for any $\beta > 1$, and so we conclude that

$$\lim_{n \to \infty} Z(n)^i_{s,t} = Z^i_{s,t}$$

exists and $Z \in G\Omega_p(V \oplus W)$. Moreover, $Z(n)$ converges to Z in p-variation topology. In fact,

$$\left| Z(n)^i_{s,t} - Z^i_{s,t} \right| \leqslant C_5 C_4 \sum_{m=n}^{\infty} \beta^{-(m-1)} \left(C_3 \varepsilon \omega(s,t) \right)^{i/p} , \qquad (6.49)$$

for all $(s,t) \in \Delta_{T_1}$ and $i = 1, \ldots, [p]$.

Therefore, Z satisfies the integration eqn (6.31) and $\pi_V(Z) = X$ on $[0, T_1]$.

Continuity

It is obvious that the map which we defined above and which sends X to the solution Z (obtained via the Picard iteration) is continuous in p-variation topology. This follows from the uniform estimate (6.49) and the fact that, for any fixed n, $X \to Z(n)$ is continuous in p-variation topology.

Uniqueness

Let $\tilde{Z}$ be another solution on some interval $[0, T_2]$. That is, $\tilde{Z} \in G\Omega_p(V \oplus W)$ is a geometric rough path such that

$$\tilde{Z} = \int \hat{f}(\tilde{Z}) \, \mathrm{d}\tilde{Z} \quad \text{and} \quad \pi_V(\tilde{Z}) = X. \tag{6.50}$$

If X is a smooth rough path, then our eqn (6.50) can be written as

$$\begin{aligned} \mathrm{d}X &= \pi_V(\mathrm{d}\tilde{Z}), \\ \mathrm{d}Y &= f(Y)\pi_V(\mathrm{d}\tilde{Z}), \quad Y_0 = y_0. \end{aligned} \tag{6.51}$$

What we are going to show is that the difference $Y - \tilde{Y} = \tilde{D}$ (where $\tilde{Y} = \pi_V(\tilde{Z})$) is zero, and therefore, as above, we consider the following equation:

$$\begin{aligned} \mathrm{d}X &= \pi_V(\mathrm{d}\tilde{Z}), \\ \mathrm{d}Y &= f(Y)\pi_V(\mathrm{d}\tilde{Z}), \quad Y_0 = y_0, \\ \mathrm{d}\tilde{Z} &= \mathrm{d}\tilde{Z}, \\ \mathrm{d}\tilde{D} &= g(Y, Y - \tilde{D})(\tilde{D})\pi_V(\mathrm{d}\tilde{Z}), \quad \tilde{D}_0 = 0. \end{aligned} \tag{6.52}$$

Hence, let us define a one-form $\tilde{\Phi}_\beta : B \to \boldsymbol{L}(B, B)$, where $B = V \oplus W \oplus (V \oplus W) \oplus W$, and

$$\tilde{\Phi}_\beta((x, y, \tilde{z}, \tilde{d}))(\xi, \eta, \tilde{\eta}, \zeta) = \left(\pi_V(\tilde{\eta}), f(y_0 + y)\pi_V(\tilde{\eta}), \tilde{\eta}, g(y, y - \beta^{-1}\tilde{d})(\tilde{d})\pi_V(\tilde{\eta})\right),$$

for any $(x, y, \tilde{z}, \tilde{d}), (\xi, \eta, \tilde{\eta}, \zeta) \in B$. Set $\tilde{\Phi} = \tilde{\Phi}_1$. Again we consider the following iteration:

$$\tilde{K}(n+1) = \int \tilde{\Phi}(\tilde{K}(n)) \, \mathrm{d}\tilde{K}(n),$$

$$\tilde{K}(1) = \int \tilde{h}((0, 0, \tilde{Z}, 0)) \, \mathrm{d}(0, 0, \tilde{Z}, 0),$$

where $\tilde{h} : B \to \boldsymbol{L}(B, B)$ is a one-form defined by

$$\tilde{h}((x, y, \tilde{z}, \tilde{d}))(\xi, \eta, \tilde{\eta}, \zeta) = \left(\pi_V(\tilde{\eta}), f(y_0)\pi_V(\tilde{\eta}), \tilde{\eta}, f(y_0)\pi_V(\tilde{\eta}) - \pi_W(\tilde{\eta})\right).$$

Here $\tilde{h}$ comes from the observation that $\tilde{D}(1) = Y(1) - \pi_W(\tilde{Z})$, so that (if $\tilde{Z}$ is a smooth rough path) we have

$$\begin{aligned} \mathrm{d}X &= \pi_V(\mathrm{d}\tilde{Z}), \\ \mathrm{d}Y(1) &= f(y_0)\pi_V(\mathrm{d}\tilde{Z}), \\ \mathrm{d}\tilde{Z} &= \mathrm{d}\tilde{Z}, \\ \mathrm{d}\tilde{D}(1) &= f(y_0)\pi_V(\mathrm{d}\tilde{Z}) - \pi_W(\mathrm{d}\tilde{Z}). \end{aligned}$$

Next we define (for any fixed $\beta > 1$, e.g. $\beta = 2$)

$$\tilde{H}(n) = \Gamma_{1,1,1,\beta^{n-1}} \tilde{K}(n),$$

and, by the same argument as before, we have

$$\left|\tilde{H}(n)^i_{s,t}\right| \leqslant C_6 \omega(s,t)^{i/p}, \tag{6.53}$$

for all $(s,t) \in \triangle_{T_3}$ and $i = 1, \ldots, [p]$, for some constant C_6, and $T_3 \leqslant T_2$ depending only on M, p, γ and an upper bound of $\max \omega$ and β. In fact, this estimate follows from the following scaling property and a similar argument, as in the proof of the existence of solutions.

Lemma 6.3.9   *(i) For any* $\tilde{K} \in G\Omega_p(V \oplus W \oplus (V \oplus W) \oplus W)$ *and any* $\tilde{\beta}, \beta \neq 0$, *we have*

$$\Gamma_{\epsilon,\epsilon,\epsilon,\beta} \int \tilde{\Phi}_{\tilde{\beta}}(\tilde{K}) \, \mathrm{d}\tilde{K} = \int \tilde{\Phi}_{\tilde{\beta}\beta}(\Gamma_{1,1,\epsilon,\beta}\tilde{K}) \, \mathrm{d}\Gamma_{1,1,\epsilon,\beta}\tilde{K}.$$

(ii) For any $n \geqslant 1$, *we have*

$$\tilde{H}(n+1) = \Gamma_{1,1,1,\beta} \int \tilde{\Phi}_{\beta^{n-1}}(\tilde{H}(n)) \, \mathrm{d}\tilde{H}(n).$$

The only thing which we should notice is that the Lipschitz constant of $\tilde{\Phi}_{\beta^{n-1}}$ is bounded by the one of $\tilde{\Phi}$, for any $\beta \geqslant 1$.

Let $P(n), Q(n) \in G\Omega_p(V \oplus W \oplus (V \oplus W) \oplus W)$, where

$$P(n) = \pi_{V \oplus W \oplus (V \oplus W)}(\tilde{K}(n)), \quad Q(n) = (P(n), 0).$$

Then, if $\tilde{Z}$ is a smooth rough path,

$$P(n) = (X, Y(n), \tilde{Z})$$

and therefore we have $\pi_{V \oplus W}(P(n)) = Z(n)$. By (6.53), we have

$$\left|Q(n)^i_{s,t} - \tilde{K}(n)^i_{s,t}\right| \leqslant C_7 \beta^{-(n-1)} \omega(s,t)^{i/p},$$

for all $(s,t) \in \triangle_{T_3}$ and $i = 1, \ldots, [p]$, where the constant C_7 depends only on M, p, γ, and an upper bound of $\max \omega$. Finally, define a map $\varphi : B \to V \oplus W$ by

$$\varphi(x, y, \tilde{z}, \tilde{d}) = (x, y - \tilde{d}).$$

Then

$$\int \mathrm{d}\varphi(\tilde{K}(n)) \, \mathrm{d}\tilde{K}(n) = \tilde{Z}$$

and

$$\int \mathrm{d}\varphi(Q(n)) \, \mathrm{d}Q(n) = Z(n).$$

Therefore, by the fact that the derivative $\mathrm{d}\varphi$ is a constant one-form,

$$\left|Z(n)^i_{s,t} - \tilde{Z}^i_{s,t}\right| \leqslant C_8 \beta^{-(n-1)} \omega(s,t)^{i/p},$$

for all $(s,t) \in \triangle_{T_3}$ and $i = 1, \ldots, [p]$, where the constant C_8 depends only on M, p, γ, and an upper bound of $\max \omega$. Hence, $Z = \tilde{Z}$.

Solution defined on the whole interval

This also follows from the uniform estimates obtained above. Therefore, we have proved the following theorem.

Theorem 6.3.1 *Let $f : W \to L(V, W)$ be a Lip $(\gamma + 1)$ function with a Lipschitz constant M. Let $\gamma > p$, and let f, together with the tensor products, be admissible. Let $X \in G\Omega_p(V)$ be a geometric rough path. Then, for any $y_0 \in W$, there exists a unique $Z \in G\Omega_p(V \oplus W)$ such that*

$$Z = \int \hat{f}(Z) \, \mathrm{d}Z \quad and \quad \pi_V(Z) = X \,. \tag{6.54}$$

Moreover, there is a constant C which depends only on M, p, γ, and C_0, such that if $\omega \leqslant C_0$ is a control, if $X, \hat{X} \in G\Omega_p(V)$, and if

$$\left| X_{s,t}^i \right|, \left| \hat{X}_{s,t}^i \right| \leqslant \omega(s,t)^{i/p} \,,$$
$$\left| X_{s,t}^i - \hat{X}_{s,t}^i \right| \leqslant \varepsilon \omega(s,t)^{i/p} \,,$$

for all $(s,t) \in \Delta$ and $i = 1, \ldots, [p]$, then

$$\left| Z_{s,t}^i \right| \leqslant \left(C\omega(s,t) \right)^{i/p}$$

and

$$\left| Z_{s,t}^i - \hat{Z}_{s,t}^i \right| \leqslant A(\varepsilon)\omega(s,t)^{i/p} \,,$$

for all $(s,t) \in \Delta$ and $i = 1, \ldots, [p]$, where $A(\varepsilon)$ is a constant which depends only on M, p, γ, and C_0 such that $\lim_{\varepsilon \to 0} A(\varepsilon) = 0$.

In this case we denote by $F(y_0, X)$ the projection $\pi_W(Z)$, and we call the map

$$F : G\Omega_p(V) \to G\Omega_p(W) \,,$$

defined by $X \to F(y_0, X)$, the Itô map defined by the differential equation

$$\mathrm{d}Y_t = f(Y_t) \, \mathrm{d}X_t \,, \quad Y_0 = y_0 \,. \tag{6.55}$$

Corollary 6.3.2 (Under the same assumptions as in Theorem 6.3.1) *For any $X \in G\Omega_p(V)$, there exists a unique solution to (6.55), and the Itô map $X \to F(y, X)$ is continuous in p-variation topology.*

6.4 Comments and notes on Chapter 6

Regarding the history of the theory of stochastic differential equations, the reader should refer to Itô and McKean (1965), Gihman and Skorohod (1972), Ikeda and Watanabe (1981), Elworthy (1982), and Malliavin (1997).

Wong and Zakai (1965a, 1965b) first proved a version of the limit theorem for solutions of Stratonovich-type SDEs, and this result was later improved by imposing a weaker condition in Ikeda and Watanabe (1981). Indeed, Ikeda and

Watanabe (1981) have observed the necessity of the Lévy area processes of Brownian motion. Stroock and Varadhan (1972) proved their famous support theorem for solutions of SDEs driven by Brownian motion using the Wong–Zakai limit theorem. For further details about the history of the limit theorem, see Malliavin (1997). Recently there have been many works extending the Wong–Zakai-type limit theorem to other stochastic processes, see, for example, Stroock and Zheng (1998), etc.

Sections 6.2 and 6.3. The present version of the universal limit theorem for $2 \leqslant p < 3$ is an improvement of that presented in Lyons (1998) and Lyons and Qian (1998) in that it applies to all rough paths, not only the geometric ones.

Section 6.4. All of the results and the ideas of their proofs are taken from Lyons (1998), with some modifications.

Recently, A. M. Davie (private communication) has made some significant progress in improving the universal limit theorem. For example, he has proved that the universal limit theorem remains true for the geometric rough paths associated with Brownian motion if the vector field is C^2, which is the best that one could expect. On the other hand, he has showed that the uniqueness is not true if we replace the Lip (r) condition by Lip $(r - 1)$. For details, see the forthcoming papers by A. M. Davie.

7

VECTOR FIELDS AND FLOW EQUATIONS

In his seminal papers Malliavin (1978a, 1978b), P. Malliavin initiated a completely new approach to Wiener functionals (functions of Brownian motion, typical examples of which are solutions to stochastic differential equations driven by Brownian motion). The main observation in his study is that most functionals in which we are interested are smooth along the directions in the Cameron–Martin space (and therefore the Cameron–Martin space can be regarded as the tangent space of the Wiener space), although in the usual sense those Wiener functionals are only measurable. Therefore, classical analysis tools (mainly functional analysis) are brought into the further study of (non)linear Wiener functionals. Since then Malliavin calculus has become one of the most active research areas in both probability and analysis. The reader may find a very exciting account in the treatise by P. Malliavin (1997). In this chapter we will exhibit a wide class of tangent directions along which it is possible to translate a path and differentiate functionals. This class is much bigger than the Cameron–Martin space usually identified as the effective tangent space. It leads to the conclusion that Itô's maps are smoother than those in the Malliavin sense. The remainder of this chapter is about Itô's vector fields. These vector fields have appeared in the recent study of analysis on Wiener spaces over curved manifolds. The main result is that there is a unique solution flow to any Itô vector field. We end this chapter with an appendix which contains a discussion of Driver's flow equation. It is Driver's geometric flow equation which motivates the introduction of Itô's vector fields and flow equations on rough path spaces.

In this chapter, all Banach spaces are assumed to be finite-dimensional, in order to avoid specifying the tensor norms involved.

7.1 Smoothness of Itô maps

In this section we study Itô maps as functionals on the space $\Omega_p(V)$ of rough paths. We not only consider the classical variations of a path, but also variations involving paths of higher degree. For simplicity, we only consider rough paths with roughness $2 \leqslant p < 3$.

Let $q_1, q_2 \geqslant 1$ be two constants such that

$$\frac{1}{p} + \frac{1}{q_1} > 1, \quad q_2 \leqslant \frac{p}{2} \tag{7.1}$$

and let

$$T^{(q_1, q_2)}\Omega_p(V) = C_{0,q_1}(\triangle, V) \oplus C_{0,q_2}(\triangle, V^{\otimes 2}).$$

We will see that $T^{(q_1, q_2)}\Omega_p(V)$ serves as the tangent space of $\Omega_p(V)$.

If $h \in C_{0,q_1}(\triangle, V)$, we will identify $h = (1, h^1)$ with the path $h_t = h^1_{0,t}$. The same notation applies to $\varphi \in C_{0,q_2}(\triangle, V^{\otimes 2})$. Given $(h, \varphi) \in T^{(q_1,q_2)}\Omega_p(V)$, $X \in \Omega_p(V)$, we use $\theta_{(h,\varphi)}X$ to denote the rough path associated with the following almost rough path:

$$(1, X^1_{s,t} + h_{s,t}, X^2_{s,t} + \varphi_{s,t}). \tag{7.2}$$

The following lemma is obvious.

Lemma 7.1.1   (i) *If* $(h, \varphi) \in T^{(q_1,q_2)}\Omega_p(V)$, $X \in \Omega_p(V)$, *then*

$$\theta_{(h,\varphi)}X = \theta_{(0,\varphi)} \circ \theta_{(h,0)}X.$$

(ii) *The map* $(X, h, \varphi) \to \theta_{(h,\varphi)}X$ *is a continuous function from* $\Omega_p(V) \times C_{0,q_1}(\triangle, V) \times C_{0,q_2}(\triangle, V^{\otimes 2})$ *to* $\Omega_p(V)$ *in variation topology.*

(iii) *If* $X \in \Omega_p(V)$ *is a geometric rough path, then so is* $\theta_h X \equiv \theta_{(h,0)}X$ *for any* $h \in C_{0,q_1}(\triangle, V)$.

Suppose α is a $\mathrm{Lip}\,(\gamma - 1)$ one-form, where $\gamma > p$, then Y is an almost rough path, where

$$Y^1_{s,t} = \alpha(X_s + h_s)(X^1_{s,t} + h^1_{s,t}) + d\alpha(X_s + h_s)(X^2_{s,t} + \varphi_{s,t}),$$
$$Y^2_{s,t} = \alpha(X_s + h_s) \otimes \alpha(X_s + h_s)(X^2_{s,t} + \varphi_{s,t}),$$

and its associated rough path is $\int \alpha(\theta_{(h,\varphi)}X)\,d\theta_{(h,\varphi)}X$. In particular,

$$\int_s^t \alpha(\theta_{(h,\varphi)}X)\,d\theta_{(h,\varphi)}X^1 = \int_s^t \alpha(\theta_h X)\,d\theta_h X^1 + \int_s^t d\alpha(\theta_h X_u)\,d\varphi_u,$$

$$\int_s^t \alpha(\theta_{(h,\varphi)}X)\,d\theta_{(h,\varphi)}X^2 = \int_s^t \alpha(\theta_h X)\,d\theta_h X^2$$
$$+ \int_s^t \alpha(\theta_h X_u) \otimes \alpha(\theta_h X_u)(d\varphi_u).$$

Therefore,

$$\lim_{\varepsilon \to 0} \frac{\int_s^t \alpha(\theta_{\varepsilon(h,\varphi)}X)\,d\theta_{\varepsilon(h,\varphi)}X^1 - \int_s^t \alpha(X)\,dX^1}{\varepsilon} = \int_s^t \alpha(X_u)\,dh_u$$
$$+ \int_s^t d\alpha(X_u)\,d\varphi_u.$$

That is, the integral $\int_s^t \alpha(X)\,dX^1$ is differentiable along any direction $(h, \varphi) \in T^{(q_1,q_2)}\Omega_p(V)$.

Definition 7.1.1 *Let B be a Banach space. A function $G : \Omega_p(V) \to B$ is differentiable at $X \in \Omega_p(V)$ along a tangent vector $(h, \varphi) \in T^{(q_1,q_2)}\Omega_p(V)$ if*

$$\lim_{\varepsilon \to 0} \frac{G(\theta_{\varepsilon(h,\varphi)}X) - G(X)}{\varepsilon}$$

exists. In this case, we use $D_{(h,\varphi)}G(X)$ to denote this limit. G is (q_1,q_2)-differentiable at X if $D_{(h,\varphi)}G(X)$ exists for all $(h, \varphi) \in T^{(q_1,q_2)}\Omega_p(V)$.

The main aim of this section is to show that the Itô map $F(y, X)$, defined by the differential equation

$$dY_t = f(Y_t)\,dX_t, \quad Y_0 = y, \tag{7.3}$$

is (q_1, q_2)-differentiable for any geometric rough path X.

If $X \in \Omega_p(V)$ and if $(h, \varphi) \in T^{(q_1,q_2)}\Omega_p(V)$, then $H \in C_{0,p}(\Delta, T^{(2)}(V \oplus V \oplus V^{\otimes 2}))$ is an almost rough path, where

$$H^1_{s,t} = (X^1_{s,t}, h_{s,t}, \varphi_{s,t}),$$
$$H^2_{s,t} = (X^2_{s,t}, 0, \ldots, 0).$$

The associated rough path of H is denoted by $[X, h, \varphi]$. It is obvious that

$$[X, h, \varphi]^1_{s,t} = (X^1_{s,t}, h_{s,t}, \varphi_{s,t}).$$

By definition, if X is geometric, then so is $[X, h, \varphi]$, and the map $(X, h, \varphi) \to [X, h, \varphi]$ is continuous in variation topology. If we define $\beta : V \oplus V \oplus V^{\otimes 2} \to V$ by

$$\beta(x, y, z) = x + y,$$

then

$$\theta_{(h,\varphi)}X^1_{s,t} = \int_s^t d\beta([X, h, \varphi])\,d[X, h, \varphi]^1$$

$$= \int_s^t d\beta([X, h, 0])\,d[X, h, 0]^1,$$

$$\theta_{(h,\varphi)}X^2_{s,t} = \int_s^t d\beta([X, h, \varphi])\,d[X, h, \varphi]^2 + \varphi_{s,t}$$

$$= \int_s^t d\beta([X, h, 0])\,d[X, h, 0]^2 + \varphi_{s,t}.$$

In the following, $F(y, X)$ denotes the Itô map defined by (7.3), $X \in G\Omega_p(V)$ is a geometric rough path, and $(h, \varphi) \in T^{(q_1,q_2)}\Omega_p(V)$. The first question is what is $F(\cdot, \theta_{(h,\varphi)}X)$? The following is the answer.

Theorem 7.1.1 *Let $f : W \to \mathbb{L}(V, W)$ be a C^4 vector field. Let $F_{\tilde{f}}(y, \tilde{X})$ be the Itô map defined by differential equation*

$$dY_t = \tilde{f}(Y_t)\,d\tilde{X}_t, \quad Y_0 = y, \tag{7.4}$$

where $\tilde{f} : W \to \mathbb{L}(V \oplus V \oplus V^{\otimes 2}, W)$,

$$\tilde{f}(y)(\xi, \eta, \zeta) = f(y)\xi + f(y)\eta + f^1(y) \circ (f(y) \otimes 1)(\zeta),$$

for all $y \in W$, $(\xi, \eta, \zeta) \in V \oplus V \oplus V^{\otimes 2}$. Then, for any $X \in G\Omega_p(V)$,

$$F(\cdot, \theta_{(h,\varphi)} X)^1 = F_{\tilde{f}}(\cdot, [X, h, \varphi])^1 \tag{7.5}$$

and $F(\cdot, \theta_{(h,\varphi)} X)^2_{s,t}$ equals

$$F_{\tilde{f}}(\cdot, [X, h, \varphi])^2_{s,t} + \int_s^t f\big(F_{\tilde{f}}(\cdot, [X, h, \varphi])_u\big) \otimes f\big(F_{\tilde{f}}(\cdot, [X, h, \varphi])_u\big)(\mathrm{d}\varphi_u). \tag{7.6}$$

Notice that if X is a smooth rough path, then $F_{\tilde{f}}(\cdot, [X, h, \varphi])^1$ is the unique solution to the following differential equation:

$$\mathrm{d}Y_t = f(Y_t)\,\mathrm{d}X_t + f(Y_t)\,\mathrm{d}h_t + f^1(Y_t) \circ (f(Y_t) \otimes 1)(\mathrm{d}\varphi_t),$$

and $F_{\tilde{f}}(\cdot, [X, h, \varphi])^2$ is the iterated path integral of $F_{\tilde{f}}(\cdot, [X, h, \varphi])^1$ since $[X, h, \varphi]$ is also a smooth rough path. Therefore, $\varepsilon \to F_{\tilde{f}}(\cdot, [X, \varepsilon h, \varepsilon \varphi])^1_{s,t}$ is differentiable and its derivative (at $\varepsilon = 0$) P satisfies the following differential equation:

$$\mathrm{d}P_t = \mathrm{d}f(Y_t)(P_t)\,\mathrm{d}X_t + f(Y_t)\,\mathrm{d}h_t + f^1(Y_t) \circ (f(Y_t) \otimes 1)(\mathrm{d}\varphi_t), \quad P_0 = 0.$$

Thus, by the continuity of Itô maps and the fact that the differential $\mathrm{d}/\mathrm{d}\varepsilon$ is a closed operator, we have the following corollary.

Corollary 7.1.1 (Under the same condition as in Theorem 7.1.1) $X \to F(y, X)^1_{s,t}$ is (q_1, q_2)-differentiable for all $(s,t) \in \triangle$. Let $\alpha : W \oplus W \to \mathbb{L}(V \oplus V \oplus V^{\otimes 2}, W \oplus W)$, where

$$\alpha(y, z)(\xi, \eta, \zeta) = \big(\tilde{f}(y)(\xi, \eta, \zeta), \mathrm{d}f(y)(z)\xi + f(y)\eta + f^1(y) \circ (f(y) \otimes 1)(\zeta)\big).$$

Let $F_\alpha(\cdot, \hat{X})$ be the Itô map defined by the differential equation

$$\mathrm{d}Y_t = \alpha(Y_t)\,\mathrm{d}\hat{X}_t.$$

Then,

$$D_{(h,\varphi)} F(y, X)^1_{s,t} = \pi_2\left(F_\alpha((y, 0), [X, h, \varphi])^1_{s,t}\right).$$

Proof First we suppose $X \in G\Omega_p(V)$ is a smooth rough path. The unique solution Z of the integral equation

$$Z = \int \hat{f}(Z)\,\mathrm{d}Z, \quad \pi_V(Z) = \theta_{(h,\varphi)} X$$

can be obtained by the following iteration:

$$Z(n) = \int \hat{f}(Z(n-1))\,\mathrm{d}Z(n-1),$$

$$Z(0) = (\theta_{(h,\varphi)} X, 0).$$

Recall that $K(n)$ is the almost rough path which defines $Z(n)$, where

$$K(n)^1_{s,t} = \hat{f}(Z(n-1)_s)(Z(n-1)^1_{s,t}) + \hat{f}^1(Z(n-1)_s)(Z(n-1)^2_{s,t}),$$

$$K(n)^2_{s,t} = \hat{f}(Z(n-1)_s) \otimes \hat{f}(Z(n-1)_s)(Z(n-1)^2_{s,t})$$

$$= \hat{f}(Z(n-1)_s) \otimes \hat{f}(Z(n-1)_s)(\theta_{(h,\varphi)} X^2_{s,t}, 0, 0, 0).$$

Thus, if we denote $K(n)_{s,t}^1 = (\theta_{(h,\varphi)}X_{s,t}^1, Y(n)_{s,t}^1)$, then

$$
\begin{aligned}
K(n)_{s,t}^1 &= \hat{f}(Z(n-1)_s)(Z(n-1)_{s,t}^1) + \hat{f}^1(Z(n-1)_s)(Z(n-1)_{s,t}^2) \\
&= \left(\theta_{(h,\varphi)}X_{s,t}^1, f(Y(n-1)_s)(\theta_{(h,\varphi)}X_{s,t}^1)\right) \\
&\qquad + \hat{f}^1(Z(n-1)_s)(Z(n-1)_{s,t}^2) \\
&\equiv \left(\theta_{(h,\varphi)}X_{s,t}^1, f(Y(n-1)_s)(X_{s,t}^1 + h_{s,t}^1)\right) \\
&\qquad + \hat{f}^1(Z(n-1)_s)(K(n-1)_{s,t}^2) \\
&\equiv \left(\theta_{(h,\varphi)}X_{s,t}^1, f(Y(n-1)_s)(X_{s,t}^1 + h_{s,t}^1)\right) \\
&\qquad + \left(0, f^1(Y(n-1)_s) \circ \left(f(Y(n-1)_s) \otimes 1\right)\right)(\theta_{(h,\varphi)}X_{s,t}^2) \\
&\equiv \left(\theta_{(h,\varphi)}X_{s,t}^1, f(Y(n-1)_s)(X_{s,t}^1 + h_{s,t}^1)\right) \\
&\qquad + \left(0, f^1(Y(n-1)_s) \circ \left(f(Y(n-1)_s) \otimes 1\right)\right)(\varphi_{s,t}),
\end{aligned}
$$

where we use '$\equiv$' to mean the difference of the two sides is controlled by $\omega(s,t)^\theta$ for some constant $\theta > 1$ and some control ω, for all $(s,t) \in \Delta$. Hence, if we set

$$
\begin{aligned}
\tilde{Y}(n)_{s,t} &= f(Y(n-1)_s)(X_{s,t}^1 + h_{s,t}^1) \\
&\qquad + f^1(Y(n-1)_s) \circ \left(f(Y(n-1)_s) \otimes 1\right)(\varphi_{s,t}),
\end{aligned}
$$

then

$$
Y(n)_{s,t}^1 = \lim_{m(D)\to 0} \sum_l \tilde{Y}(n)_{t_{l-1},t_l}.
$$

That is, if X is a smooth rough path, then $F(\cdot, \theta_{(h,\varphi)}X)_{s,t}^1 = \lim_{n\to\infty} Y(n)_{s,t}^1$, where $Y(n)^1$ is the following iteration:

$$
\begin{aligned}
\tilde{Y}(n)_{s,t} &= f(Y(n-1)_s)(X_{s,t}^1 + h_{s,t}^1) \\
&\qquad + f^1(Y(n-1)_s) \circ \left(f(Y(n-1)_s) \otimes 1\right)(\varphi_{s,t}), \\
Y(n)_{s,t}^1 &= \lim_{m(D)\to 0} \sum_l \tilde{Y}(n)_{t_{l-1},t_l}.
\end{aligned}
$$

Notice that the above system is exactly the Picard iteration of the following ordinary differential equation:

$$
dY_t = f(Y_t)\, dX_t + f(Y_t)\, dh_t + f^1(Y_t) \circ (f(Y_t) \otimes 1)(d\varphi_t).
$$

Therefore, we have (7.5) in the case when X is a smooth rough path. By the continuity of Itô maps, we may conclude (7.5) holds for any geometric rough path X. The proof of (7.6) is similar. If X is a smooth rough path, then we have

$$
\begin{aligned}
K(n)_{s,t}^2 &= \hat{f}(Z(n-1)_s) \otimes \hat{f}(Z(n-1)_s)(\theta_{(h,\varphi)}X_{s,t}^2, 0, 0, 0) \\
&\equiv \hat{f}(Z(n-1)_s) \otimes \hat{f}(Z(n-1)_s)(\varphi_{s,t}, 0, 0, 0),
\end{aligned}
$$

so that

$$Y(n)^2_{s,t} = \lim_{m(D)\to 0} \sum_l \hat{K}(n)^2_{t_{l-1},t_l} + \sum_l Y(n)^1_{s,t_{l-1}} \otimes Y(n)^1_{t_{l-1},t_l},$$

where

$$\hat{K}(n)^2_{s,t} = f(Y(n-1)_s) \otimes f(Y(n-1)_s)(\varphi_{s,t}).$$

However, by (7.5) and the fact that $[X, h, \varphi]$ is a smooth rough path, we know that

$$\lim_{m(D)\to 0} \sum_l Y(n)^1_{s,t_{l-1}} \otimes Y(n)^1_{t_{l-1},t_l} \to F_{\tilde{f}}(\cdot, [X, h, \varphi])^2_{s,t}, \quad \text{as} \quad n \to \infty,$$

and moreover,

$$\lim_{m(D)\to 0} \sum_l \hat{K}(n)^2_{t_{l-1},t_l} = \int_s^t f(Y(n-1)_u) \otimes f(Y(n-1)_u)(\mathrm{d}\varphi_u)$$

which goes to $\int_s^t f(Y_u) \otimes f(Y_u)(\mathrm{d}\varphi_u)$ as $n \to \infty$ since $Y(n)^1$ tends to Y^1 uniformly. Then, by the continuity of Itô maps again, we therefore conclude that (7.6) remains true for any geometric rough path. ∎

7.2 Itô's vector fields

In this section, we introduce a class of vector fields along which we can differentiate Itô functionals. In the next section we will show that there exists a unique solution flow on the space of geometric rough paths for such a vector field. In order to illustrate the main idea, in this section we will consider the space $\Omega_1(V)$ of Lipschitz paths.

Let us first explain the meaning of a vector field on a path space. Consider the space $\Omega(M)$ of all continuous and piecewise-smooth paths starting at a fixed point o in a smooth manifold M. Then the tangent space at a path $X \in \Omega(M)$ is the vector space of all continuous piecewise-smooth vector fields A along the path X such that $A_0 = 0$ (as we fix the initial point), that is $A_t \in T_{X_t}M$ and $A_0 = 0$ (T_qM is the tangent space at $q \in M$). In particular, if $M = V$ is a vector space, then we may identify T_qV with V itself, and therefore a tangent vector along the path X is again a path in V with initial point zero. Therefore, a vector field A on the path space $\Omega(V)$ can be regarded as a map (with the corresponding regular property) from the path space $\Omega(V)$ to itself.

When a smooth manifold is endowed with an affine connection, then we may pull back a vector field into a vector field on the path space over a vector space via an Itô map (see the appendix to this chapter). Moreover, the pull-back itself can be obtained by solving some differential equations. This last point leads us to the following consideration.

Let $(h, \varphi) \in T^{(q_1, q_2)}\Omega_p(V)$. Then $[X, h, \varphi] \in \Omega_p(V \oplus V \oplus V^{\otimes 2})$ for any $X \in \Omega_p(V)$. Let $A : V \to \mathbb{L}(V \oplus V \oplus V^{\otimes 2}, V)$ be a smooth vector field. Let G be the Itô map defined by the differential equation

$$dY_t = A(Y_t) \, d\hat{X}_t, \quad Y_0 = y, \tag{7.7}$$

and define $M : \Omega_p(V) \to \Omega_p(V)$ by

$$M(X) = G(0, [X, h, \varphi]), \quad \forall X \in \Omega_p(V). \tag{7.8}$$

Such a map M is called an Itô vector field on $\Omega_p(V)$.

In this section we only consider the case where $q_1 = q_2 = p = 1$, and M, defined by (7.8), is regarded as a vector field on the space $\Omega_1(V)$ of Lipschitz paths in V. We will leave the general case to the next section. We aim to solve the following flow equation:

$$\frac{d\Phi^t}{dt} = M(\Phi^t), \quad \Phi^0 = X, \tag{7.9}$$

for any initial value $X \in \Omega_1(V)$. The essential property which we will use is described in the following definition.

Definition 7.2.1 *A map* $M : \Omega_1(V) \to \Omega_1(V)$ *is called a Lipschitz vector field on* $\Omega_1(V)$ *(in variation topology) if, for any control* ω_1*, there exists a control* $\omega_2 \geqslant \omega_1$ *and a constant* C *dependent only on* T *and* $\max \omega_1$*, such that if* $X, \hat{X} \in \Omega_1(V)$ *satisfy*

$$\left| X_{s_1, s_2} \right|, \left| \hat{X}_{s_1, s_2} \right| \leqslant \omega_2(s_1, s_2), \qquad \forall (s_1, s_2) \in \Delta,$$
$$\left| X_{s_1, s_2} - \hat{X}_{s_1, s_2} \right| \leqslant \varepsilon \omega_2(s_1, s_2), \qquad \forall (s_1, s_2) \in \Delta,$$

then

$$\left| M(X)_{s_1, s_2} \right| \leqslant C\omega_2(s_1, s_2), \qquad \forall (s_1, s_2) \in \Delta,$$
$$\left| M(X)_{s_1, s_2} - M(\hat{X})_{s_1, s_2} \right| \leqslant C\varepsilon \omega_2(s_1, s_2), \qquad \forall (s_1, s_2) \in \Delta.$$

An Itô vector field defined via a differential equation is a Lipschitz vector field on the space of Lipschitz paths.

In the following we assume that M is a Lipschitz vector field on $\Omega_1(V)$, and let $X \in \Omega_1(V)$. Let us solve the following initial value problem:

$$\frac{d\Phi^t}{dt} = M(\Phi^t), \quad \Phi^0 = X. \tag{7.10}$$

By definition, for any $X \in \Omega_1(V)$, we may choose a control ω and a constant C such that the following conditions hold.

- $|X_{s_1, s_2}|, |M(X)_{s_1, s_2}| \leqslant \frac{1}{2}\omega(s_1, s_2)$, for all $(s_1, s_2) \in \Delta$.

- If $Y, \hat{Y} \in \Omega_1(V)$ and if

$$\left|Y_{s_1,s_2}\right|, \left|\hat{Y}_{s_1,s_2}\right| \leqslant \omega(s_1,s_2), \qquad \forall (s_1,s_2) \in \Delta,$$

$$\left|Y_{s_1,s_2} - \hat{Y}_{s_1,s_2}\right| \leqslant \varepsilon\omega(s_1,s_2), \qquad \forall (s_1,s_2) \in \Delta,$$

then

$$\left|M(Y)_{s_1,s_2}\right| \leqslant C\omega(s_1,s_2), \qquad \forall (s_1,s_2) \in \Delta,$$

and

$$\left|M(Y)_{s_1,s_2} - M(\hat{Y})_{s_1,s_2}\right| \leqslant C\varepsilon\omega(s_1,s_2), \qquad \forall (s_1,s_2) \in \Delta.$$

We use the following iteration:

$$\Phi^t(n) = X + \int_0^t M(\Phi^u(n-1))\,du,$$

$$\Phi^t(0) = X. \tag{7.11}$$

Lemma 7.2.1 *Let $T_1 = (1/2C) \wedge T$. Then, for all $|t|, |s| \leqslant T_1$,*

$$\left|\Phi^t(n)_{s_1,s_2}\right| \leqslant \omega(s_1,s_2),$$

$$\left|\Phi^t(n)_{s_1,s_2} - \Phi^s(n)_{s_1,s_2}\right| \leqslant C|t-s|\omega(s_1,s_2),$$

and

$$\left|\Phi^t(n)_{s_1,s_2} - \Phi^t(n-1)_{s_1,s_2}\right| \leqslant \frac{(C|t|)^n}{n!}\omega(s_1,s_2), \tag{7.12}$$

for all $(s_1,s_2) \in \Delta$.

Therefore, $\Phi^t(n)$ tends to a limit $\Phi^t \in \Omega_1(V)$ in variation topology, which solves the flow eqn (7.10). Moreover, by (7.12), we have, for all $|t| \leqslant T_1$,

$$\left|\Phi^t(n)_{s_1,s_2} - \Phi^t_{s_1,s_2}\right| \leqslant \sum_{k=n}^{\infty} \frac{(C|t|)^k}{k!}\omega(s_1,s_2), \qquad \forall (s_1,s_2) \in \Delta. \tag{7.13}$$

Lemma 7.2.1 follows from a very simple induction argument. Let us use $\Phi^t(X)$ to denote the solution to the initial problem (7.10).

Next suppose $X, \hat{X} \in \Omega_1(V)$ are two Lipschitz paths, and choose a control ω such that the above conditions are satisfied for both X and $\hat{X}$. Suppose that

$$\left|X_{s_1,s_2} - \hat{X}_{s_1,s_2}\right| \leqslant \frac{1}{2}\varepsilon\omega(s_1,s_2), \qquad \forall (s_1,s_2) \in \Delta.$$

Then by a similar argument we have, for all $|t| \leqslant T_1$,

$$\left|\Phi^t(X)_{s_1,s_2} - \Phi^t(\hat{X})_{s_1,s_2}\right| \leqslant \varepsilon\omega(s_1,s_2), \qquad \forall (s_1,s_2) \in \Delta.$$

Notice that all of the above estimates are uniform, so that we may extend the solution to the whole interval. Therefore, we have the following theorem.

Theorem 7.2.1 *There exists a unique family $\{\Phi^t(X)\}$ for any Lipschitz path $X \in \Omega_1(V)$ such that*

(i) The map $t \to \Phi^t(X)$ is Lipschitz continuous in variation topology, namely

$$\left|\Phi^t(X)_{s_1,s_2} - \Phi^s(X)_{s_1,s_2}\right| \leqslant C_1|t - s|\omega(s_1,s_2), \quad \forall(s_1,s_2) \in \Delta.$$

(ii) There is a constant C_2, dependent only on an upper bound of ω, which controls the variation of X such that

$$\left|\Phi^t(X)_{s_1,s_2}\right| \leqslant C_2\omega(s_1,s_2), \quad \forall(s_1,s_2) \in \Delta.$$

(iii) For all $|t| \leqslant T$,

$$\Phi^t(X)_{s_1,s_2} = X_{s_1,s_2} + \int_0^t M\left(\Phi^u(X)\right)_{s_1,s_2} du, \quad \forall(s_1,s_2) \in \Delta.$$

Next, we consider the application of the Euler iteration to the flow eqn (7.10) which gives a finite-dimensional algorithm for the unique solution of this flow equation. It is this method which we will adopt to solve the flow eqn (7.10) on $G\Omega_p(V)$ for $p \geqslant 2$.

Let us continue to use the same notation and the same assumptions as above. The Euler iteration is defined as follows:

$$\Psi^t(0) = X,$$

$$\Psi^t(n) = X + \frac{t}{n}\sum_{l=1}^n M\left(\Psi(n-1)^{((l-1)/n)t}\right).$$

First, by an induction argument we have, for all n and all $|t| \leqslant T_1$,

$$\left|\Psi^t(n)_{s_1,s_2}\right| \leqslant \omega(s_1,s_2), \quad \forall(s_1,s_2) \in \Delta. \tag{7.14}$$

Lemma 7.2.2 *For all n and $|s|, |t| \leqslant T_1$, we have*

$$\left|\Psi^t(n)_{s_1,s_2} - \Psi^s(n)_{s_1,s_2}\right| \leqslant C|t - s|\omega(s_1,s_2), \tag{7.15}$$

for all $(s_1,s_2) \in \Delta$, where C is a constant which is dependent on T_1 and C_1.

Proof The above is obvious for $n = 0, 1$. Notice that

$$\Psi^t(n)_{s_1,s_2} - \Psi^s(n)_{s_1,s_2}$$

$$= \frac{t}{n}\sum_{l=1}^n M\left(\Psi^{((l-1)/n)t}(n-1)\right)_{s_1,s_2} - \frac{s}{n}\sum_{l=1}^n M\left(\Psi^{((l-1)/n)s}(n-1)\right)_{s_1,s_2}$$

$$= \frac{t-s}{n}\sum_{l=1}^n M\left(\Psi^{((l-1)/n)t}(n-1)\right)_{s_1,s_2}$$

$$+ \frac{s}{n}\sum_{l=1}^n \left[M\left(\Psi^{((l-1)/n)t}(n-1)\right)_{s_1,s_2} - M\left(\Psi^{((l-1)/n)s}(n-1)\right)_{s_1,s_2}\right].$$

Therefore, by (7.15) and the induction assumption, we have

$$\left| \Psi^t(n)_{s_1,s_2} - \Psi^s(n)_{s_1,s_2} \right|$$
$$\leqslant C_1 |t - s| \omega(s_1, s_2)$$
$$+ \frac{|s|}{n} \sum_{l=1}^{n} \left| M\big(\Psi^{((l-1)/n)t}(n-1)\big)_{s_1,s_2} - M\big(\Psi^{((l-1)/n)s}(n-1)\big)_{s_1,s_2} \right|$$

$$\leqslant C_1 |t - s| \omega(s_1, s_2) + \frac{|s|}{n} \sum_{l=1}^{n} 2C_1 \frac{l-1}{n} |t - s| \omega(s_1, s_2)$$

$$\leqslant C |t - s| \omega(s_1, s_2).$$

This proves the estimate. ∎

Lemma 7.2.3 *For all* $|t| \leqslant T_1$*, we have*

$$\left| \Phi^t(n)_{s_1,s_2} - \Psi^t(n)_{s_1,s_2} \right| \leqslant \frac{C}{n} t^2 \omega(s_1, s_2), \quad \forall (s_1, s_2) \in \Delta. \tag{7.16}$$

Therefore,

$$\left| \Phi^t(X)_{s_1,s_2} - \Psi^t(n)_{s_1,s_2} \right| \leqslant \left[\sum_{k=n}^{\infty} \frac{(C|t|)^k}{k!} + \frac{C}{n} t^2 \right] \omega(s_1, s_2), \tag{7.17}$$

for all $(s_1, s_2) \in \Delta$*, for some constant* C *which is dependent on* C_1 *and* T_1*.*

Proof We will use induction to prove (7.16). It is trivial for $n = 0, 1$ as

$$\Phi^t(0) = \Psi^t(0), \quad \Phi^t(1) = \Psi^t(1). \tag{7.18}$$

Now

$$\Phi^t(n)_{s_1,s_2} - \Psi^t(n)_{s_1,s_2}$$
$$= \sum_{l=1}^{n} \int_{((l-1)/n)t}^{(l/n)t} \left(M\left(\Phi^u(n-1)\right)_{s_1,s_2} - M\big(\Psi^{((l-1)/n)t}(n-1)\big)_{s_1,s_2} \right) du$$
$$= \sum_{l=1}^{n} \int_{((l-1)/n)t}^{(l/n)t} \left(M\left(\Phi^u(n-1)\right)_{s_1,s_2} - M\left(\Psi^u(n-1)\right)_{s_1,s_2} \right) du$$
$$+ \sum_{l=1}^{n} \int_{((l-1)/n)t}^{(l/n)t} \left(M\left(\Psi^u(n-1)\right)_{s_1,s_2} - M\big(\Psi^{((l-1)/n)t}(n-1)\big)_{s_1,s_2} \right) du,$$

so that (we may assume $t > 0$) by the induction assumption

$$\left| \Phi^t(n)_{s_1,s_2} - \Psi^t(n)_{s_1,s_2} \right| \leqslant C\omega(s_1, s_2) \sum_{l=1}^{n} \int_{((l-1)/n)t}^{(l/n)t} \frac{2C_1}{n-1} u^2 \, du$$
$$+ C\omega(s_1, s_2) \sum_{l=1}^{n} \int_{((l-1)/n)t}^{(l/n)t} \left| u - \frac{l-1}{n} t \right| du$$

$$\leqslant \frac{C}{n} t^2 \omega(s_1, s_2) \,.$$

Thus we have completed the proof by induction. ∎

Notice that the previous estimates only involve an upper bound of a control ω which can be chosen according to the vector field M and the initial path X. Therefore, we have the following theorem.

Theorem 7.2.2 *Let $X \in \Omega_1(V)$ and let M be a Lipschitz vector field. Then the Euler iteration $\Psi^t(n)$, where $n = 0, 1, 2, \ldots$, converges to the unique solution $\Phi^t(X)$ of the flow eqn (7.10) in one-variation topology. Therefore,*

$$\Psi^t(n)^i_{s_1, s_2} \to \Phi^t(X)^i_{s_1, s_2}, \quad as \quad n \to \infty \,,$$

for all $(s_1, s_2) \in \Delta$, where $\Psi^t(n)^i_{s_1, s_2}$ and $\Phi^t(X)^i_{s_1, s_2}$ denote the ith iterated path integrals over $[s_1, s_2]$ corresponding to $\Psi^t(n)$ and $\Phi^t(X)$, respectively.

7.3 Flows of Itô vector fields

In this section we are going to use the Euler iteration to solve the following flow equation:

$$\frac{\mathrm{d}\Phi^t}{\mathrm{d}t} = M(\Phi^t) \,, \quad \Phi^0 = X \,, \tag{7.19}$$

for all geometric rough paths $X \in G\Omega_p(V)$, where M is an Itô vector field, namely $M(X)$ is the Itô map defined by the following differential equation:

$$\mathrm{d}(M_t, Y_t) = A(M_t, Y_t) \, \mathrm{d}\hat{X}_t \,, \quad M_0 = 0 \,, \tag{7.20}$$

for some smooth map

$$A : V \oplus W \to \mathbb{L}(V \oplus V \oplus V^{\otimes 2}, V \oplus W) \,,$$

where $(h, \varphi) \in T^{(q_1, q_2)}\Omega_p(V)$ is fixed. For a smooth rough path X, $M(X)$ is the solution of the following differential equation:

$$\begin{aligned} \mathrm{d}M_t &= f_1(M_t, Y_t) \, \mathrm{d}X_t + f_2(M_t, Y_t) \, \mathrm{d}H_t \,, \quad M_0 = 0 \,, \\ \mathrm{d}Y_t &= g_1(M_t, Y_t) \, \mathrm{d}X_t + g_2(M_t, Y_t) \, \mathrm{d}H_t \,, \end{aligned} \tag{7.21}$$

where $H = [h, \varphi]$,

$$\begin{aligned} f_1 &: V \oplus W \to \mathbb{L}(V, V) \,, \\ f_2 &: V \oplus W \to \mathbb{L}(V \oplus V^{\otimes 2}, V) \,, \\ g_1 &: V \oplus W \to \mathbb{L}(V, W) \,, \end{aligned}$$

and

$$g_2 : V \oplus W \to \mathbb{L}(V \oplus V^{\otimes 2}, W) \,.$$

For simplicity, we will assume that in this section $V = \mathbb{R}^d$ and $W = \mathbb{R}^m$ are finite-dimensional vector spaces, and that f_1, f_2, g_1, g_2 are smooth with bounded derivatives up to degree $[p] + 1$. Therefore, we may assume that

$$\left| D^i f_1 \right|, \left| D^i f_2 \right|, \left| D^i g_1 \right|, \left| D^i g_2 \right| \leqslant C_1, \tag{7.22}$$

for $i = 0, \ldots, [p] + 1$.

Let $X \in G\Omega_p(V)$ be a smooth rough path. Then the Euler iteration is the following sequence:

$$\Psi(0)^t = X,$$

$$\Psi(n)^t = X + \frac{t}{n} \sum_{l=1}^{n} M\left(\Psi(n-1)^{((l-1)/n)t}\right), \quad \forall n \geqslant 1, \tag{7.23}$$

and all of the $\Psi(n)^t$ are smooth rough paths too. If we want to emphasize the dependence on X, we will write $\Psi(n)^t$ by $\Psi^t(n, X)$. The same notation applies to Φ^t as well. The above iteration can be considered to be the iteration for the traces of those rough paths, and their higher-order terms are determined by iterated path integrals.

Define $M_n^t = M\left(\Psi(n)^t\right)$ and $Y_n^t = Y\left(\Psi(n)^t\right)$. Then

$$M_n^t = M\left(X + \frac{t}{n} \sum_{l=1}^{n} M_{n-1}^{((l-1)/n)t}\right), \quad n \geqslant 1,$$

$$Y_n^t = Y\left(X + \frac{t}{n} \sum_{l=1}^{n} M_{n-1}^{((l-1)/n)t}\right), \quad n \geqslant 1,$$

and therefore, by definition,

$$dM_n^t = f_1(M_n^t, Y_n^t) \, dX + f_2(M_n^t, Y_n^t) \, dH + \frac{t}{n} \sum_{l=1}^{n} f_1(M_n^t, Y_n^t) \, dM_{n-1}^{((l-1)/n)t} \tag{7.24}$$

and

$$dY_n^t = g_1(M_n^t, Y_n^t) \, dX + g_2(M_n^t, Y_n^t) \, dH + \frac{t}{n} \sum_{l=1}^{n} g_1(M_n^t, Y_n^t) \, dM_{n-1}^{((l-1)/n)t}.$$

For fixed t and n, we define

$$M^{(k_j, \ldots, k_1)} = M_{n-j}^{((k_j-1)/(n-j+1)) \times \cdots \times ((k_1-1)/n)t},$$

$$Y^{(k_j, \ldots, k_1)} = Y_{n-j}^{((k_j-1)/(n-j+1)) \times \cdots \times ((k_1-1)/n)t},$$

for $j = 1, \ldots, n$ and $k_i = 1, \ldots, n - i + 1$. By (7.24), we have

$$dM^{(k_j,\ldots,k_1)} = f_1\big(M^{(k_j,\ldots,k_1)}, Y^{(k_j,\ldots,k_1)}\big)\,dX + f_2\big(M^{(k_j,\ldots,k_1)}, Y^{(k_j,\ldots,k_1)}\big)\,dH$$

$$+\frac{t}{n-j}\prod_{l=1}^{j}\frac{k_l-1}{n-(l-1)}$$

$$\times \sum_{k_{j+1}=1}^{n-j} f_1\big(M^{(k_j,\ldots,k_1)}, Y^{(k_j,\ldots,k_1)}\big)\,dM^{(k_{j+1},\ldots,k_1)}$$

$$(7.25)$$

and

$$dY^{(k_j,\ldots,k_1)} = g_1\big(M^{(k_j,\ldots,k_1)}, Y^{(k_j,\ldots,k_1)}\big)\,dX + g_2\big(M^{(k_j,\ldots,k_1)}, Y^{(k_j,\ldots,k_1)}\big)\,dH$$

$$+\frac{t}{n-j}\prod_{l=1}^{j}\frac{k_l-1}{n-(l-1)}$$

$$\times \sum_{k_{j+1}=1}^{n-j} g_1\big(M^{(k_j,\ldots,k_1)}, Y^{(k_j,\ldots,k_1)}\big)\,dM^{(k_{j+1},\ldots,k_1)},$$

$$(7.26)$$

for $j = 1,\ldots, n-1$. Moreover, for all k_i, we have

$$dM^{(k_n,\ldots,k_1)} = f_1\big(M^{(k_n,\ldots,k_1)}, Y^{(k_n,\ldots,k_1)}\big)\,dX + f_2\big(M^{(k_n,\ldots,k_1)}, Y^{(k_n,\ldots,k_1)}\big)\,dH$$

$$(7.27)$$

and

$$dY^{(k_n,\ldots,k_1)} = g_1\big(M^{(k_n,\ldots,k_1)}, Y^{(k_n,\ldots,k_1)}\big)\,dX + g_2\big(M^{(k_n,\ldots,k_1)}, Y^{(k_n,\ldots,k_1)}\big)\,dH\,.$$

$$(7.28)$$

Now, repeating (7.24)–(7.28), we have

$$dM^{(k_{n-j},\ldots,k_1)} = f_1\big(M^{(k_{n-j},\ldots,k_1)}, Y^{(k_{n-j},\ldots,k_1)}\big)\,dX$$

$$+ f_2\big(M^{(k_{n-j},\ldots,k_1)}, Y^{(k_{n-j},\ldots,k_1)}\big)\,dH$$

$$+\sum_{q=1}^{j}\frac{t^q}{j(j-1)\cdots(j-q+1)}\sum_{k_{n-j+1}=1}^{j}\cdots\sum_{k_{n-j+q}=1}^{j-q+1}$$

$$\prod_{l=1}^{n-j}\frac{k_l-1}{n-l+1}\cdots\prod_{l=1}^{n-j+q-1}\frac{k_l-1}{n-l+1}$$

$$\times f_1\big(M^{(k_{n-j},\ldots,k_1)}, Y^{(k_{n-j},\ldots,k_1)}\big)\circ\cdots$$

$$\circ f_1\big(M^{(k_{n-j+q-1},\ldots,k_1)}, Y^{(k_{n-j+q-1},\ldots,k_1)}\big)$$

$$\circ \big[f_1\big(M^{(k_{n-j+q},\ldots,k_1)}, Y^{(k_{n-j+q},\ldots,k_1)}\big)\,dX$$

$$+ f_2\big(M^{(k_{n-j+q},\ldots,k_1)}, Y^{(k_{n-j+q},\ldots,k_1)}\big)\,dH\big] \quad (7.29)$$

and

$$dY^{(k_{n-j},\ldots,k_1)} = g_1\big(M^{(k_{n-j},\ldots,k_1)}, Y^{(k_{n-j},\ldots,k_1)}\big)\, dX$$
$$+ g_2\big(M^{(k_{n-j},\ldots,k_1)}, Y^{(k_{n-j},\ldots,k_1)}\big)\, dH$$
$$+ \sum_{q=1}^{j} \frac{t^q}{j(j-1)\cdots(j-q+1)} \sum_{k_{n-j+1}=1}^{j} \cdots \sum_{k_{n-j+q}=1}^{j-q+1}$$
$$\prod_{l=1}^{n-j} \frac{k_l-1}{n-l+1} \cdots \prod_{l=1}^{n-j+q-1} \frac{k_l-1}{n-l+1}$$
$$\times g_1\big(M^{(k_{n-j},\ldots,k_1)}, Y^{(k_{n-j},\ldots,k_1)}\big)$$
$$\circ f_1\big(M^{(k_{n-j},\ldots,k_1)}, Y^{(k_{n-j},\ldots,k_1)}\big) \circ \cdots$$
$$\circ f_1\big(M^{(k_{n-j+q-1},\ldots,k_1)}, Y^{(k_{n-j+q-1},\ldots,k_1)}\big)$$
$$\circ \big[f_1\big(M^{(k_{n-j+q},\ldots,k_1)}, Y^{(k_{n-j+q},\ldots,k_1)}\big)\, dX$$
$$+ f_2\big(M^{(k_{n-j+q},\ldots,k_1)}, Y^{(k_{n-j+q},\ldots,k_1)}\big)\, dH \big], \quad (7.30)$$

for $j = 1,\ldots,n-1$ and $k_j = 1,\ldots,n-j+1$. In particular,

$$dM^{(k_1)} = f_1\big(M^{(k_1)}, Y^{(k_1)}\big)\, dX + f_2\big(M^{(k_1)}, Y^{(k_1)}\big)\, dH$$
$$+ \sum_{q=1}^{n-1} \frac{t^q}{(n-1)\cdots(n-q)} \sum_{k_2=1}^{n-1} \cdots \sum_{k_{q+1}=1}^{n-q}$$
$$\prod_{l=1}^{1} \frac{k_l-1}{n-l+1} \cdots \prod_{l=1}^{n-q} \frac{k_l-1}{n-l+1}$$
$$\times f_1\big(M^{(k_1)}, Y^{(k_1)}\big) \circ \cdots \circ f_1\big(M^{(k_q,\ldots,k_1)}, Y^{(k_q,\ldots,k_1)}\big)$$
$$\circ \big[f_1\big(M^{(k_{q+1},\ldots,k_1)}, Y^{(k_{q+1},\ldots,k_1)}\big)\, dX$$
$$+ f_2\big(M^{(k_{q+1},\ldots,k_1)}, Y^{(k_{q+1},\ldots,k_1)}\big)\, dH \big],$$

and therefore

$$\mathrm{d}\Psi(n)^t = \mathrm{d}X + \frac{t}{n} \sum_{k_1=1}^{n} \mathrm{d}M^{(k_1)}$$

$$= \mathrm{d}X + \frac{t}{n} \sum_{k_1=1}^{n} \left[f_1\big(M^{(k_1)}, Y^{(k_1)}\big)\, \mathrm{d}X + f_2\big(M^{(k_1)}, Y^{(k_1)}\big)\, \mathrm{d}H \right]$$

$$+ \frac{t}{n} \sum_{k_1=1}^{n} \sum_{q=1}^{n-1} \frac{t^q}{(n-1)\cdots(n-q)} \sum_{k_2=1}^{n-1} \cdots \sum_{k_{q+1}=1}^{n-q}$$

$$\prod_{l=1}^{1} \frac{k_l - 1}{n - l + 1} \cdots \prod_{l=1}^{n-q} \frac{k_l - 1}{n - l + 1}$$

$$\times f_1\big(M^{(k_1)}, Y^{(k_1)}\big) \circ \cdots \circ f_1\big(M^{(k_q,\ldots,k_1)}, Y^{(k_q,\ldots,k_1)}\big)$$

$$\circ \big[f_1\big(M^{(k_{q+1},\ldots,k_1)}, Y^{(k_{q+1},\ldots,k_1)}\big)\, \mathrm{d}X$$

$$+ f_2\big(M^{(k_{q+1},\ldots,k_1)}, Y^{(k_{q+1},\ldots,k_1)}\big)\, \mathrm{d}H \big]. \tag{7.31}$$

Let

$$U = V^0 \oplus \sum_{\substack{1 \leqslant j \leqslant n \\ 1 \leqslant k_i \leqslant n-i+1}} \oplus \big(V^{(k_j,\ldots,k_1)} \oplus W^{(k_j,\ldots,k_1)} \big),$$

where V^0, $V^{(k_j,\ldots,k_1)}$ are copies of V and $W^{(k_j,\ldots,k_1)}$ are copies of W. For fixed t, n, we introduce the following functions:

$$F_0 : U \to \mathbb{L}\big(V, V^0\big),$$

$$F^{(k_{n-j},\ldots,k_1)} : U \to \mathbb{L}\big(V, V^{(k_{n-j},\ldots,k_1)}\big),$$

$$G_1^{(k_{n-j},\ldots,k_1)} : U \to \mathbb{L}\big(V, W^{(k_{n-j},\ldots,k_1)}\big),$$

$$G_2^{(k_{n-j},\ldots,k_1)} : U \to \mathbb{L}\big(V \oplus V^{\otimes 2}, W^{(k_{n-j},\ldots,k_1)}\big),$$

and

$$h_0 : U \to \mathbb{L}\big(V \oplus V^{\otimes 2}, V^0\big),$$

$$h^{(k_{n-j},\ldots,k_1)} : U \to \mathbb{L}\big(V \oplus V^{\otimes 2}, V^{(k_{n-j},\ldots,k_1)}\big),$$

which are defined as the following. For

$$\bar{x} = \big(x^0, x^{(k_{n-j},\ldots,k_1)}\big)_{\substack{0 \leqslant j \leqslant n-1 \\ 1 \leqslant k_i \leqslant n-i+1}},$$

$$\bar{y} = \big(y^0, y^{(k_{n-j},\ldots,k_1)}\big)_{\substack{0 \leqslant j \leqslant n-1 \\ 1 \leqslant k_i \leqslant n-i+1}},$$

and $\xi \in V$, $\eta \in V \oplus V^{\otimes 2}$,

$$F_0(\bar{x},\bar{y})\xi = \xi + \frac{t}{n}\sum_{k_1=1}^{n} f_1\big(x^{(k_1)},y^{(k_1)}\big)\xi$$

$$+ \frac{t}{n}\sum_{k_1=1}^{n}\sum_{q=1}^{n-1}\frac{t^q}{(n-1)\cdots(n-q)}\sum_{k_2=1}^{n-1}\cdots\sum_{k_{q+1}=1}^{n-q}$$

$$\prod_{l=1}^{1}\frac{k_l-1}{n-l+1}\cdots\prod_{l=1}^{n-q}\frac{k_l-1}{n-l+1}$$

$$\times f_1\big(x^{(k_1)},y^{(k_1)}\big)\circ\cdots\circ f_1\big(x^{(k_q,\ldots,k_1)},y^{(k_q,\ldots,k_1)}\big)$$

$$\circ f_1\big(x^{(k_{q+1},\ldots,k_1)},y^{(k_{q+1},\ldots,k_1)}\big)\xi,$$

$$h_0(\bar{x})\eta = \frac{t}{n}\sum_{k_1=1}^{n} f_2\big(x^{(k_1)},y^{(k_1)}\big)\eta$$

$$+ \frac{t}{n}\sum_{k_1=1}^{n}\sum_{q=1}^{n-1}\frac{t^q}{(n-1)\cdots(n-q)}\sum_{k_2=1}^{n-1}\cdots\sum_{k_{q+1}=1}^{n-q}$$

$$\prod_{l=1}^{1}\frac{k_l-1}{n-l+1}\cdots\prod_{l=1}^{n-q}\frac{k_l-1}{n-l+1}$$

$$\times f_1\big(x^{(k_1)},y^{(k_1)}\big)\circ\cdots\circ f_1\big(x^{(k_q,\ldots,k_1)},y^{(k_q,\ldots,k_1)}\big)$$

$$\circ f_2\big(x^{(k_{q+1},\ldots,k_1)},y^{(k_{q+1},\ldots,k_1)}\big)\eta,$$

$$F^{(k_{n-j},\ldots,k_1)}(\bar{x},\bar{y})\xi = f_1\big(x^{(k_{n-j},\ldots,k_1)},y^{(k_{n-j},\ldots,k_1)}\big)\xi$$

$$+\sum_{q=1}^{j}\frac{t^q}{j(j-1)\cdots(j-q+1)}\sum_{k_{n-j+1}=1}^{j}\cdots\sum_{k_{n-j+q}=1}^{j-q+1}$$

$$\prod_{l=1}^{n-j}\frac{k_l-1}{n-l+1}\cdots\prod_{l=1}^{n-j+q-1}\frac{k_l-1}{n-l+1}$$

$$\times f_1\big(x^{(k_{n-j},\ldots,k_1)},y^{(k_{n-j},\ldots,k_1)}\big)\circ\cdots$$

$$\circ f_1\big(x^{(k_{n-j+q-1},\ldots,k_1)},y^{(k_{n-j+q-1},\ldots,k_1)}\big)$$

$$\circ f_1\big(x^{(k_{n-j+q},\ldots,k_1)},y^{(k_{n-j+q},\ldots,k_1)}\big)\xi,$$

and

$$h^{(k_{n-j},\ldots,k_1)}(\bar{x},\bar{y})\eta = f_2\big(x^{(k_{n-j},\ldots,k_1)},y^{(k_{n-j},\ldots,k_1)}\big)\eta$$

$$+ \sum_{q=1}^{j} \frac{t^q}{j(j-1)\cdots(j-q+1)} \sum_{k_{n-j+1}=1}^{j} \cdots \sum_{k_{n-j+q}=1}^{j-q+1}$$

$$\prod_{l=1}^{n-j} \frac{k_l-1}{n-l+1} \cdots \prod_{l=1}^{n-j+q-1} \frac{k_l-1}{n-l+1}$$

$$\times f_1\big(x^{(k_{n-j},\ldots,k_1)},y^{(k_{n-j},\ldots,k_1)}\big) \circ \cdots$$

$$\circ f_1\big(x^{(k_{n-j+q-1},\ldots,k_1)},y^{(k_{n-j+q-1},\ldots,k_1)}\big)$$

$$\circ f_2\big(x^{(k_{n-j+q},\ldots,k_1)},y^{(k_{n-j+q},\ldots,k_1)}\big)\eta\,.$$

Also

$$G_1^{(k_{n-j},\ldots,k_1)}(\bar{x},\bar{y})\xi = g_1\big(x^{(k_{n-j},\ldots,k_1)},y^{(k_{n-j},\ldots,k_1)}\big)\xi$$

$$+ \sum_{q=1}^{j} \frac{t^q}{j(j-1)\cdots(j-q+1)} \sum_{k_{n-j+1}=1}^{j} \cdots \sum_{k_{n-j+q}=1}^{j-q+1}$$

$$\prod_{l=1}^{n-j} \frac{k_l-1}{n-l+1} \cdots \prod_{l=1}^{n-j+q-1} \frac{k_l-1}{n-l+1}$$

$$\times g_1\big(x^{(k_{n-j},\ldots,k_1)},y^{(k_{n-j},\ldots,k_1)}\big)$$

$$\circ f_1\big(x^{(k_{n-j+1},\ldots,k_1)},y^{(k_{n-j+1},\ldots,k_1)}\big) \circ \cdots$$

$$\circ f_1\big(x^{(k_{n-j+q-1},\ldots,k_1)},y^{(k_{n-j+q-1},\ldots,k_1)}\big)$$

$$\circ f_1\big(x^{(k_{n-j+q},\ldots,k_1)},y^{(k_{n-j+q},\ldots,k_1)}\big)\xi$$

and

$$G_2^{(k_{n-j},\ldots,k_1)}(\bar{x},\bar{y})\eta = g_2\big(x^{(k_{n-j},\ldots,k_1)},y^{(k_{n-j},\ldots,k_1)}\big)\eta$$

$$+ \sum_{q=1}^{j} \frac{t^q}{j(j-1)\cdots(j-q+1)} \sum_{k_{n-j+1}=1}^{j} \cdots \sum_{k_{n-j+q}=1}^{j-q+1}$$

$$\prod_{l=1}^{n-j} \frac{k_l-1}{n-l+1} \cdots \prod_{l=1}^{n-j+q-1} \frac{k_l-1}{n-l+1}$$

$$\times g_1\big(x^{(k_{n-j},\ldots,k_1)},y^{(k_{n-j},\ldots,k_1)}\big)$$

$$\circ f_1\big(x^{(k_{n-j+1},\ldots,k_1)},y^{(k_{n-j+1},\ldots,k_1)}\big) \circ \cdots$$

$$\circ f_1\big(x^{(k_{n-j+q-1},\ldots,k_1)},y^{(k_{n-j+q-1},\ldots,k_1)}\big)$$

$$\circ f_2\big(x^{(k_{n-j+q},\ldots,k_1)},y^{(k_{n-j+q},\ldots,k_1)}\big)\eta\,.$$

Finally, set

$$F_{(n,t)} = \left(F_0, F^{(k_{n-j},\ldots,k_1)}, G_1^{(k_{n-j},\ldots,k_1)}\right)_{\substack{0\leqslant j\leqslant n-1 \\ 1\leqslant k_i\leqslant n-i+1}}$$

and

$$h_{(n,t)} = \left(h_0, h^{(k_{n-j},\ldots,k_1)}, G_2^{(k_{n-j},\ldots,k_1)}\right)_{\substack{0\leqslant j\leqslant n-1 \\ 1\leqslant k_i\leqslant n-i+1}} .$$

Then

$$\bar{X} = \left(\Psi(n)^t, M^{(k_{n-j},\ldots,k_1)}, Y^{(k_{n-j},\ldots,k_1)}\right)_{\substack{0\leqslant j\leqslant n-1 \\ 1\leqslant k_i\leqslant n-i+1}}$$

is the unique solution to the following differential equation:

$$d\bar{X} = F_{(n,t)}(\bar{X})\,dX + h_{(n,t)}(\bar{X})\,dH . \tag{7.32}$$

The essential fact which we need about this equation is contained in the following proposition.

Proposition 7.3.1 *Consider the space*

$$U = V^0 \oplus \sum_{\substack{1\leqslant j\leqslant n \\ 1\leqslant k_i\leqslant n-i+1}} \oplus\left(V^{(k_j,\ldots,k_1)} \oplus W^{(k_j,\ldots,k_1)}\right)$$

(note that V^0 and each of $V^{(k_j,\ldots,k_1)}$ is a copy of V and each $W^{(k_j,\ldots,k_1)}$ is a copy of W) and suppose that it is endowed with the maximum norm

$$|\bar{x}| = \max\left\{\left|x^{(k_{n-j},\ldots,k_1)}\right|, \left|x^0\right|, \left|y^{(k_{n-j},\ldots,k_1)}\right|\right. :$$
$$\left. 0\leqslant j\leqslant n-1, 1\leqslant k_i\leqslant n-i+1\right\}.$$

Then, for any $p\geqslant 1$ and $0 < C_0 < 1$, there is a constant $C(p)$ depending only on C_0, C_1, and p such that, for all n and t, if $|tC_1|\leqslant C_0$, then

$$\left|D^i g_{(n,t)}\right|, \left|D^i h_{(n,t)}\right| \leqslant C(p),$$

for $i = 1,\ldots,[p]+1$.

Proof It is obvious that

$$|F_0(\bar{x},\bar{y})| \leqslant 1 + (tC_1) + \sum_{q=1}^{n-1}(tC_1)^{q+1}$$

$$\leqslant 1 + \frac{1}{1-C_0}$$

and

$$\left|F^{(k_{n-j},\ldots,k_1)}(\bar{x},\bar{y})\right| \leqslant C_1 + \sum_{q=1}^{j}(tC_1)^q C_1$$

$$\leqslant \frac{C_1}{1-C_0} .$$

Similar estimates are true for h_0, $h^{(k_{n-j},\ldots,k_1)}$, and G_i. Next we consider their derivatives. A simple calculation shows that

$$\frac{\partial F^{(k_{n-j},\ldots,k_1)}}{\partial x^{(l_{n-i},\ldots,l_1)}}, \ \frac{\partial F^{(k_{n-j},\ldots,k_1)}}{\partial y^{(l_{n-i},\ldots,l_1)}} = 0, \quad \text{if} \quad i > j,$$

$$\frac{\partial F^{(k_{n-j},\ldots,k_1)}}{\partial x^{(k_{n-j},\ldots,k_1)}} \quad \text{or} \quad \frac{\partial F^{(k_{n-j},\ldots,k_1)}}{\partial y^{(k_{n-j},\ldots,k_1)}}$$

$$= Df_1\big(x^{(k_{n-j},\ldots,k_1)}, y^{(k_{n-j},\ldots,k_1)}\big)$$

$$+ \sum_{q=1}^{j} \frac{t^q}{j(j-1)\cdots(j-q+1)} \sum_{k_{n-j+1}=1}^{j} \cdots \sum_{k_{n-j+q}=1}^{j-q+1}$$

$$\prod_{l=1}^{n-j} \frac{k_l-1}{n-l+1} \cdots \prod_{l=1}^{n-j+q-1} \frac{k_l-1}{n-l+1}$$

$$\times Df_1\big(x^{(k_{n-j},\ldots,k_1)}, y^{(k_{n-j},\ldots,k_1)}\big)$$

$$\circ f_1\big(x^{(k_{n-j+1},\ldots,k_1)}, y^{(k_{n-j+1},\ldots,k_1)}\big) \circ \cdots$$

$$\circ f_1\big(x^{(k_{n-j+q},\ldots,k_1)}, y^{(k_{n-j+q},\ldots,k_1)}\big),$$

$$\frac{\partial F^{(k_{n-j},\ldots,k_1)}}{\partial x^{(l_{n-i},\ldots,l_1)}}, \ \frac{\partial F^{(k_{n-j},\ldots,k_1)}}{\partial y^{(l_{n-i},\ldots,l_1)}} = 0,$$

$$\text{if} \quad i \leqslant j \quad \text{and} \quad (k_{n-j},\ldots,k_1) \neq (l_{n-j},\ldots,l_1),$$

and, if $i < j$,

$$\frac{\partial F^{(k_{n-j},\ldots,k_1)}}{\partial x^{(l_{n-i},\ldots,l_{n-j+1},k_{n-j},\ldots,k_1)}} \quad \text{or} \quad \frac{\partial F^{(k_{n-j},\ldots,k_1)}}{\partial y^{(l_{n-i},\ldots,l_{n-j+1},k_{n-j},\ldots,k_1)}}$$

$$= \sum_{\substack{n-j+q\geqslant n-i \\ 1\leqslant q\leqslant j}} \frac{t^q}{j(j-1)\cdots(j-q+1)}$$

$$\sum_{k_{n-j+1}=1}^{j} \cdots \sum_{k_{n-j+q}=1}^{j-q+1} \delta^{(l_{n-i},\ldots,l_{n-j+1})}_{(k_{n-i},\ldots,k_{n-j+1})}$$

$$\prod_{l=1}^{n-j} \frac{k_l-1}{n-l+1} \cdots \prod_{l=1}^{n-j+q-1} \frac{k_l-1}{n-l+1}$$

$$\times f_1 \circ \cdots$$

$$\circ Df_1\big(x^{(l_{n-i},\ldots,l_{n-j+1},k_{n-j},\ldots,k_1)}, y^{(l_{n-i},\ldots,l_{n-j+1},k_{n-j},\ldots,k_1)}\big) \circ \cdots$$

$$\circ f_1.$$

Therefore, if $|tC_1| \leqslant C_0$, then we have

$$\left| D^1 F^{(k_{n-j}, \ldots, k_1)} \right|$$

$$\leqslant \sum_{(l_{n-i}, \ldots, l_1)} \left| \frac{\partial F^{(k_{n-j}, \ldots, k_1)}}{\partial x^{(l_{n-i}, \ldots, l_1)}} \right| + \left| \frac{\partial F^{(k_{n-j}, \ldots, k_1)}}{\partial y^{(l_{n-i}, \ldots, l_1)}} \right|$$

$$= \sum_{i \leqslant j} \left| \frac{\partial F^{(k_{n-j}, \ldots, k_1)}}{\partial x^{(l_{n-i}, \ldots, l_{n-j+1}, k_{n-j}, \ldots, k_1)}} \right| + \left| \frac{\partial F^{(k_{n-j}, \ldots, k_1)}}{\partial y^{(l_{n-i}, \ldots, l_{n-j+1}, k_{n-j}, \ldots, k_1)}} \right|$$

$$\leqslant 2 \sum_{i \leqslant j} \sum_{\substack{n-j+q \geqslant n-i \\ 1 \leqslant q \leqslant j}} \frac{i \cdots (j - q + 1) C_0^q}{j(j-1) \cdots (j - q + 1)}$$

$$= 2 \sum_{i=0}^{j} \sum_{j - i \leqslant q \leqslant j} \frac{C_0^q}{j(j-1) \cdots i}$$

$$\leqslant 2 \left(1 + \frac{1}{j} \right) \frac{1}{1 - C_0} \leqslant C_2 \,,$$

where C_2 is a constant depending only on C_0. A similar argument applies to F_0. Therefore,

$$\left| D^1 F^{(k_{n-j}, \ldots, k_1)} \right| + \left| D^1 F_0 \right| \leqslant C_3 \,,$$

for some constant C_3 depending only on C_0. The same inequality holds for other functions h, G_i. Therefore, as the norm on W is the maximum norm, we have

$$\left| D^1 F_{(n,t)} \right| \leqslant \max_{(k_{n-j}, \ldots, k_1)} \left| D^1 F^{(k_{n-j}, \ldots, k_1)} \right| + \left| D^1 F_0 \right|$$

$$\leqslant C_3$$

and the same inequality holds for $h_{(n,t)}$. By repeating the same argument, we may complete the proof.   ■

From the above section, we know that for any smooth rough path X we have a unique solution Φ^t (also denoted by $\Phi^t(X)$), and $\Psi^t(n, X)^i_{\bullet} \to \Phi^t(X)^i_{\bullet}$ uniformly, where $\Psi^t(n, X)^i_{\bullet}$ and $\Phi^t(X)^i_{\bullet}$ denote the corresponding ith iterated integrals. Then, by applying Corollary 6.3.2 to eqn (7.32) and Proposition 7.3.1, we have the following theorem.

Theorem 7.3.1 *Let $T_0 > 0$ such that $|T_0 C_1| < 1$. Then there is a constant $K(\varepsilon)$ depending only on an upper bound $\max \omega$, C_1, $|T_0 C_1|$, q_1, q_2, p, and $\varepsilon > 0$, such that the following conditions hold.*

(i) $\lim_{\varepsilon \to 0} K(\varepsilon) = 0$.

(ii) *If $X, \hat{X} \in G\Omega_p(V)$ are two smooth rough paths, and ω is a control such that*

$$\left| X^i_{s_1, s_2} \right|, \left| \hat{X}^i_{s_1, s_2} \right| \leqslant \omega(s_1, s_2), \quad \forall (s_1, s_2) \in \Delta$$

and

$$\left| X^i_{s_1,s_2} - \hat{X}^i_{s_1,s_2} \right| \leqslant \varepsilon \omega(s_1,s_2), \quad \forall (s_1,s_2) \in \Delta,$$

for $i = 1, \ldots, [p]$, *then, for all* $|t| \leqslant T_0$, *we have*

$$\left| \Phi^t(X)^i_{s_1,s_2} - \Phi^t(\hat{X})^i_{s_1,s_2} \right| \leqslant K(\varepsilon) \omega(s_1,s_2), \quad \forall (s_1,s_2) \in \Delta,$$

for $i = 1, \ldots, [p]$, *where* $\Phi^t(X)$ *is the unique solution flow established in the previous section and* $\Phi^t(X)^i_{s_1,s_2}$ *are its* i*th iterated integrals over the interval* (s_1, s_2).

By the above theorem, the map $X \to \Phi^t(X)$, which is defined for any Lipschitz path (and so also for any smooth rough path) X, can be uniquely extended to be a continuous map in p-variation topology on $G\Omega_p(V)$, denoted again by $\Phi^t(X)$, and $\{\Phi^t\}$, where $\Phi^t : X \to \Phi^t(X)$, for any $X \in G\Omega_p(V)$, is called the solution flow of the Itô vector field M.

7.4 Appendix: Driver's flow equation

This appendix contains a derivation of Driver's flow equation. We will assume that (M, g) is a compact, d-dimensional manifold with an affine connection ∇, and ω is its connection one-form on the frame bundle $L(M)$.

Denote by $W_o^\infty(I, N)$ the space of all continuous and piecewise-smooth paths in a manifold N with time running interval $I = [0, 1]$ and initial value $o \in N$.

Let $\sigma \in W_o^\infty(I, M)$. Then a tangent vector W at σ, denoted by $W \in T_\sigma W_o^\infty(I, M)$, is by definition a continuous and piecewise-smooth vector field along the path σ with initial point zero, that is $W_s \in T_{\sigma_s} M$, $W_0 = 0$.

With the affine connection ∇ on $L(M)$, and a frame $u_o \in L(M)$ satisfying $\pi(u_o) = o$, we have the Itô development map, denoted by J, which is the map from $W_o^\infty(I, \mathbb{R}^d)$ to $W_o^\infty(I, M)$. By definition, if $z \in W_o^\infty(I, \mathbb{R}^d)$ and $\sigma = \pi(\gamma)$, where γ is the unique solution of the stochastic differential equation

$$d\gamma_t = B(\gamma_t) \, dz_t, \quad \gamma_0 = u_o$$

and B denotes the standard horizontal vector field on $L(M)$ of the affine connection ω, then $J(z) = \sigma$. In this case γ is called the development of z, denoted by $I(z)$. Therefore, $J = \pi \circ I$. There is a unique inverse map J^{-1} of the Itô map J. In fact, if $\sigma \in W_o^\infty(I, M)$ and

$$z_t = \int_0^t \theta(d\gamma_s),$$

where $\gamma = H(\sigma)$ is the unique horizontal lift of σ at $u_o \in L(M)$, then $J^{-1}(\sigma) = z$, that is $J^{-1} = \theta \circ H$.

Given a path $\sigma \in W_o^\infty(I, M)$ and a tangent vector $W \in T_\sigma W_o^\infty(I, M)$, let $z = J^{-1}(\sigma)$ and $M = J_*^{-1} W$, where J_*^{-1} denotes the tangent map of J^{-1}, so that

$$J_*^{-1} : T_\sigma W_o^\infty(I, M) \to T_z W_0^\infty(I, \mathbb{R}^d), \quad z = J^{-1}(\sigma).$$

Let us deduce the formula for $M = J_*^{-1} W$, where W is a vector field along σ. This formula is essentially obtained by Driver (1992).

Let $\sigma^t \in W_o^\infty(I, M)$ such that $\sigma^0 = \sigma$ and

$$\left.\frac{d\sigma_s^t}{dt}\right|_{t=0} = W_s, \quad \forall s \in I.$$

For example, we can choose $\sigma_s^t = \exp_{\sigma_s}(tW_s)$. Then, by definition,

$$M_s = \left.\frac{dz_s^t}{dt}\right|_{t=0}, \quad z^t = J^{-1}(\sigma^t).$$

Denote by γ^t the unique horizontal lift of σ^t at u_o. Then clearly

$$\pi_* \left.\left(\frac{\partial \gamma_s^t}{\partial s}\right)\right|_{t=0} = W_s, \quad z_s^t = \int_0^s \theta(d\gamma_u^t). \tag{7.33}$$

In particular, $\partial z_s^t / \partial s = \theta\left(\partial \gamma_s^t / \partial s\right)$. Let $T = \partial \gamma_s^t / \partial t$ and $S = \partial \gamma_s^t / \partial s$. Then $[S, T] = 0$ and, since S is horizontal, $\omega(S) = 0$.

It is obvious that

$$\frac{\partial}{\partial t} \frac{\partial z_s^t}{\partial s} = T\theta(S) = \frac{\partial}{\partial s} \frac{\partial z_s^t}{\partial t}.$$

Integrating the two sides of this equation over $[0, s]$, we thus obtain

$$M_s = \int_0^s T\theta(S)(\tau)|_{t=0}\, d\tau. \tag{7.34}$$

By the first structure equation and the fact that $[S, T] = 0$, we have

$$T\theta(S) = S\theta(T) - \omega(T)\theta(S) + \Theta(hT, S). \tag{7.35}$$

Then, since

$$\theta(T)|_{t=0}(s) = (\gamma_s^t)^{-1}\pi_* \left.\left(\frac{\partial \gamma_s^t}{\partial t}\right)\right|_{t=0}$$

$$= \gamma_s^{-1} W_s,$$

we have

$$S\theta(T)|_{t=0} = \frac{d}{ds}\gamma_s^{-1} W_s. \tag{7.36}$$

By the second structure equation, $S\omega(T) = \Omega(S, hT)$, hence

$$\omega(T)|_{t=0} = \int_0^s \Omega(S, hT)|_{t=0}\, d\tau. \tag{7.37}$$

On the other hand,

$$S|_{t=0}(\tau) = \dot{\gamma}_\tau,$$

$$hT|_{t=0}(s) = B_{\gamma_s}(\gamma_s^{-1} W_s),$$

so that, by eqn (7.37),

$$\omega(T)|_{t=0}(s) = \int_0^s \Omega(\dot{\gamma}_\tau, B_{\gamma_\tau}(\gamma_\tau^{-1} W_\tau)) \, d\tau$$
$$= \int_0^s \Omega(d\gamma_\tau, B(\gamma_\tau)(\gamma_\tau^{-1} W_\tau)) \,.$$

Inserting this equation into eqn (7.35), we obtain

$$T\theta(S)|_{t=0}(s) = S\theta(T) - \omega(T)\theta(S) + \Theta(hT, S)|_{t=0}$$
$$= \frac{d}{ds}\gamma_s^{-1} W_s - \int_0^s \Omega(d\gamma_\tau, B_{\gamma_\tau}(\gamma_\tau^{-1} W_\tau))\theta(S)|_{t=0} - \Theta(S, hT)$$
$$= \frac{d}{ds}\gamma_s^{-1} W_s - \int_0^s \Omega(d\gamma_\tau, B_{\gamma_\tau}(\gamma_\tau^{-1} W_\tau))(\circ \dot{z}_s)$$
$$- \Theta(\dot{\gamma}_s, B_{\gamma_s}(\gamma_s^{-1} W_s)) \,,$$

and thus

$$M_t = \gamma_t^{-1} W_t - \int_0^t \int_0^s \Omega(d\gamma_\tau, B_{\gamma_\tau}(\gamma_\tau^{-1} W_\tau)) \, dz_s$$
$$- \int_0^t \Theta(d\gamma_s, B_{\gamma_s}(\gamma_s^{-1} W_s)) \,. \qquad (7.38)$$

Theorem 7.4.1 *Let* $\sigma \in W_o^\infty(I, M)$, $z = J^{-1}(\sigma)$, $W \in T_\sigma W_o^\infty(I, M)$, *and* $M = J_*^{-1} W$. *Then*

$$M_t = K_t - \int_0^t \int_0^s \Omega(B_{\gamma_\tau} \, dz_\tau, B_{\gamma_\tau} K_\tau) \, dz_s - \int_0^t \Theta(B_{\gamma_s} \, dz_s, B_{\gamma_s} K_s) \,, \qquad (7.39)$$

where $K_t = \gamma_t^{-1} W_t$ *and*

$$d\gamma_t = B(\gamma_t) \, dz_t \,, \quad \gamma_0 = u_o \,.$$

Important examples of vector fields W are the parallel translations of elements in the Cameron–Martin space $\mathbb{H}_0^1$. Given $h \in \mathbb{H}_0^1$, we define $X^h(\sigma)_t = \gamma_t h_t$, where γ is the horizontal lift of σ, namely the solution of the stochastic differential equation

$$d\gamma_s = B(\gamma_s)\gamma_s^{-1} \circ d\sigma_s \,, \quad \gamma_0 = u_o \,.$$

Take

$$\varphi_t = \int_0^t \circ dz_s \otimes h_s - h_s \otimes \circ dz_s \,,$$

and consider the stochastic differential equation

$$d\gamma_t^\varepsilon = B(\gamma_t^\varepsilon) \circ dz_t + \varepsilon B(\gamma_t^\varepsilon) \, dh_t + \frac{\varepsilon}{2}[B, B](\gamma_t^\varepsilon) \circ d\varphi_t \,,$$
$$\gamma_0^\varepsilon = u_o \,. \qquad (7.40)$$

Then

$$X^h(\sigma)_t = \left.\frac{\partial \sigma_t^\varepsilon}{\partial \varepsilon}\right|_{\varepsilon=0},$$

where $\sigma_t^\varepsilon = \pi(\gamma_t^\varepsilon)$.

7.5   Comments and notes on Chapter 7

Regarding Malliavin calculus, one should look at the excellent monograph by Malliavin (1997).

The classical theorem of Cameron and Martin (1944) states that the Wiener measure is quasi-invariant under translations along directions in the Cameron–Martin space. This fact allows us to differentiate Wiener functionals (i.e. solutions of stochastic differential equations) along these directions, and thus the Cameron–Martin space can be regarded as the tangent space of the classical Wiener space. The analysis of the Wiener space, following this line of study, has developed into a very important subject over the last few decades, lead by L. Gross, T. Hida, P. Malliavin, etc., and accumulated as Malliavin calculus.

From the late 1980s, with attempts to understand the Bismut formula for the derivative of the heat kernel, attention has turned to a nonlinear version of the Cameron–Martin formula, and Driver's flow equation was proposed in order to prove the Cameron–Martin formula for the law of Brownian motion on a compact Riemannian manifold. For more details, see Malliavin (1997) and Hsu (1998). The main idea developed in this chapter comes from the observation that Wiener functionals are more 'smooth' than has been revealed before, and thus it might be helpful to work on the space of paths together with the Lévy area. The material provided here is only a first attempt, and we hope to provide further developments in the future. The main results are taken from Lyons and Qian (1997c).

BIBLIOGRAPHY

Adler, R. J. (1990). *An Introduction to Continuity, Extrema and Related Topics for General Gaussian Processes*. Lecture Notes—Monograph Series Vol. 12. Institute of Math. Statistics, Hayward, California.

Arnold, L. (1973). *Stochastic Differential Equations: Theory and Applications*. Wiley, New York.

Arnold, L. and Kliemann, W. (1983). Qualitative theory of stochastic systems. In *Probabilistic Analysis and Related Topics 3* (ed. A. T. Bharucha-Reid), pp. 1–79. Academic Press, New York.

Azencott, R. (1982). Formule de Taylor stochastique et développement asymptotique d'intégrales de Feynman. In *Seminar on Probability, XVI, Supplement*, pp. 237–85. Lecture Notes in Mathematics 921. Springer, Berlin.

Bachelier, L. (1992). *Calcul des Probabilités*. Reprint of the 1912 original. Les Grands Classiques Gauthier–Villars. Éditions Jacques Gabay, Sceaux.

Bachelier, L. (1995). *Théorie de la Spéculation. Théorie Mathématique du Jeu*. Reprint of the 1900 original. Les Grands Classiques Gauthier–Villars. Éditions Jacques Gabay, Sceaux.

Barlow, M. T. (1982). One-dimensional stochastic differential equations with no strong solution. *J. London Math. Soc.*, **26**, 335–47.

Bass, R. F. (1988). Probability estimates for multiparameter Brownian processes. *Ann. Prob.*, **16**, 251–64.

Bass, R. F., Hambly, B. M., and Lyons, T. J. (1998). Extending the Wong–Zakai theorem to reversible Markov processes via Lévy area. *J. Euro. Math. Soc.* In press.

Ben Arous, G. (1989). Flots et séries de Taylor stochastiques. *Prob. Theory Related Fields*, **81**, 29–77.

Bernštein, S. (1934). Principles de la théorie des équations différentielles stochastiques. *Trudy Fiz.-Mat., Steklov Inst., Akad. Nauk.*, **5**, 95–124.

Bertion, J. (1989). Sur une intégrale pour les processus à α-variation bornée. *Ann. Prob.*, **17**, 1521–35.

Brzeźniak, Z. (1995). Stochastic partial differential equations in M-type 2 Banach spaces. *Potential Anal.*, **4**, 1–45.

Brzeźniak, Z. (1997). On stochastic convolution in Banach spaces and applications. *Stoch. Stoch. Rep.*, **61**, 245–95.

Brzeźniak, Z. and Carroll, A. (2000). Approximations of the Wong–Zakai type for stochastic differential equations in M-type 2 Banach spaces with applications to loop spaces. *IHP*. In press.

Brzeźniak, Z. and Elworthy, D. (2000). Stochastic differential equations on Banach manifolds. *Methods Funct. Anal. Topology*, **6**, 43–84.

Burdzy, K. (1993). Variation of iterated Brownian motion. In *Workshop and Conference on Measure-Valued Processes, Stochastic Partial Differential Equations and Interacting Systems*. CRM Proceedings and Lecture Notes.

Cameron, R. H. and Martin, W. T. (1944). Transformation of Wiener integrals under translations. *Ann. Math.*, **45**, 386–96.

Cameron, R. H. and Martin, W. T. (1945). Transformations of Wiener integrals under a general class of linear transformations. *Trans. Amer. Math. Soc.*, **58**, 184–219.

Capitaine, M. and Donati-Martin, C. (2001). The Lévy area process for the free Brownian motion. *J. Funct. Anal.*, **179**, 153–69.

Castell, F. and Gaines, J. G. (1996). The ordinary differential equation approach to asymptotically efficient schemes for solution of stochastic differential equations. *Ann. Inst. H. Poincaré, Prob. Stat.*, **32**, 231–50.

Chen, K. T. (1954). Iterated integrals and exponential homomorphisms. *Proc. London Math. Soc.*, **4**, 502–12.

Chen, K. T. (1957). Integration of paths, geometric invariants and a generalized Baker–Hausdorff formula. *Ann. Math.*, **65**, 163–78.

Chen, K. T. (1958). Integration of paths—a faithful representation of paths by non-commutative formal power series. *Trans. Amer. Math. Soc.*, **89**, 395–407.

Chen, K. T. (1961). Formal differential equations. *Ann. Math.*, **73**, 110–33.

Chen, K. T. (1967). Iterated path integrals and generalized paths. *Bull. Amer. Math. Soc.*, **73**, 935–8.

Chen, K. T. (1968). Algebraic paths. *J. Algebra*, **10**, 8–36.

Chen, K. T. (1971). Algebras of iterated path integrals and fundamental groups. *Trans. Amer. Math. Soc.*, **156**, 359–79.

Chen, K. T. (1973). Iterated integrals of differential forms and loop space homology. *Ann. Math.*, **97**, 215–46.

Chen, K. T. (1977). Iterated path integrals. *Bull. Amer. Math. Soc.*, **83**, 831–79.

Chung, K. L. and Williams, R. J. (1983). *Introduction to Stochastic Integration*. Birkhäuser, Boston.

Cohen, S. and Estrade, A. (2000). Non-symmetric approximations for manifold-valued semi-martingales. *Ann. Inst. H. Poincaré, Prob. Stat.*, **36**, 45–70.

Coutin, L. and Qian, Z. (2000). Stochastic differential equations driven by fractional Brownian motions. *C. R. Acad. Sci. Paris*, **331**, Série I, 75–80.

Coutin, L. and Qian, Z. (2002). Stochastic analysis, rough path analysis and fractional Brownian motions. *Prob. Theory Related Fields*, **122**, 108–40.

Da Prato, G. and Zabczyk, J. (1992). *Stochastic Equations in Infinite Dimensions*. Encyclopedia of Mathematics and Its Applications 44. Cambridge University Press.

Dellacherie, C. and Meyer, P.-A. (1978). *Probabilities and Potential.* North-Holland Mathematics Studies 29. North-Holland, Amsterdam.

Dettweiler, E. (1985). Stochastic integration of Banach space-valued functions. In *Stochastic Space–Time Models and Limit Theorem*, pp. 33–79. D. Reidel Publishing.

Dettweiler, E. (1989). On the martingale problem for Banach space-valued SDEs. *J. Theoret. Prob.*, **2**, 159–97.

Dettweiler, E. (1991). Stochastic integration relative to Brownian motion on a general Banach space. *Boga-Tr. J. Math.*, **15**, 6–44.

Doob, J. L. (1942). The Brownian movement and stochastic equations. *Ann. Math.*, **43**, 351–69.

Doob, J. L. (1953). *Stochastic Processes.* Wiley, New York.

Doob, J. L. (1984). *Classical Potential Theory and Its Probabilistic Counterpart.* Springer–Verlag, New York.

Doss, H. (1977). Liens entre équations différentielles stochastiques et ordinaires. *Ann. Inst. H. Poincaré*, **13**, 99–125.

Driver, B. K. (1992). A Cameron–Martin type quasi-invariance theorem for Brownian motion on a compact Riemannian manifold. *J. Funct. Anal.*, **110**, 272–376.

Dudley, R. M. (1967). The sizes of compact subsets of Hilbert space and continuity of Gaussian processes. *J. Funct. Anal.*, **1**, 290–330.

Dudley, R. M. (1973). Sample functions of the Gaussian process. *Ann. Prob.*, **1**, 66–103.

Dudley, R. M. and Norvaiša, R. (1999). *Differentiability of Six Operators on Non-Smooth Functions and p-Variation.* Lecture Notes in Mathematics 1703. Springer–Verlag, Berlin.

Elworthy, K. D. (1982). *Stochastic Differential Equations on Manifolds.* London Mathematical Society Lecture Notes in Mathematics 70. Cambridge University Press.

Elworthy, K. D. (1988). Geometric aspects of diffusions on manifolds. In *École d'Été de Probabilités de Saint-Flour XV–XVII, 1985–1987* (ed. P. L. Hennequin), pp. 276–425. Lecture Notes in Mathematics 1362. Springer, Berlin.

Emery, M. (1989). *Stochastic Calculus in Manifolds.* Springer, Berlin.

Fernique, X. (1975). Régularité des trajectoires des fonctions aléatoires Gaussiennes. In *École d'Été de Probabilités de Saint-Flour 1974*, pp. 1–96. Lecture Notes in Mathematics 480. Springer–Verlag.

Fliess, M. and Normand-Cyrot, D. (1982). Algèbres de Lie nilpotentes, formule de Baker–Campbell–Hausdorff et intégrales itérées de K. T. Chen. In *Séminaire de Probabilités XVI*, pp. 257–67. Lecture Notes in Mathematics 920. Springer, Berlin.

Föllmer, H. (1981a). Dirichlet processes. In *Stochastic Integrals* (ed. D. Williams), pp. 476–8. Proceedings of LMS Durham Symposium, University of Durham, July 7–17, 1980. Lecture Notes in Mathematics 851. Springer, Berlin.

Föllmer, H. (1981b). Calcul d'Itô sans probabilités. In *Séminaire de Probabilités XV*, pp. 143–50. University of Strasbourg, 1979–1980. Lecture Notes in Mathematics 850. Springer, Berlin.

Föllmer, H. (1986). Time reversal on Wiener space. In *Stochastic Processes— Mathematics and Physics*, pp. 119–29. Bielefeld, 1984. Lecture Notes in Mathematics 1158. Springer, Berlin.

Fukushima, M. (1980). *Dirichlet Forms and Markov Processes*. North-Holland, Amsterdam.

Fukushima, M., Oshima, Y., and Takeda, M. (1994). *Dirichlet Forms and Symmetric Markov Processes*. De Gruyter Studies in Mathematics 19. Walter de Gruyter, Berlin.

Gaines, J. G. and Lyons, T. J. (1997). Variable step size control in the numerical solution of stochastic differential equations. *SIAM J. Appl. Math.*, **51**, 1455–84.

Gihman, I. I. and Skorohod, A. V. (1972). *Stochastic Differential Equations*. Springer–Verlag, Berlin.

Goodman, V. and Kuelbs, J. (1991). Rates of clustering for some Gaussian self-similar processes. *Prob. Theory Related Fields*, **88**, 47–75.

Gross, L. (1965). Abstract Wiener spaces. *Math. Stat. Prob.*, **2**, 31–42. Proceedings 5th Berkeley Symposium.

Gross, L. (1970). Abstract Wiener measure and infinite-dimensional potential theory. In *Lectures in Modern Analysis and Applications, II*, pp. 84–116. Lecture Notes in Mathematics 140. Springer, Berlin.

Gross, L. (1993). Uniqueness of ground states for Schrödinger operators over loop groups. *J. Funct. Anal.*, **112**, 373–441.

Gross, L. (1998). Harmonic functions on loop groups. Séminaire Bourbaki. Vol. 1997–1998. *Astérisque*, **252**, Exp. No. 846, 5, 271–86.

Gyöngy, I. (1988). On the approximation of stochastic differential equations. *Stochastics*, **23**, 331–52.

Hackbusch, W. (1998). From classical numerical mathematics to scientific computing. In *Proceedings of the International Congress of Mathematicians*, Vol. I. Berlin. *Doc. Math.*, Extra Vol. I, 235–54.

Hain, R. M. (1984). Iterated integrals and homotopy periods. *Mem. Amer. Math. Soc.*, **47**.

Hambly, B. M. and Lyons, T. J. (1998). Stochastic area for Brownian motion on the Sierpinski gasket. *Ann. Prob.*, **26**, 132–48.

Haussmann, U. (1979). On the integral representation of functionals of Itô processes. *Stochastics*, **3**, 17–28.

Haussmann, U. and Pardoux, E. (1986). Time reversal of diffusions. *Ann. Prob.*, **14**, 1188–205.

He, S. W., Wang, J. G., and Yan, J. A. (1992). *Semi-martingale Theory and Stochastic Calculus*. Kexue Chubanshe (Science Press), Beijing.

Hirsch, M. W. (1984). The dynamical systems approach to differential equations. *Bull. (New Series) Amer. Math. Soc.*, **11**, 1–64.

Hsu, E. P. (1995). Quasi-invariance of the Wiener measure on the path space over a compact Riemannian manifold. *J. Funct. Anal.*, **134**, 417–50.

Ikeda, N. and Watanabe, S. (1981). *Stochastic Differential Equations and Diffusion Processes*. North-Holland, Kodansha, Tokyo.

Itô, K. (1944). Stochastic integral. *Proc. Imp. Acad. Tokyo*, **20**, 519–24.

Itô, K. (1946). On a stochastic integral equation. *Proc. Imp. Acad. Tokyo*, **22**, 32–5.

Itô, K. (1950). Stochastic differential equations in differential manifolds. *Nagoya Math. J.*, **1**, 35–47.

Itô, K. (1951*a*). On stochastic differential equations. *Mem. Amer. Math. Soc.*, **4**.

Itô, K. (1951*b*). On a formula concerning stochastic differentials. *Nagoya Math. J.*, **3**, 55–65.

Itô, K. and McKean, H. P. (1965). *Diffusion Processes and their Sample Paths*. Springer–Verlag, Berlin.

Jacod, J. (1979). *Calcul Stochastique et Problèmes de Martingales*. Lecture Notes in Mathematics 714. Springer, Berlin.

Jones, J. D. S. and Léandre, R. (1991). L^p Chen forms on loop spaces. In *Stochastic Analysis* (ed. M. Barlow and I. Bingham), pp. 104–62. Cambridge University Press.

Karatzas, I. and Shreve, S. E. (1988). *Brownian Motion and Stochastic Calculus*. Graduate Texts in Mathematics 113. Springer–Verlag, New York.

Kloeden, P. E. and Platen, E. (1991). Stratonovich and Taylor expansions. *Math. Nachr.*, **151**, 33–50.

Kunita, H. (1984). Stochastic differential equations and stochastic flows of diffeomorphisms. In *École d'Été de Probabilités de Saint-Flour XII, 1982* (ed. P. L. Hennequin), pp. 143–303. Lecture Notes in Mathematics 1097. Springer, Berlin.

Kunita, H. (1990). *Stochastic Flows and Stochastic Differential Equations*. Cambridge Studies in Advanced Mathematics 24. Cambridge University Press.

Kunita, H. and Watanabe, S. (1967). On square-integrable martingales. *Nagoya Math. J.*, **30**, 209–45.

Kuo, H. H. (1975). *Gaussian Measures in Banach Spaces.* Lecture Notes in Mathematics 463. Springer–Verlag, Berlin.

Kurtz, T. G. and Protter, P. (1991). Weak limit theorems for stochastic integrals and differential equations. *Ann. Prob.*, **19**, 1035–70.

Kurtz, T. G., Pardoux, E., and Protter, P. (1995). Stratonovich stochastic differential equations driven by general semi-martingales. *Ann. Inst. H. Poincaré*, **31**, 351–78.

Landau, H. J. and Shepp, L. A. (1970). On the supremum of a Gaussian process. *Sankhyà A*, **32**, 369–78.

Léandre, R. (1997). Invariant Sobolev calculus on free loop space. *Acta Appl. Math.*, **46**, 267–350.

Ledoux, M. (1996). Isoperimetry and Gaussian analysis. In *Lectures on Probability Theory and Statistics*, pp. 165–294. Saint-Flour, 1994. Lecture Notes in Mathematics 1648. Springer, Berlin.

Ledoux, M. and Talagrand, M. (1991). *Probability in Banach Spaces. Isoperimetry and Processes.* Ergebnisse der Mathematik und ihrer Grenzgebiete (3), 23. Springer–Verlag, Berlin.

Ledoux, M., Lyons, T. J., and Qian, Z. (2000). Lévy area of Wiener processes. *Ann. Prob.* In press.

Lévy, P. (1937). *Théorie de l'Addition des Variables Aléatoires.* Gauthier–Villars, Paris.

Lévy, P. (1948). *Processus Stochastiques et Mouvement Brownien.* Gauthier–Villars, Paris.

Lyons, T. J. (1991). On the nonexistence of path integrals. *Proc. Roy. Soc. London Ser. A*, **432**, 281–90.

Lyons, T. J. (1994). Differential equations driven by rough signals (I): An extension of an inequality of L. C. Young. *Math. Res. Lett.*, **1**, 451–64.

Lyons, T. J. (1995). The interpretation and solution of ordinary differential equations driven by rough signals. *Proc. Symp. Pure Math.*, **57**, 115–28.

Lyons, T. J. (1998). Differential equations driven by rough signals. *Rev. Mat. Iberoamer.*, **14**, 215–310.

Lyons, T. J. and Qian, Z. (1996). Calculus for multiplicative functionals, Itô's formula, and differential equations. In *Itô's Stochastic Calculus and Probability Theory*, pp. 233–50. Springer, Tokyo.

Lyons, T. J. and Qian, Z. (1997a). Flow equations on spaces of rough paths. *J. Funct. Anal.*, **149**, 135–59.

Lyons, T. J. and Qian, Z. (1997b). Calculus of variation for multiplicative functionals. In *New Trends in Stochastic Analysis*, pp. 348–74. Charingworth, 1994. World Scientific, River Edge, NJ.

Lyons, T. J. and Qian, Z. (1997c). A class of vector fields on path spaces. *J. Funct. Anal.*, **145**, 205–23.

Lyons, T. J. and Qian, Z. (1997*d*). Stochastic Jacobi fields and vector fields induced by varying area on path spaces. *Prob. Theory Related Fields*, **109**, 539–70.

Lyons, T. J. and Qian, Z. (1998). Flows of diffeomorphisms induced by geometric multiplicative functionals. *Prob. Theory Related Fields*, **112**, 91–119.

Lyons, T. J. and Zeitouni, O. (1999). Conditional exponential moments for iterated Wiener integrals. *Ann. Prob.*, **27**, 1738–49.

Lyons, T. J. and Zheng, W. A. (1988). A crossing estimate for the canonical process on a Dirichlet space and a tightness result. *Astérisque 157–8*, 249–71. Colloque Paul Lévy sur les Processus Stochastiques. Palaiseau, 1987.

Ma, Z.-M. and Röckner, M. (1992). *Introduction to the Theory of (Non-Symmetric) Dirichlet Forms*. Springer–Verlag, Berlin.

Malliavin, P. (1978*a*). Stochastic calculus of variation and hypoelliptic operators. In *Proceedings International Symposium SDE Kyoto, 1976* (ed. K. Itô), pp. 195–263. Kinokuniya, Tokyo.

Malliavin, P. (1978*b*). C^k-hypoellipticity with degeneracy. In *Stochastic Analysis* (ed. A. Friedman and M. Pinsky), pp. 199–214, 327–40. Academic Press, New York.

Malliavin, P. (1984). Analyse différentielle sur l'espace de Wiener. In *Proceedings of the International Congress of Mathematicians*, Vol. 1 and 2, pp. 1089–96. Warsaw, 1983. PWN, Warsaw.

Malliavin, P. (1993). Infinite dimensional analysis. *Bull. Sci. Math.*, **117**, 63–90.

Malliavin, P. (1997). Stochastic Analysis. Grundlehren der Mathematischen Wissenschaften 313. Springer–Verlag, Berlin.

McShane, E. J. (1974). *Stochastic Calculus and Stochastic Models*. Academic Press, New York.

Norris, J. R. (1995). Twisted sheets. *J. Funct. Anal.*, **132**, 273–334.

Nualart, D. (1995). *The Malliavin Calculus and Related Topics*. Springer–Verlag, New York.

Pardoux, E. (1990). Applications of anticipating stochastic calculus to stochastic differential equations. In *Stochastic Analysis and Related Topics II* (ed. H. Korezlioglu and A. S. Üstünel), pp. 63–105. Silivri, 1988. Lecture Notes in Mathematics 1444. Springer, Berlin.

Paršin, A. N. (1958). On cohomologies of spaces of paths. *Mat. Sb.*, **44**, 3–52. English translation: *Amer. Math. Soc.*, 187–96, 1969.

Pisier, G. (1986). Probabilistic methods in the geometry of Banach spaces. In *Probability and Analysis*, pp. 167–241. Varenna, 1985. Lecture Notes in Mathematics 1206. Springer, Berlin.

Protter, P. (1977). On the existence, uniqueness, convergence and explosions of solutions of systems of stochastic differential equations. *Ann. Prob.*, **5**, 243–61.

Ree, R. (1958). Lie elements and an algebra associated with shuffles. *Ann. Math.*, **68**, 210–20.

Ree, R. (1960). Generalized Lie elements. *Canad. J. Math.*, **12**, 493–502.

Ren, J. (1990). Analyse quasi-sûre des équations différentielles stochastiques. *Bull. Sci. Math.*, **114**, 187–214.

Reutenauer, C. (1993). *Free Lie Algebras.* Oxford University Press.

Revuz, D. and Yor, M. (1991). *Continuous Martingales and Brownian Motion.* Grundlehren der Mathematischen Wissenschaften 293. Springer–Verlag, Berlin.

Russo, F. and Vallois, P. (1993). Forward, backward symmetric stochastic integration. *Prob. Theory Related Fields*, **97**, 403–21.

Schwartz, L. (1989). La convergence de la série de Picard pour les EDS (équations différentielles stochastiques). In *Séminaire de Probabilités XXIII*, pp. 343–54. Lecture Notes in Mathematics 1372. Springer, Berlin.

Silverstein, M. L. (1974). *Symmetric Markov Processes.* Lecture Notes in Mathematics 426. Springer, Berlin.

Silverstein, M. L. (1976). *Boundary Theory for Symmetric Markov Processes.* Lecture Notes in Mathematics 516. Springer, Berlin.

Sipilainen, E.-M. (1993). A pathwise view of solutions of stochastic differential equations. Ph.D. thesis. University of Edinburgh.

Skorohod, A. V. (1961). Stochastic equations for diffusion processes in a bounded region. *Theory Prob. Appl.*, **6**, 264–74.

Skorohod, A. V. (1965). *Studies in the Theory of Random Processes.* Addison–Wesley, Reading, Mass. [Reprinted by Dover Publications, New York.]

Stein, E. M. (1970). *Singular Integrals and Differentiability Properties of Functions.* Princeton University Press.

Strichartz, R. S. (1987). The Campbell–Baker–Hausdorff–Dynkin formula and solutions of differential equations. *J. Funct. Anal.*, **72**, 320–45.

Stroock, D. W. and Taniguchi, S. (1994). Diffusions as integral curves, or Stratonovich without Itô. In *The Dynkin Festschrift*, pp. 333–69. Birhäuser Boston, MA.

Stroock, D. W. and Varadhan, S. R. S. (1972). On the support of diffusion processes with application to the strong maximum principle. In *Proceedings of Sixth Berkeley Symposium Math. Stat. Prob. III*, pp. 333–68. University of California Press.

Stroock, D. W. and Varadhan, S. R. S. (1979). *Multidimensional Diffusion Processes.* Grundlehren der Mathematischen Wissenschaften 233. Springer–Verlag, Berlin.

Stroock, D. W. and Zheng, W. A. (1998). Markov chain approximations to symmetric diffusions. *Ann. Inst. H. Poincaré, Prob. Stat.*, **33**, 619–49.

Sugita, H. (1992). Various topologies in the Wiener space and Lévy's stochastic area. *Prob. Theory Related Fields*, **91**, 283–96.

Sussmann, H. J. (1978). On the gap between deterministic and stochastic ordinary differential equations. *Ann. Prob.*, **6**, 19–41.

Talagrand, M. (1987). Regularity of Gaussian processes. *Acta Math.*, **159**, 99–149.

Üstünel, A. S. (1995). *An Introduction to Analysis on Wiener Space.* Lecture Notes in Mathematics 1610. Springer–Verlag, Berlin.

Varadarajan, V. S. (1974). *Lie Groups, Lie Algebras and their Representations.* Prentice–Hall, Englewood Cliffs, NJ.

Walsh, J. B. (1986). An introduction to stochastic partial differential equations. In *École d'Été de Probabilités de Saint-Flour XIV, 1984*, pp. 265–439. Lecture Notes in Mathematics 1180. Springer, Berlin.

Watanabe, S. (1984). *Lectures on Stochastic Differential Equations and Malliavin Calculus.* Notes by M. Gopalan Nair and B. Rajeev. Published for the Tata Institute of Fundamental Research, Bombay. Springer–Verlag, Berlin.

Whitehead, J. H. C. (1947). An expression of Hopf's invariant as an integral. *Proc. Nat. Acad. Sci. USA*, **33**, 117–23.

Williams, D. R. E. (1997). Solutions of differential equations driven by làdlàg paths of finite p-variation ($1 \leqslant p < 2$). Ph.D. thesis. Imperial College London.

Wong, E. and Zakai, M. (1965a). On the relationship between ordinary and stochastic differential equations. *Int. J. Eng. Sci.*, **3**, 213–29.

Wong, E. and Zakai, M. (1965b). On the convergence of ordinary integrals to stochastic integrals. *Ann. Math. Stat.*, **36**, 1560–4.

Yamato, Y. (1979). Stochastic differential equations and nilpotent Lie algebras. *Z. Wahr. und verw. Gebiete*, **47**, 213–29.

Young, L. C. (1936). An inequality of Hölder type, connected with Stieltjes integration. *Acta Math.*, **67**, 251–82.

INDEX

214